Linguistic and Cultural Online Communication Issues in the Global Age

Kirk St.Amant
Texas Tech University, USA

Information Science REFERENCE

INFORMATION SCIENCE REFERENCE
Hershey · London · Melbourne · Singapore

Acquisitions Editor:	Kristin Klinger
Development Editor:	Kristin Roth
Senior Managing Editor:	Jennifer Neidig
Managing Editor:	Sara Reed
Assistant Managing Editor:	Sharon Berger
Copy Editor:	Julie LeBlanc
Typesetter:	Sharon Berger
Cover Design:	Lisa Tosheff
Printed at:	Yurchak Printing Inc.

Published in the United States of America by
Information Science Reference (an imprint of IGI Global)
701 E. Chocolate Avenue, Suite 200
Hershey PA 17033
Tel: 717-533-8845
Fax: 717-533-8661
E-mail: cust@idea-group.com
Web site: http://www.idea-group-ref.com

and in the United Kingdom by
Information Science Reference (an imprint of IGI Global)
3 Henrietta Street
Covent Garden
London WC2E 8LU
Tel: 44 20 7240 0856
Fax: 44 20 7379 0609
Web site: http://www.eurospanonline.com

Library of Congress Cataloging-in-Publication Data

Linguistic and cultural online communication issues in the global age / Kirk St. Amant, editor.

p. cm.

Summary: "This book provides readers with in-depth information on the various linguistic, cultural, technological, legal, and other factors that affect interactions in online exchanges. It provides information that implements effective decisions related to the uses and designs of online media when interacting with individuals from other cultures"--Provided by publisher.

Includes bibliographical references and index.

ISBN 978-1-59904-213-8 (hardcover) -- ISBN 978-1-59904-215-2 (ebook)

1. Intercultural communication. 2. Information technology--Social aspects. 3. Communication and technology. 4. Communication--Social aspects. 5. Distance education. I. St. Amant, Kirk, 1970-

HM1211.L56 2007

306.44'602854678--dc22

2006033769

British Cataloguing in Publication Data
A Cataloguing in Publication record for this book is available from the British Library.

Table of Contents

Section V
International Aspects of Online Education

Detailed Table of Contents

In this chapter, the author examines three challenges that can result in conflict and miscommunication in international online exchanges. Each factor, in turn, relates to a different aspect of what the author calls "Internet multilinguality"—or the increasing number of languages and dialects used by individuals in cyberspace. These three factors are language barriers, cultural differences, and technological difficulties. In the chapter, the author also provides recommendations for effectively addressing these factors in international online exchanges.

This chapter introduces the concept of embodied metaphor as a framework for studying the multifaceted context of cross-cultural interactions via computer-mediated communication (CMC). In the chapter, the author argues that the embodied metaphor approach offers a more reliable method for understanding, explaining and framing the user's adaptive response to CMC. The author also examines how culture has the potential to shape online interactions by influencing the way metaphors are enacted. By examining how an embodied metaphor approach can be used to examine such situations, the chapter provides readers with a new means for understanding of the interplay between CMC and culture.

At present, two-thirds to three-fourths of the world's Internet users are not native speakers of English. Within this context, translation becomes important, for it would allow companies to tap a global—and multilingual—customer base via e-commerce. This chapter examines how machine translation—or the use of software programs to convert texts from one language to another—should be viewed within this global e-commerce context. In this chapter, the author provides an overview of how machine translation works and examines the benefits and limitations of using such systems to "customize" e-commerce Web sites for different linguistic groups. Through this examination, readers gain a better understanding of how to integrate machine translation use into global e-commerce strategies.

The authors of this chapter present an individual-based approach to understanding cultural similarities and differences in the borderless environment created by online media. In addressing this topic, the authors also present a typology of individual cultural value orientations that readers can use to understand how cultural factors might affect discourse in global online contexts. This typology, in turn, emphasizes the need for making distinctions at the individual level before group level comparisons in order to grasp the complexity of today's global culture. The authors then use the results of an empirical study to demonstrate the usefulness of this approach in international online contexts.

This chapter presents the results of a study that examined the effects of culture and computer-mediated communication (CMC) in relation to Chinese college students' use of English in online exchanges with U. S. college students. In presenting this research, the author uses an ethnography of communication approach to uncover four language patterns in Chinese college students' use of English. These patterns were others-oriented talk, mentor-mentee talk with limited relationship development, adoption of others' talk, and icebreaker talk. The experiment also revealed that Chinese cultural values have a strong effect on Chinese college students' use of English in CMC. The cultural values that seemed to have the greatest effects on these exchanges were Chinese significance of personal relationships, Chinese collectivism, Chinese understanding of authority, and Chinese use of CMC.

Section II
Global Virtual Communities and International Online Collaboration

Globalization has brought about or re-invents the concept of workspaces such that work is no longer confined to a traditional space in a specific geographical setting whereby individuals are restricted to face-to-face meetings and interactions. Within this context, virtual workspaces bring issues of culture and its effect on behaviors to the forefront of discourse in global cyberspace. To examine this situation, the author of this chapter explores the role of culture and its implications for virtual workspaces. In so doing, the author addresses how culture also plays a role in how behaviors are adapted from traditional to virtual workspaces. Through this examination, the author reveals how virtual workspaces provide organizations with a mechanism for fostering effective international relationships in a variety of professional contexts.

Virtual teams and their leaders are key players in global organizations, yet the dispersed nature of such teams has changed how people work in groups. It has also redefined the nature of teamwork. Within this framework, leadership issues in computer-mediated communication are vital because of the increasing prevalence of the virtual organization, the flattening of organizational structures, and the corresponding interest in managing virtual groups and teams. The authors of this chapter examine the communication behaviors of participants in two different case studies to determine if number, length, and content of messages are sufficient criteria to identify emergent leaders in asynchronous and synchronous environments.

Within a single business day, employees might be required to work with colleagues located in a variety of nations in a series of online intercultural dialogs (OIDs). Preparing individuals to work effectively within such contexts, however, is no easy task. The author of this essay examines the importance of doing research on such situations. The author also presents a methodology for examining and assessing such situations and discusses how the results of such research can be applied to business and academic practices worldwide.

Section IV
Governance and Legal Factors in Global Cyberspace

Early discussions of the Internet often advocated the idea that nation states would have either limited or no control over this exclusively electronic domain. The subsequent need for protocols and standardization of the Internet, however, led to the development of formal institutions such as the Internet Corporation for Assigned Names and Numbers (ICANN). It also led more nations to adopt policies designed to promote, control, and mold the Internet culturally and politically within their jurisdictional boundaries. The author of this chapter outlines the characteristics and intrinsic qualities of the Internet which make it a global entity and examines how it has been deemed ungovernable in conventional terms. In so doing, the author examines how different legal and political developments shaped public policy related to online access in different regions and examines the tensions between local and international interests in attempting to control cyberspace.

By adhering to cultural expectations, a Web site can build credibility—or ethos—with an intercultural audience. In this regard, laws create certain expectations in audiences, and when a seller violates such laws, or appears to violate such laws, it can create a loss of consumer confidence. The authors of this chapter examine certain legal and cultural issues that will instantly brand a Web site as unreliable to users from certain backgrounds. In presenting these ideas, the authors wish to provide a method for raising awareness of diverse legal cultures. Such increased awareness, in turn, can help organizations and individuals reduce the need to make costly changes as a result of legal advice given outside of an appropriate cultural context.

The conceptualization of privacy has been the subject of much debate for more than a century in scholarship ranging from social philosophy and sociology to law, finance, and medicine. Legal issues concerning the right to privacy, moreover, are part of a complex web of state and national laws. To understand these factors from a more focused perspective, the authors of this chapter examine the international legal issues resulting from the EU's 1998 Data Protection and Privacy Directive and its effects on online interactions in a global context. Through such a focus, the authors hope to provide readers with the knowledge needed to understand and to adapt to varying cultural and national perspectives of privacy when interacting in international online exchanges.

Section V
International Aspects of Online Education

More than half of all institutions of higher education in the U.S. now offer entirely online or blended courses. Within this context, a central question emerges—how well suited are these contexts for students from different cultures participate in such courses? The answer to this question can have pronounced implications for the future of online education and to teaching in general. The authors of this chapter provide a first step toward answering this complex question. In this chapter, the authors explore the nature of cyberculture and describe how cyberculture affects the online learning experiences of international students. Through the exploration of cyberculture and online education, the authors reveal what increased global online access could mean for participants in online educational contexts.

New technologies and computer-mediated communication (CMC) in general seem inherently suited to create constructive cross-cultural communication. As a result, the new possibilities for cross-cultural teaching and learning through dialogue and negotiation in the online environment compel us to reconceptualize the traditional role of the instructor. To this end, the author of this chapter seeks to answer the question, "What does it mean to teach collaboratively, interactively, and open-endedly within a global context?" To address this question, the author first examines several other questions related to such educational contexts. The author then provides an overview of the dialogue-enabling properties of the online environment and its potential to support of multiple learning styles.

Workplaces are transforming in the global age. Workers are increasingly participating in a global workforce comprised of people who are socially and demographically diverse. Further complicating this situation is the fact that e-learning is increasingly gaining in prominence as a method for training such a workforce. The author of this chapter examines these new trends by first providing an overview of the changing nature of the workplace and then discusses the emergence of e-learning within this context. In providing this overview, the author also examines the social and cultural implications such developments have for adult education and training in the global age.

The author of this chapter examines the impact of commodification on education when institutions of higher education sell knowledge as a commercial good. In addressing this topic, the author provides insights on the impact multiple sociocultural and communicative norms have on virtual communication and does so through research results that examine a case of distance education related to an Australian university. Through this case study, the author presents ideas for informing the change in policy, the administration, and the delivery of distance education and virtual communication in global environments. The author also discusses the kinds of skills leaders and participants in such educational environments need in order to interact effectively with each other in such situations.

Foreword

This volume is of interest and value on several levels. First, for individuals beginning their exploration of the complex, interlocking issues entailed by cross-cultural communication online, the essays collected here both serve as a helpful introduction *and* provide important new insights and inspiration for further research. Second, for researchers and practitioners already working in these domains, several chapters (e.g., Chapters IV, X, and XI) expand on significant contributions in these fields. In particular, as these and additional essays (e.g., Chapter VIII) explicitly connect the research and insight gathered here with the larger literatures of intercultural communication and of online discourse, they provide helpful context and orientation to understanding culture and communication online (see also Macfadyen, Roche, & Doff, 2004; Sudweeks, Hrachovec, & Ess, 2006).

Most importantly, the volume provides crucial insight for a much broader audience—namely, anyone who uses the Internet and the World Wide Web to communicate across cultures. As several chapters in this volume exemplify, much extant research in these domains has focused on the very practical work of cross-cultural collaboration for the sake of business, research, and distance learning. The contributions of such research to more effective cross-cultural communication online are critically important, of course—and not simply to a relatively small number of academics and professionals collaborating in virtual teams, but to anyone who undertakes such communication. Simply put, this increasingly means a sizable majority of Internet users *per se.* That is, as the Internet continues its dramatic expansion across the globe—so as to include, at the time of this writing, nearly one-sixth of the world's population—more of us will inevitably communicate across multiple cultures. Of course, the great danger here, especially for those with no previous experience in cross-cultural communication, is the naïve presumption that "everyone else is just like me," an ethnocentricism whose consequences only begin with communication *failures*, because we fail to understand and respect "the Other"—first of all, as we fail to understand the critical *differences* at work in cross-cultural communication, much less take these differences into account in our own communicative efforts (a failure I have called "computer-mediated colonization"). At worst, such naïve ethnocentrism ends in cultural and political imperialism. Hence in a *communicative* world increasingly interconnected across multiple cultures by the Internet and World Wide Web, what is desperately needed is a guide for effective cross-cultural communication online. Such a text, to my knowledge, remains to be written; but the chapters and insights collected here constitute much needed and significant contributions towards developing a resource of this sort. This text will thus be of interest and value to researchers in particular and to anyone more broadly who desires to communicate more effectively in cross-cultural exchanges that take place online.

REFERENCES

Macfadyen, L. P., Roche, J., & Doff, S. (2004). *Communicating across cultures in cyberspace: A bibliographical review of online intercultural communication.* Hamburg, Germany: Lit-Verlag.

Sudweeks, F., Hrachovec, H., & Ess, C. (2006). *Proceedings of the 5th International Conference on Cultural Attitudes Towards Technology and Communication 2006.* Murdoch, WA: Murdoch University.

Charles Ess

Drury University, USA, and Norwegian University of Science and Technology, Norway

***Charles Ess** is professor of philosophy and religion and distinguished professor of interdisciplinary studies, Drury University (USA); and professor II in the Program for Applied Ethics, Norwegian University of Science and Technology. Dr. Ess has received awards for teaching excellence and scholarship, and published extensively in comparative (East-West) philosophy, applied ethics, discourse ethics, history of philosophy, feminist Biblical studies, and computer-mediated communication (CMC). With Fay Sudweeks, Dr. Ess co-chairs the biennial conferences, Cultural Attitudes Towards Technology and Communication (CATaC). He has served as a visiting professor at IT-University, Copenhagen (2003), and a Fulbright senior scholar at the University of Trier (Denmark, 2004).*

Preface

In his book *The Wired Neighborhood*, Stephen Doheny-Farina (1996) examines how online media are changing the ways in which we think about communities. That is, our ability to interact and form networks of interpersonal connections is no longer limited by traditional barriers of geographic distance. Rather, online media now allow individuals from across the world to come together in online spaces in order to discuss their mutual interests. Or, as Doheny-Farina puts it, virtual communities "offer us the opportunity to construct utopian collectives—communities of interest, education, tastes, beliefs, and skills. In cyberspace, we can remake the world out of an unsettled landscape" (p. 16). Online media are thus global media that permit us to participate in a growing international community where we interact based on interest of mind rather than convenience of proximity. And with each passing day, online communities are becoming more international and intercultural in nature.

THE GLOBAL GROWTH OF ONLINE ACCESS

As of this writing, roughly one billion persons worldwide have ready access to the Internet (Internet Usage Statistics, 2006). While this number is but one-sixth of the world's current population, international online access continues to grow with amazing speed. For example, the number of global Internet users increased by almost 2.8 million between June and July of 2004 alone (ClickZ, 2004b). Moreover, certain trends indicate much of this growth will occur in areas outside of the United States.

The number of Australian Internet users increased by some 400,000 individuals between June and August of 2004 (ClickZ, 2004a), while Canadian spending on IT infrastructure is expected to grow by US$4 billion between 2004 and 2008 (Insurance-Canada, 2004). Western Europe has experienced similar kinds of growth in the areas of wireless local area networks (WLANs) and subscriptions increasing markedly—particularly in the Scandinavian countries (eMarketer, 2004c). In Japan, over 50% of the adult population is online (AsiaBiz Tech, 2003), while in South Korea, 76% of all citizens now have a broadband connection (up to 20 megabits per second is some cases) that are relatively inexpensive (Forsberg, 2005; Borland & Kanellos, 2004).

The most astounding growth, however, is taking place in developing nations. Thanks to a series of government and industry programs, Internet use is booming in India and China. As a result, India has become a leading location for software programming and information technology (IT) production (The Economist, 2000a; Kripalani & Engardio, 2003), and the number of Internet connections in India has increased by over 50% between 2004 and the end of 2005 with some 38.5 million individuals now online (Burns, 2006). China, likewise, has seen impressive increases as Internet access grew from 2.1 million users in 1999 to almost 100 million by the end of 2004 (The Economist, 2000b; eMarketer, 2004d), and some experts expect this number to balloon to some 200 million users by 2007 (eMarketer, 2004d). This rapid expansion has also allowed Chinese to become the second most common language in online interactions (Internet World Stats, 2006a).

Other developing regions are also seeing impressive growth. In Africa, public- and private-sector organizations have undertaken initiatives to increase online access across the continent (The Economist, 2000c; Kalia, 2001). These factors perhaps explain why Africa's number of Internet users has grown by almost 430% between

2000 and 2005, and they could also account for why sales of laptop computers remain strong across the continent (Internet World Stats, 2005; eMarketer, 2004b). In Latin America, the number of Internet users in Brazil has increased by 430,000 in recent months, and Global Crossings Ltd. has created a high-speed communication network that interlinks most major cities in the region (ClickZ, 2004b; NYSE, 2001). Additionally, Latin American has the world's lowest Wi-Fi costs and a fast-growing market for mobile handsets—factors that could result in rapid growth in online access in the region (eMarketer, 2004a; Buddle.com, 2006). Similarly, in Eastern Europe, the number of individuals going online has grown almost exponentially in the last five years. Lithuania, Latvia, Poland, and Russia have all experienced 300+% growth in online access since 2000 (Burns, 2005), and some expect online access in the region to climb from 17 to 27% by 2006 (NUA, 2003).

As a result of these trends, the prospects for online media to create truly global communities increases every day. Yet the ability to make contact with others does not necessarily mean international online interactions (IOIs) will be effective. Nor does it mean truly global communities will arise in online environments. Rather, access is only one part of forming international online communities. The other part is communication, and this component is far more complex.

CULTURE, COMMUNICATION, AND CYBERSPACE

The first and perhaps greatest obstacle to international online discourse is language. While roughly 30% of individuals who interact online communicate in English, more speakers of other languages are creating their own discourse areas in cyberspace (Internet World Stats, 2006a). According to Internet World Stats (2006a), roughly 14% of Internet users are Chinese speakers, 7.5% are Spanish speakers, and 2.3% are Russian speakers. These statistics, however, reflect only a small percentage of the actual number of speakers of each language—11%, 18%, and 17% respectively (Internet World Stats, 2006a). Thus, as online access increases in more developing nations, these percentages will surely grow to reflect a "less-Anglo" Internet. Moreover, as more non-English speakers populate cyberspace and find increasing options to interact in languages other than English, then language could become a barrier to international online discourse. Language differences, however, are only one obstacle to effective international exchanges in online environments.

Cultural communication expectations can also affect the nature and success of IOIs. That is, cultures can have differing expectations of what is an acceptable way to discuss a topic. They can also vary on what topics are acceptable for discussion (Barnum & Li, 2006; Yunker, 2003; Keegan & Green, 2003). Areas of communication difference between cultures can include everything from the structure of overall messages to the use of individual words. As Kaplan (1966, 2001), Campbell (1998a), and Barnum and Li (2006) note, members of Western cultures tend to prefer a more direct, linear structure of presentation, with each point connecting to the next (Barnum & Li, 2006; Campbell, 1998a, 1998b; Hall 1981a, 1981b). Many Asian cultures, however, often use a "spiral" approach in which the speaker or poster circles around key issues without ever addressing them directly (Barnum & Li, 2006; Ulijn & St.Amant, 2001; Campbell, 1998a, 1998b; Hall, 1981a, 1981b). Such structures often reflect deep-seated cultural values that, while unknown to the observer, influence a great deal of external actions (Hofstede & Hofstede, 2005; Varner & Beamer, 2004; Hu & Grove, 1999). As a result, behavior by members of one culture can often be misunderstood or misinterpreted by individuals from another culture (Ulijn & St.Amant, 2001; Campbell, 1998b; Hu & Grove, 1999).

Similarly, Hall (1981a, 1981b, 1983) notes that cultural groups can have markedly different perspectives on how much information needs to be explicitly vs. implicitly conveyed in the same setting—a perspective re-enforced by Ulijn and St.Amant's (2001) findings related to differences in perception of events by individuals from five different cultural groups (French, Dutch, German, Italian, and Chinese). Moreover, as both Weiss (1998) and Grundy (1998) note, persons from different cultures can have varying expectations of both what subjects to address in the same genre of communication (e.g., a letter vs. an e-mail) and how those subjects should be addressed in that genre. Li's (1999) and Li and Koole's (1998) research takes this idea of culture and communication differences one step further by examining how such variations can even occur at the individual word level. In such cases, each culture might associate a different connotative meaning with the use of particular words within certain contexts. Such assumptions lead to each party acting in a way that the other considers inappropriate or disrespectful.

Research in cross-cultural online discourse indicates such differences can affect the perceptions cultures can have of the same online exchange (Callahan, 2005; Hewling, 2005; Wurtz, 2005; Ma, 1996). These differences also have implications for how members of different cultures evaluate and react to both online communiqués and the design/visual aspects of online media (Yunker, 2003; Honold, 1999; Forslund, 1996). Moreover, research in translation reveals that individuals use their native culture's communication expectations to evaluate the acceptability and the effectiveness of a message—even if that message is presented in another language (Ulijn, 1996; Campbell, 1998). As a result, effective IOIs will involve more than just a common language. They will also require a deeper understanding of how cultures use language and visuals to convey information.

Cultural values can also give rise to different legal systems with varying perspectives on the same topic (Swire & Litan, 1998). Such differences, moreover, can affect IOIs in unexpected ways. While online media allows individuals to exchange information across national borders, such exchanges involve actual persons who reside in nations with existing laws. These laws can place real or perceived restrictions—in the form of legal penalties—on what individuals say and do online (Rosenthal, 2005; Doyle, 2004). They can also affect what kinds of information individuals in different nations can legally exchange or receive via online media (Swire & Litan, 1998). As a result, the development of international online communities can involve an intricate legal dance of who can say what, and when information can and cannot be exchanged.

At the same time, various factors are expanding the international spread of online access in a way that focuses on increasing IOIs. More companies, for example, are adopting international outsourcing as a core part of their business plans (Baily & Farrell, 2004; Rosenthal, 2004). In many cases, such outsourcing practices require effective online access to be successful, and the result is a variety of corporate and government programs designed to get workers in more nations—particularly developing nations—online (Warschauer, 2003; Kalia, 2001; The Economist, 2004). Simultaneously, the recent deregulation of international education via the General Agreement on Trade in Services (GATS) has prompted various public and private educational institutions to develop online courses and curricula targeted at an international student base (eMarketer, 2005; The Economist, 2002). The success of such online educational ventures has even prompted certain national governments to develop online curricula targeted at a global student body (Commonwealth of Learning, 2005; Daniel, 2005).

Within this complicated framework of culture, communication, and cyberspace, two central questions emerge:

1. How will factors of culture (and related aspects of law) affect international online interactions?
2. What steps can individuals and organizations take to address such cultural factors effectively in order to interact successfully in international cyberspace?

The essays in this collection provide readers with a foundation for answering these questions. Readers should not, however, consider this text as a comprehensive guide or a handbook to understanding IOIs. Rather, they should view this book as a resource for understanding how cultural factors can affect communication in cyberspace. Through the information provided in the various chapters, readers should gain the knowledge base and the insights needed to make more informed choices when interacting in global cyberspace. These entries also provide readers with the initial perspectives needed to investigate factors of culture and cyberspace further, and in a way that meets the needs of particular individuals or organizations.

ORGANIZATION OF THE TEXT

This book examines the complex nature of international online interactions from five primary perspectives:

1. Language and culture
2. Virtual communities and online collaboration
3. Culture and online design
4. Governance and legal factors
5. Online education

The chapters in each section, in turn, provide readers with a range of perspectives on a particular subject related to culture and online exchanges.

The chapters in the book's first section, "Language, Culture, and Cyberspace," examine how factors of language and culture affect online discourse and present methods for addressing these factors effectively. The initial essay by Shaoyi He examines the proliferation of different languages online and presents recommendations for overcoming barriers created by linguistic and cultural differences. Next, Rosanna Tarsiero expands upon He's ideas by exploring how using different metaphors when discussing online communication can lead to better ways of understanding international online exchanges. Nicole St. Germaine-Madison, in turn, offers a perspective on the uses of computer programs to help users overcome language barriers in global exchanges, while Wei-Na Lee and Sejung Marina Choi present a typology for classifying and understanding Web users according to culture. The section then concludes with Yun Xia's presentation of a study involving the effects of online media on interactions between U.S. and Chinese students.

The book's second section, "Global Virtual Communities and International Online Collaboration," examines how cultural factors affect interaction in online groups. In this section, Bolanle A. Olaniran examines how shifts to virtual workgroups and online teams make the nature of work increasingly global in nature. Simeon J. Simoff and Fay Sudweeks take the examination of virtual teams one step further and present a framework for identifying potential leaders from within such teams. Pamela Estes Brewer then completes this section by discussing the importance of studying exchanges in such online environments and presenting a method for performing such studies.

The book's third section, "Aspects of Online Design in International Environments," provides various perspectives and approaches related to examining visual aspects of international online exchanges. The section begins with Barry Thatcher et al.'s study of how cultural factors seem to affect Web site design and how design factors seem to reflect cultural rhetorical traditions. Next, Anthony Faiola and Sorin Adam Matei use the psychological concept of cognitive styles to examine how culture affects the ways in which individuals use and assess visual media such as Web sites. In the section's concluding essay, Marc Hermeking ties together ideas from the previous two chapters by examining how cultural expectations related to visual design can affect responses to online marketing materials.

The book's fourth section, "Governance and Legal Factors in Global Cyberspace," begins with Yasmin Ibrahim's discussion of how the global nature of online media affects the ways in which different governments try to regulate its use. Kendall Kelly and Jennifer Jones take a more focused approach to online governance by discussing the legal and cultural issues that affect the successes of online media in international contexts. Vinita Agarwal and Charles J. Stewart conclude this section by exploring one particular legal aspect—privacy—and how different national regulations related to privacy can affect IOIs.

The book's final section, "International Aspects of Online Education," contains four chapters that review the prospects of using online media to develop global educational environments. In this section, Chun-Min Wang and Thomas C. Reeves examine the prospects of using online media to create international online classrooms and how such classrooms involve balancing existing national cultures with cybercultures created in such contexts. Tatjana Chorney moves the discussion from meeting spaces to information transmission in an essay that overviews how aspects of culture need to be considered when addressing the learning styles of different online students. Karim A. Remtulla then expands this discussion by exploring how online media have implications for training globally dispersed workforces. Finally, Angela T. Ragusa shifts the discussion to a more macro level by presenting an overview of how to view overall curricula and program administration in a manner that addresses online education in international environments.

As cyberspace becomes increasingly international, users will need to understand how factors of culture can affect online exchanges. By presenting various perspectives of five areas related to such exchanges, the essays in this collection provide a foundational understanding of what these factors are and how such factors can be addressed in cross-cultural interactions. Readers can then use this information to make more informed decisions and to create online communities that are increasingly international in nature.

REFERENCES

AsiaBiz Tech. (2003, March 12). *Over 50 percent of Japanese population online.* Retrieved June 24, 2003, from http://www.nua.com/surveys/index.cgi?f=VS&art_id=905358740&rel=true

Baily, M. N., & Farrell, D. (2004, July). Exploding the myths of offshoring. *The McKinsey Quarterly.* Retrieved August 15, 2006, from http://www.mckinseyquarterly.com/article_print.aspx?L2=7&L3=10&ar=1453

Barnum, C., & Li, H. (2006). Chinese and American technical communication: A cross-cultural comparison of differences. *Technical Communication, 53,* 143-166.

Borland, J., & Kanellos, M. (2004, July 28). *South Korea leads the way.* Retrieved November 2, 2005, from http://news.com.com/South+Korea+leads+the+way/2009-1034_3-5261393.html

Buddle.com. (2006). *Latin America—mobile market—overview & statistics.* Retrieved August 20, 2006, from http://www.budde.com.au/Reports/Contents/Latin-America-Mobile-Market-Overview-Statistics-2172.html

Burns, E. (2006, February 7). *India's Internet users spiked in '05.* Retrieved August 12, 2006, from http://www.clickz.com/showPage.html?page=3583391

Burns, E. (2005, November 3). *Baltic states Internet population.* Retrieved November 3, 2005, from http://www.clickz.com/stats/sectors/geographics/print.php/3561451

Callahan, E. (2005). Cultural similarities and differences in the design of university Web sites. *Journal of Computer-Mediated Communication, 11. Retrieved from* http://jcmc.indiana.edu/vol11/issue1/callahan.html

Campbell, C. P. (1998a). *Beyond language: Cultural predispositions in business communication.* Retrieved from http://infohost.nmt.edu/~cpc/internationalethos.html

Campbell, C. P. (1998b). Rhetorical ethos: A bridge between high-context and low-context cultures? In S. Niemeier, C. P. Campbell, & R. Dirven (Eds.), *The cultural context in business communication* (pp. 31-47). Philadelphia: John Benjamins.

ClickZ. (2004a, August 25). *Active Internet users by country, July 2004.* Retrieved October 6, 2004, from http://www.clickz.com/stats/sectors/geographics/article.php/3397231

ClickZ. (2004b, September 22). *Active Internet users by country, August 2004.* Retrieved October 6, 2004, from http://www.clickz.com/stats/big_picture/geographics/article.php/3410261

Commonwealth of Learning. (2005). *Commonwealth executive MBA/MPA.* Retrieved October 3, 2005, from http://www.col.org/cemba/

Daniel, J. (2005). *Collaboration in the time of competition.* Retrieved October 3, 2005, from http://www.col.org/speeches/JD_0511ICDEDelhi.htm

Doheny-Farina, S. (1996). *The wired neighborhood.* New Haven, CT: Yale University Press.

Doyle, J. F. (2004). *Avoiding outsourcing pitfalls.* Retrieved December 12, 2004, from http://www.outsourcing-requests.com/center/jsp/requests/print/story.jsp?id=4626

Economist, The. (2000a, July 22). When India wires up. *The Economist,* 39-40.

Economist, The. (2000b, July 22). Wired China. *The Economist,* 24-28.

Economist, The. (2000c, September 9). Tapping in to Africa. *The Economist,* 49.

Economist, The. (2002, October 7). Offsite learning: On target? Part II. *The Economist*. Retrieved December 5, 2005, from http://www.economist.com/displaystory.cfm?story_id=1377324

Economist, The. (2004, March 13). Beyond the digital divide. *The Economist: Technology Quarterly Supplement,* 8.

eMarketer. (2004a). *11 trends to watch in 2004.* Retrieved January 21, 2004, from http://www.emarketer.com/news/article.php?1002633&format=printer_friendly

eMarketer. (2004b, August 20). *Laptop sales continue to climb in third world.* Retrieved October 27, 2004, from www.emarketer.com/Article.aspx?1003006&printerFriendly=yes

eMarketer. (2004c, September 20). *Mobility in the frozen north.* Retrieved September 29, 2004, from http://www.emarketer.com/Article.aspx?1003048&printerFriendly =yes

eMarketer. (2004d). *China's coming of age online.* (2004, November 16). Retrieved November 17, 2004, from http://www.emarketer.com/Article.aspx?1003139&printerFriendly=yes

eMarketer. (2005, October 24). *Online degrees gain acceptance.* Retrieved October 24, 2005, from http://www.emarketer.com/Article.aspx?1003647&printerFriendly=yes

Forsberg, B. (2005). The future is South Korea: Technology firms try out latest in world's most wired society. *San Francisco Chronicle,* (March 13). Retrieved November 3, 2005, from http://www.sfgate.com/cgi-bin/article.cgi?f=/c/a/2005/03/13/BROADBAND.TMP

Forslund, C. J. (1996). Analyzing pictorial messages across cultures. In D. C. Andrews (Eds.), *International dimensions of technical communication* (pp. 45-58). Arlington, VA: Society for Technical Communication.

Grundy, P. (1998). Parallel texts and diverging cultures in Hong Kong: Implications for intercultural communication. In S. Niemeier, C. P. Campbell, & R. Dirven (Eds.), *The cultural context in business communication* (pp. 167-183). Philadelphia: John Benjamins.

Hall, E. T. (1981a). *Beyond culture.* New York: Anchor Books.

Hall, E. T. (1981b). *The silent language.* New York: Anchor Books.

Hall, E. T. (1983). *The dance of life: The other dimension of time.* New York: Anchor Books.

Hewling, A. (2005). Culture in the online class: Using message analysis to look beyond nationality-based frames of reference. *Journal of Computer-Mediated Communication, 11.* Retrieved from http://jcmc.indiana.edu/vol11/issue1/hewling.html

Hofstede, G., & Hofstede, G. J. (2005). *Cultures and organizations: Software of the mind.* New York: McGraw-Hill.

Honold, P. (1999). Learning how to use a cellular phone: Comparison between German and Chinese users. *Technical Communication, 46,* 196-205.

Hu, W., & Grove, C. (1999). *Encountering the Chinese: A guide for Americans* (2nd ed.). Yarmouth, ME: Intercultural Press.

Insurance-Canada. (2004, September 29). *Canada gets with IT.* Retrieved February 5, 2005, from http://www.insurance-canada.ca/ebusiness/canada/eMarketer-Canada-IT-409.php

Internet World Stats. (2005). *Internet usage statistics for Africa.* Retrieved December 5, 2005, from http://www.internetworldstats.com/stats1.htm

Internet World Stats. (2006a). *Internet users by language.* Retrieved August 28, 2006, from http://www.internetworldstats.com/stats7.htm

Internet World Stats. (2006b). *Internet user statistics—the big picture.* Retrieved August 28, 2006, from http://www.internetworldstats.com/stats.htm

Kalia, K. (2001, July/August). Bridging global digital divides. *Silicon Alley Reporter,* 52-54.

Kaplan, R. B. (1966). Cultural thought patterns in intercultural education. *Language Learning, 16,* 1-20.

Kaplan, R. B. (2001). What in the world is contrastive rhetoric? In C.G. Panetta (Ed.), *Contrastive rhetoric revisited and redefined* (pp. vii-xx). Mahwah, NJ: Lawrence Erlbaum.

Keegan, W. J., & Green , M. C. (2003). *Global marketing* (3rd ed.). Upper Saddle River, NJ: Prentice-Hall.

Kripalani, M., & Engardio, P. (2003, December 8). The rise of India. *BusinessWeek,* 66-76.

Li, X. (1999). *Chinese-Dutch business negotiations: Insights from discourse.* Atlanta: Rodopi.

Li, X., & Koole, T. (1998). Cultural keywords in Chinese-Dutch business negotiations. In S. Niemeier, C. P. Campbell, & R. Dirven (Eds.), *The cultural context in business communication* (pp. 185-213). Philadelphia: John Benjamins.

Ma, R. (1996). Computer-mediated conversations as a new dimension of intercultural communication between East Asian and North American college students. In S. Herring (Ed.), *Computer-mediated communication: Linguistic, social, and cross-cultural perspectives* (pp. 173-185). Amsterdam: John Benjamins.

NUA. (2003, February 19). *IDC research: Net usage up in Central and Eastern Europe.* Retrieved June 23, 2003, from http://www.nua.com/surveys/index.cgi?f=VS&art_id=905358723&rel=true

NYSE. (2001, Summer). Tying Latin American together. *NYSE Magazine,* 9.

Rosenthal, B. E. (2005). *New outsourcing risks in 2005 and how to mitigate them.* Retrieved January 2, 2005, from http://www.outsourcing-requests.com/center/jsp/requests/print/story.jsp?id=4721

Rosenthal, B. E. (2004). *META predicts offshoring will continue to grow at 20 percent clips through 2008.* Retrieved August 16, 2006, from http://www.outsourcing-requests.com/center/jsp/requests/print/story.jsp?id=4714

Swire, P. P., & Litan, R. E. (1998). *None of your business: World data flows, electronic commerce, and the European privacy directive.* Washington, DC: Brookings Institution Press.

Ulijn, J. M. (1996). Translating the culture of technical documents: Some experimental evidence. In D. C. Andrews (Ed.), *International dimensions of technical communication* (pp. 69-86), Arlington, VA: Society for Technical Communication.

Ulijn, J. M., & St.Amant, K. (2001). Mutual intercultural perception: How does it affect technical communication—some data from China, The Netherlands, Germany, France, and Italy. *Technical Communication, 47,* 220-237.

Varner, I., & Beamer, L. (2004). *Intercultural communication in the global workplace.* Boston: McGraw-Hill/ Irwin.

Warschauer, M. (2003). *Technology and social inclusion: Rethinking the digital divide.* Cambridge, MA: MIT Press.

Weiss, S. E. (1998). Negotiating with foreign business persons: An introduction for Americans with propositions on six cultures. In S. Niemeier, C. P. Campbell, & R. Dirven (Eds.), *The cultural context in business communication* (pp. 51-118). Philadelphia: John Benjamins.

Wurtz, E. (2005). A cross-cultural analysis of Web sites from high-context cultures and low-context cultures. *Journal of Computer-Mediated Communication, 11.* Retrieved from http://jcmc.indiana.edu/vol11/issue1/wuertz.html

Yunker, J. (2003). *Beyond borders: Web globalization strategies.* Boston: New Riders.

Kirk St.Amant

Editor

Texas Tech University, USA

Acknowledgments

I would like to thank all involved in the creation of this collection, including the reviewers who dedicated their time and expertise to this project and the editorial production staff at Idea Group Inc., and Kristin Roth in particular, for their assistance and professionalism throughout this project.

My thanks also goes out to all of the individuals who contributed to this collection; their intelligence, insights, and commitment to examining culture and online communication are greatly appreciated.

Finally, I wish to thank my daughter, Lily Catherine St.Amant, for being a continual source of inspiration in all that I do, my wife, Dori St.Amant, for her unwavering patience and understanding during this project, and my grandparents, Albert and Lily St.Amant, for teaching me the importance of culture and language.

Kirk St.Amant

Lubbock, Texas, USA
September 2006

Section I
Language, Culture, and Cyberspace

Chapter I
Internet Multilinguality:
Challenges, Dimensions, and Recommendations

Shaoyi He
California State University San Marcos, USA

ABSTRACT

In the hay-days of the Internet boom, most software packages used on the Internet were in English, as were most Web sites and search engines. Since 2000, however, there have been more non-English-speaking users than English-speaking users on the Internet. Besides this language issue, there are other important issues involved in multilinguality on the Internet: cultural, technological, political, and legal issues. This chapter examines three challenges of Internet multilinguality: language barriers, cultural differences, and technological difficulties. This chapter also provides three recommendations for overcoming barriers created by language, culture, and technology.

INTRODUCTION

Since its creation in the United States in 1969, the Internet has evolved from the then monolingual DARPANET into a World Wide Web, with an increasing multilinguality of more than 1,000 languages (Crystal, 2001a). In the hay-days of the Internet boom, most software packages used on the Internet were in English, the vast majority of Web sites were in English, and search engines were also mostly in English (Crystal, 2001b). Nowadays, multilinguality has become an important aspect of the Internet, which would have an indispensable impact on anyone that is involved in online communication over the Internet and the Web. As the result, the Internet has made it possible for Web users to access more information in various languages more efficiently than ever before. Since 2000, there have been more non-English-speaking users than English-speaking users on the Internet (Global Reach, 2004). This trend towards multilingualization of the Internet is not surprising at all, because 92% of the world's population speaks a primary language other than English. Besides the language issue, there are other important issues involved in multilinguality on the Internet: cultural, technological, political, and legal issues (Hillier, 2003; Hopkins, 2000; Yunker, 2003). This chapter will first address three challenges of Internet multilinguality—language barriers, cultural differences, and technological difficulties—and then address three vital dimensions of Internet multilinguality—multilingual Web sites, multilingual domain names, and multilingual search engines. Also, this chapter will outline three recommendations to overcome the barriers that are created by language, culture, and technology: (1) Web site glocalization; (2) multilingual search engines, and (3) multilingual domain name. In addition, this chapter will outline the future trends for Internet multilinguality, because the triangular relationship among language, culture, and technology would continue to create new challenges for multilingual online communication.

BACKGROUND

Although it is very difficult to know the exact number of languages spoken on earth, there are 6,912 languages cataloged in the 15th edition of *Ethnologue: Languages of the World*, an encyclopedic reference work that tries to organize all of the world's known living languages by continent and country (Ethnologue, 2005; Language Facts, 2006). It is estimated that 96% of these languages (around 6,600) are spoken by a mere 4% of the world's population, and the eight most popular languages are spoken by more than half of the world's population. These eight languages are English, Mandarin Chinese, Hindi (with Urdu), Spanish, Russian, Arabic and Bengali, and Portuguese (National Virtual Translation Center, 2006; UNESCO WebWorld, 2006). English as the most widely spoken and written language with some 300-400 million native speakers is also the most widely learned second language with 1.9 billion speakers, nearly a third of the world's population (Wikipedia, 2006). English was also a dominant language on the Internet around 1997 when 80-85% of the Web sites were in English (Jenkins, 1997; Wallraff, 2000). In 1999, 96% of e-commerce Web sites in the *.com* domain were in English (The Default Language, 1999). The dominance of English on the Internet and the Web came to an end in 2000 when the number of non-English-speaking Internet users surpassed that of English-speaking Internet users (Global Reach, 2004). In 2001, the State of the Internet Report produced by the U.S. Internet Council and the International Technology & Trade Associates (ITTA) found that native-English speakers represented approximately 45% of the online population. As a matter of fact, more and more non-English-language Web sites were developed to meet the rising needs of the majority of the non-English-speaking customers on the Internet. For example, the Yahoo! France and Yahoo! Japan Web directories have been credited as the early successful efforts that resulted in the increasing of the numbers of French and Japanese-speaking Internet users (Jenkins, 1997). At the end of 2002, more then half of America's 100 largest companies had developed their multilingual Web sites (Tongues of the World, 2002). Since then, major companies have been moving steadily to multilingualizing their Web sites. For instance, Siemens, which sells products in more than 190 countries, has more than three dozen Web sites in more than 40 languages. DaimlerChrysler sells in more than 200 countries, with 28 Web sites in more than 20 languages. Furthermore, at the end of 2005, the English-speaking Internet population dropped to 30.6%, while the non-English-speaking Internet population increased to 69.4% (Internet World Stats, 2006). Besides English, the other most popular languages on the Internet are Chinese (13.0%), Japanese (8.5%), Spanish (6.3%), German (5.6%), French (4.0%), Korean (3.3%), Portuguese (3.2%), Italian (2.8%), and Russian (2.3%). Obviously, nearly half (49%) of all Internet users are from a small number of non-English-speaking countries in East Asia and Europe, and many of them might be limited to the native language for accessing information on the Internet (Internet World Stats, 2006). Such a rapid increase in the number of non-English-speaking Internet users strongly implies that multilinguality is becoming more and more crucial and indispensable in a marketplace that called for a good practice in computer-mediated communication on the Internet and the Web (Appleby, 2003; Perrault & Gregory, 2000; Seilheimer, 2004; Starr, 2005).

Multilinguality was early defined as "the possibility to use different languages according to different countries' needs" when the project of DATAMED (DATA capturing in MEDiterranean countries) was carried out within the European Strategic Program for Research and Development in Information Technologies (Commission of the European Communities, 1998). Another concise definition was provided by Kay (1998): "Multilinguality is a characteristic of tasks that involve the use of more than one natural language" (p. 245). He then immediately pointed out: "In working environments where more than one language is in use, the problem of storing and retrieving information acquires a multilingual dimension" (Kay, 1998, p. 245). The following issues related to multilinguality were defined by Picchi and Peters (1997) for the purpose of analyzing the topics of multilingual language recognition and representation, and multilingual search and retrieval:

- **Internationalization:** Enabling worldwide communication, no matter what the language.
- **Localization:** Adapting to local needs.
- **Multilingual digital library:** Containing documents in more than one language.
- **Multilingual document:** Containing text in more than one language.
- **Cross-language retrieval:** Retrieving any type of text composed or indexed in one language via a query formulated in another language.

Besides the above issues that represent some vital dimensions of multilinguality on the Internet, there are some issues representing other important dimensions of Internet multilinguality, as defined below from *Wikipedia, the Free Encyclopedia* (http://en.wikipedia.org/wiki/Wikipedia):

- **Multilingualization:** The practice of adapting a product for multiple languages (http://en.wiktionary.org/wiki/multilingualization).
- **Globalization:** The worldwide phenomenon of technological, economic, political, and cultural exchanges, brought about by modern communication, transportation, and legal infrastructure as well as the political choice to consciously open cross-border links in international trade and finance (http://en.wikipedia.org/wiki/Globalization).
- **Glocalization:** The creation or distribution of products or services intended for a global or trans-regional market, but customized to suit local laws or culture (http://en.wikipedia.org/wiki/Glocalization).
- **Machine translation:** A sub-field of computational linguistics that investigates the use of computer software to translate text or speech between natural languages (http://en.wikipedia.org/wiki/Machine_Translation).
- **Search engine optimization:** A set of methods aimed at improving the ranking of a Web site in search engine listings (http://en.wikipedia.org/wiki/Search_Engine_Optimization).
- **Internationalized domain name (IDN):** An Internet domain name that (potentially) contains non-ASCII characters. Such domain names could contain letters with diacritics, as required by many European languages,

or characters from non-Latin scripts such as Arabic or Chinese (http://en.wikipedia.org/wiki/Internationalized_Domain_Name).

- **Language technology:** Is often called human language technology (HLT) and consists of computational linguistics (or CL) and speech technology as its core, but includes also many application-oriented aspects of them (http://en.wikipedia.org/wiki/Language_Technology).

In the following sections, firstly, three main challenges in Internet multilinguality will be identified for online communication. Secondly, three important dimensions of Internet multilinguality will be discussed with examples. Finally, three practical recommendations for Internet multilinguality will be outlined for further consideration.

INTERNET MULTILINGUALITY: THREE MAIN CHALLENGES

Challenges for multilinguality on the Internet are mainly from three sources: language, culture, and technology, which have been observed and studied by numerous scholars (Appleby, 2003; He, 1999; Hillier, 2003; Pecchi & Peters, 1997; Schneider, 2005; Starr, 2005; Wallraff, 2000; Woldering, 2006). The following is a brief description of these challenges.

Challenge of Language Barriers

The language barriers to online communication not only refer to the historical dominance of the English language, but also refer to the presently increasing multilinguality on the Internet for global electronic commerce (Crystal, 2001b; He, 2005). The Web sites that are only in English are already a natural communication barrier for non-English-speaking Internet users, not to mention all the other Web sites that are developed in various local languages. This has contributed to an information accessibility gap between English-speaking and non-English-speaking Internet users. Many researchers regarded language barriers as being one of the major problems for searching, accessing, and retrieving multilingual information and knowledge on the Internet, and looked at the role that language played in creating difficulties for online communication as well as the impact of language diversity on the development and maintenance of multilingual Web sites (Berner, 2003; Cheon, 2005; MacLeod, 2000; Perrault & Gregory, 2000; Seilheimer, 2004; Spethman, 2003). Furthermore, the language barriers would also create the problem of multilingual searching on the Web. Bilingual or multilingual Internet users often found it difficult to choose the appropriate language for their search. Some users enter queries in their native language and prefer to retrieve documents in a foreign language, while others prefer the opposite. Some users enter queries in one language but want to retrieve the documents in multiple languages (Rieh & Rieh, 2005). The emergence of multilinguality on the Internet not only bought in the challenges and technologies regarding language barriers, but also called for collective efforts to overcome language barriers, thus enabling all the users from different countries and regions to effectively browse, navigate, access, filter, process, and retrieve multilingual information on the Internet and the Web (Appleby, 2003; He, 2005; Yunker, 2003; Tixier, 2005).

Challenge of Cultural Differences

The cultural differences may cause misinterpretation in online communication. Something that is appropriate and acceptable in one culture may not be appropriate and acceptable in another culture. Therefore, on a multilingual Web site, the use of symbols, icons, images, colors, values, rituals, idioms, metaphors, salutations, geography, and time zones needs be appropriately chosen according to cultural contexts, thus avoiding being misunderstood or offensive (Hillier, 2003; Singh, Zhao, & Hu, 2005; Tsikriktsis, 2002). An article published in *Forbes* (February 21, 2000) entitled "Cultural Web Faux Pas" reported that "it doesn't matter that your site sells diapers or offers tips on American quilting patterns—disregard cultural differences and your site will be doomed." Another article entitled "Are We All Speaking the Same Language?", published in *The Times* of London (April 20, 2000), pointed out that "many companies that elect to rely purely on a Web site to trade internationally are failing to take linguistic and cultural differences sufficiently into account." Therefore, it is very important to have a better understanding of the cultural differences when developing a multilingual Web site. What is user friendly on the Web site for visitors in one culture may not be as user friendly for visitors in another culture. Although the Web site interfaces targeted to a different audience in different countries or different cultures may not be necessarily completely different from each other, there must be some features that allow the targeted audience to feel comfortable when visiting the Web site. The cultural markers that signify a cultural affiliation (e.g., the native language, a national symbol, or color) would greatly help localizing a global Web site. For example, CNN uses cultural icons such as national flags to attract people from different countries to visit its Web site (Chau et al., 2002).

Challenge of Technological Difficulties

The technological difficulties mainly refer to the problems with character sets—there is no ASCII analogue for the non-Latin characters used by the Chinese, Japanese, Korean, Arabic, Hebrew, and other non-English languages. Take Chinese for example. It is difficult for users of Chinese to browse, navigate, filter, and process the Chinese information on the Web because the Internet protocols, technologies, and applications are English oriented and not Chinese friendly. There are two major difficulties in searching Chinese information on the Internet: (1) the internal coding schemes for processing Chinese language are different from those for processing English language on the computers, and (2) there are incompatibilities between the character set coding schemes used in Mainland China, Hong Kong, Singapore, and Taiwan. Firstly, the Chinese language uses the double byte character set (DBCS) while English language uses the single byte character set (SBCS). Because of this difference, most Internet applications are incapable of identifying DBCS encoding, thus causing problems for Chinese information retrieval on the Web. An English browser or search engine must either have a built-in or add-on capability to recognize Chinese coding schemes in order to correctly display Chinese characters (Lin, 1997; Liu, 1997). Secondly, the Web sites created in Mainland China and Singapore are usually in GB code, the popular coding scheme for the simplified Chinese characters, while the Web sites created in Hong Kong and Taiwan are usually in BIG5 code, the popular coding scheme for the traditional Chinese characters. If a browser or search engine has inappropriately selected the GB code for a BIG5-coded Web page, it would not correctly display the traditional Chinese characters. Also, due to the complexity of encoding and decoding Chinese characters, it is likely that some Chinese characters would still be incorrectly displayed, even if the correct coding scheme is used (He, 1999). In recent years, Unicode, as a new character set coding scheme, has been implemented to include both of the simplified Chinese and traditional Chinese characters, thus reducing problems caused by compiling and translating Chinese characters coded in different character sets (Emerson, 2000).

INTERNET MULTILINGUALITY: THREE MAJOR DIMENSIONS

Although there are multiple dimensions of multilinguality on the Internet, the sections below only focus on three of them that are closely related to the challenges from language, culture, and technology.

Multilingual Web Sites

A multilingual Web site provides the same information in different languages. The first multilingual Web site in Europe, EuroNews, was launched January 1, 1993, to simultaneously cover world news from a European perspective in seven languages: English, French, German, Italian, Portuguese, Russian, and Spanish. (EuroNews, 2005). A typical multilingual Web site should contain a mixture of global content and local content. The global content need be presented in many languages for customers all over the world, and usually covers product information, technical support documents, tutorials, corporate profiles, worldwide branding messages, and the design of the Web itself. The local content has to be written for each target language audience, and usually includes locally available products, local promotions, sales and advertising campaigns, and local points-of-purchase indices. While the global content is applicable everywhere and is relatively insensitive to national or cultural differences, the local content provides the most relevant information to convince the users that the Web site fits in their culture (Hopkins, 2000; Schneider, 2005; Yunker, 2003).

A well-designed multilingual Web site for a given audience would adequately meet their information needs. For instance, a Chinese version of a corporate Web site will be helpful to promote information relevant to a Chinese audience. It would also help to overcome potential cultural barriers when users are able to navigate, understand, and interact on the Web site in the native language. According to WorldLingo, a leading innovator and provider of translation and localization services and technology, Web users will not only stay twice as long, but will also be up to four times more likely to purchase from a multilingual Web site that communicates in their native language (WorldLingo, 2006). In addition, a properly designed multilingual Web site has the potential for an increase in sales. If an e-commerce Web site is translated into a few of the major world languages (i.e., Spanish, French, German, and Italian), there is potentially a 400% increase in sales. Multilingual Web sites will continue to remain a necessity for businesses and organizations that are engaged in global e-commerce (Hillier, 2003; Payne, 2005; Schneider, 2005; Starr, 2005).

Multilingual Domain Names

Multilingual domain names, also known as internationalized domain names (IDNs), are Web addresses in non-ASCII character-based languages, such as Chinese, Korean, Japanese, Arabic, and Russian. Traditionally, domain names were written in English, which is an ASCII-based language (HostLibrary, 2004). A non-English-speaking user of the Internet had to type an English domain name in order to access a Web site. So, there have been complaints about the English-only domain name systems (DNSs), calling for the use of multilingual characters to represent Web addresses. Efforts to upgrade the monolingual domain name systems to support non-ASCII characters began when there were public technical discussions of internationalized top-level domain names (TLDs) in the domain name system (DNS) in 1996. The formation of the Multilingual Internet Names Consortium (MINC) in 2000 made it possible to promote domain name internationalization to include multilingual and multiscript support for Web addresses (Wee, 2002). With the availability in more than 350 languages, IDNs would enable Internet users around the world to overcome language barriers in using non-ASCII characters from the desired language in Web addresses. In November 2000, Chinese characters could be used to register .com, .org, and .net domain names at the Web site of VeriSign. In November 2001, multilingual domain names were available for sale on major domain name registrars' Web sites (Boyarsk et al., 2002). With a rapid Internet penetration in non-English-speaking countries, more and more domain names in Chinese, Japanese, Korean, Vietnamese, Arabic, Hebrew, and Russian have been registered in many domain name registrars. For example, 19% of the .com and .net Web addresses registered in Korea were IDN domain names at the end of 2003 (HostLibrary, 2004).

The significant development of IDNs not only enabled the support of multilingual domain names, but also brought up some interesting issues that need to be addressed in terms of the implementation. First, many non-ASCII character-based languages have common characters that may be represented differently after conversion to ASCII. Chinese, Japanese, and Korean all have similar characters, as do Latin, Cyrillic, and Greek (Wielansky, 2002). Second, there exist more complicated aspects of IDN support for some special languages, such as dealing with simplified and traditional versions of the same Chinese characters, and handling bidirectional text in Arabic or Hebrew, which may need some additional time for full implementation (Ishida, 2006; Wee, 2002; Wu, 2003). Third, the domain name of a global brand has to be registered in as many languages as possible for avoiding potential cybersquatting. For example, Coca-Cola as a global brand involving multiple languages may need to be registered in all of these languages to prevent future cybersquatting. Such a practice could put a burden in time and cost on those organizations that have already spent tremendous time and effort developing an online presence (Wielansky, 2002). In addition, the registered IDN domain names may be unevenly distributed in recent years. According to statistics from two sources, HostLibrary.com and webhosting.info, at least 74% of the IDN domain registration was done by companies or organizations in three countries: the United States, Korea, and Japan (HostLibrary, 2004; Cheon, 2005).

Multilingual Search Engines

In order to meet the increasing needs in accessing multilingual information on the Internet, a number of Web search engines (e.g., Google and AlltheWeb) have been offering search capabilities in different languages, so that users are able to restrict their search to Web sites in one or several pre-selected language(s). As early as in 1997, a multilingual search engine, called TITAN, was developed. Based on a bilingual dictionary, it allowed translation of queries from Japanese to English and English to Japanese. TITAN helps Japanese users to search the Web using their own language (Hayashi et al., 1997). MULINEX is a multilingual search engine that extracts information about documents by making use of language identification, thematic classification, and automatic summarization (Capstick et al., 1998). The ITEM multilingual search engine permits querying and retrieving documents in three languages (Spanish, Catalan, and English); the user is allowed to select from a number of Natural Language Processing options, and to refine the results of such lexical processing (Verdejo et al., 2000). AltaVista, Excite, and Lycos are in the first group of search engines with multilingual capabilities. AltaVista (http://www.altavista.com) provided the first multilingual search capabilities in June 1997, and now has search capability in 36 languages. Lycos (http://www.lycos.com) began its international penetration in May 1997, and can search 10 European languages and two Asian languages. Excite (http://www.excite.com) created an international division in April 1997, and now searches nine European languages and two Asian languages.

Figure 1. Search engines and languages supported

Search Engine	Languages
AllTheWeb (http://www.alltheweb.com)	44
Yahoo! (http://www.yahoo.com)	37
AltaVista (http://www.altavista.com)	36
Google (http://www.google.com)	35
MSN Search (http://search.msn.com)	25

As a matter of fact, Web search engines with multilingual capabilities have become indispensable for accessing multilingual information on the Internet. Although the number of languages supported vary between different search engines, there are a number of major search engines, namely *AltaVista*, *AllTheWeb*, *Ask Jeeves*, *EuroSeek*, *Google*, *Excite*, *HotBot*, *Lycos*, *MSN Search*, and *Yahoo!*, that altogether have search capabilities in more than 40 languages, such as Chinese, Czech, Danish, Dutch, Estonian, French, Finnish, German, Greek, Italian, Japanese, Korean, Latvian, Lithuanian, Norwegian, Portuguese, Spanish, and Swedish (Sullivan, 2004). Five popular search engines with the number of languages they support can be seen in Figure 1.

Besides the coverage of languages, multilingual search engines should also have the following capabilities that were identified by Stepanov (2000):

- Search materials in different languages.
- Retrieve documents in defined language only.
- Choose interface in desirable language.
- Work correctly with multicoding languages.
- Translate query.
- Translate search results.
- Translate document itself.

Multilingual capabilities of search engines are always improving because the need for multilingual and cross-language search on the Internet will be permanently increasing.

INTERNET MULTILINGUALITY: THREE PRACTICAL RECOMMENDATIONS

Based on the above main challenges and important dimensions that are identified for multilinguality on the Internet, three practical recommendations are outlined as follows.

Web Site Glocalization

There have been different approaches to develop and implement a multilingual Web site, and two have been historically adopted for many of the Web sites. One is Web site globalization and the other Web site localization. The former globalizes the content for broadening a businesses appeal to a general international audience, and the latter localizes the content to appeal to customers of a particular nation or culture (Cyr & Trevor-Smith, 2004; Knoppers, 1998; Seilheimer, 2004; Spethman, 2003; Tixier, 2005). As the combination of globalization and localization, glocalization is "a term that was invented in order to emphasize that the globalization of a product is more likely to succeed when the product or service is adapted specifically to each locality or culture it is marketed in" (Whatis.com, 2005). Glocalization is a process of adopting specific elements from other cultures without losing the original identity, and blending the ideas, brands, and practices from different cultures in such a balanced way that one is not overwhelmed by the other. For instance, glocalization has led to the emergence of phenomena such as world music, gourmet cooking, and ethnic body adornment (Shen, Woolley, & Prior, 2006). According to Tixier (2005): "Glocalization is the implementation of the constraints of globalization in accordance with the constraints of the local environment and requested by the final user" (p. 16).

Here, glocalization is regarded as the third path, besides globalization and localization, to develop major Internet Web sites for various sectors of industry in the United States and Europe. While the Web sites' structure could remain the same, the content of the Web sites could be adapted to the cultural, political, legal, behavioral, and technical aspects in the different markets or countries (Tixier, 2005).

There are many examples of glocalization. For example, McDonald's in China have changed the menu to serve rice to local Chinese customers, and McDonald's in France have replaced the familiar Ronald McDonald mascot with Asterix the Gaul, a popular French cartoon character. When applied to global e-commerce, Web sites glocalization means thinking global and acting local. In a study on the Chinese Web sites of the 100 top global brands, Maynard and Tian (2004) found that a glocal strategy, as opposed to a standardized global strategy, was being practiced in cyberspace by many of the companies with top brands. These companies glocalized their Chinese Web pages by integrating local cultural characteristics into their brands' strategies, thus presenting themselves as being socially accommodating to the local market.

Web-Based Machine Translation

Machine translation systems connected by the Internet may allow cross-language communication with anyone, from anywhere, and at anytime. The Web-based translation makes it possible to instantly interpret regional dialects and idiomatic expressions in various languages. With the machine translation systems, the products and services can be obtained or provided in a local language, thus helping international business and markets to grow, and allowing local cultures and languages to thrive (Lehman-Wilzig, 2001).

There are various companies that provide software and service for Web-based machine translation. The following are some of the more popular ones:

- **Babel Fish:** (http://babelfish.altavista.com) Provided by AltaVista, Babel Fish enables the translation of short passages and Web sites to and from English in 10 languages.
- **FreeTranslation.com:** (http://www.freetranslation.com) Powered by SDL's Enterprise Translation Server, FreeTranslation offers free translations of text or Web pages in one of seven languages.
- **IBM's alphaWorks:** (http://www.alphaworks.ibm.com/aw.nsf/html/mt) A machine translation demo lets you get the gist of your favorite Web pages in one of seven languages.
- **InterTran:** (http://www.tranexp.com/intertran) A free Web translation service that can translate single words, phrases, sentences, and entire Web pages in one of 21 languages.
- **SYSTRAN:** (http://www.systransoft.com) Founded in 1968, SYSTRAN develops leading MT technologies and provides free online translation in one of five languages.
- **TRADOS:** (http://www.trados.com) A suite of applications that enable users to master the many facets of translation work, and address the requirements of multiple frontends, terminology management, alignment of legacy translations, and project management. It supports more than 60 languages.

Human Language Technology

The rapid growth of the Internet and the Web poses new challenges to human language technology, while the increasing multilinguality of the Internet constitutes an additional challenge for online communication. Therefore, we need human language technology to enable people to communicate with machines using natural communication and overcome the language barriers for e-commerce, education, and international cooperation. Human language technology enables multilingual support of global e-commerce Web sites that goes beyond Web site localization or text translation, improving Web site usability in terms of natural languages. Such a support implements cross-language information retrieval technologies that are able to handle searches made in the user's local language across knowledge bases coded in any number of other languages. The language technologies on HLT-enabled Web sites cover applications of multilingual user interfaces and services, multilingual content management, multilingual trading, and multilingual customer relationship management, hence making multilingual Web sites more competitive in a global marketplace (Khan, 2002; Uszkoreit, 2006).

In addition, Bodomo (2005) and Khan (2002) identified some major problems imposed by HLT that have to be addressed as follows:

- Exchanging data in different languages
- Supporting different languages in the user-interface
- Retrieving multilingual information and documents
- Ensuring consistency in terminology, and in writing and translation style
- Supporting the character sets and encoding used to represent the information
- Providing localization of user interfaces
- Supporting cross-language information retrieval systems
- Providing language identification techniques
- Supporting multilingual text generation and summarization
- Providing Web-based tools: Web-based translators, search engines, multilingual dictionaries

As more and more massive amounts of multilingual information becomes available on the Internet and the Web, it is becoming increasingly crucial that human language technologies are appropriately used for achieving global communication in a multilingual multilinguality information society (Bodomo, 2005).

FUTURE TRENDS

In the information age represented by the Internet, multilingual Web sites have become a common ground for online communication for peoples across national boundaries. On the Web, users are 70% more likely to stay on a Web site that speaks their native languages. Also, their willingness to buy something online increases by at least 40% if the Web site is localized (Vogt, 2003). Every year in the next 10 years or longer, there will be more than 100 million non-English-speaking people going onto the Internet and the Web (Mosquera, 2001). With such a steady growing online population, Internet multilinguality would remain an important aspect of the Internet and the Web, just as pointed out by Crystal (2001): "The future looks good for Web multilingualism." As mentioned before, the triangular relationship among language, technology, and culture would continue to create new challenges for global e-commerce; therefore, we need to make sure that the Internet as a multilingual community will allow every individual, every culture and every language, and every technology to contribute to and benefit from a multilingual online world.

CONCLUSION

Internet multilinguality as an important topic involves issues of language, culture, technology, and law "that could markedly affect international interactions in online environments" (St.Amant, 2005). This chapter has provided an in-depth discussion on some interesting aspects of Internet multilinguality, focusing on three main challenges and three major dimensions. Three practical recommendations were also discussed in detail. All these would lead to a better understanding of multilinguality on the Internet, so that: (1) the obstacles created by language barriers could be removed, (2) the misunderstanding brought by cultural differences would be avoided, and (3) the problems caused by technological difficulties could be solved.

REFERENCES

Appleby, S. (2003). Multilingual information technology. *BT Technology Journal, 21*(1), 75-83.

Bodomo, A. (2005). *Human language technology in multilingual perspectives: Institutions and applications.* Retrieved January 19, 2007, from http://www.idea-group.com/files/prefaces/ijthi%20preface%202(1).pdf

Berner, S. (2003). "Lost in translation": Cross-lingual communication, and virtual academic communities. In *Proceedings of the 5th Annual Conference on World Wide Web Applications.* Retrieved January 19, 2007, from http://general.rau.ac.za/infosci/www2003/Papers/Berner,%20S%20Lost%20in%20Translation.pdf

Boyarsk, J., et al. (2002). Seminar addresses multilingual domain names. *Intellectual Property & Technology Law Journal, 14*(3), 32.

Capstick, J., et al. (1998, August 18-21). Mulinex: Multilingual Web search and navigation. In *Proceedings of the International Conference on Natural Language Processing and Industrial Applications (NLP+IA'98),* Moncton, New-Brunswick, Canada.

Chau, P., et al. (2002). Cultural differences in the online behavior of consumers. *Communications of the ACM, 45*(10), 138-143.

Cheon, K. (2005). Multilingualism and the domain name system. In *Reforming Internet governance: Perspectives from the Working Group on Internet governance* (pp. 67-72). Retrieved January 19, 2007, from http://www.wgig.org/docs/book/Kangsik_Cheon%20.pdf

Commission of the European Communities. (1998). *Report on users' requirements.* Retrieved January 19, 2007, from http://petral.istat.it/datamed/docs/man/d8d0055o.doc

Crystal, D. (2001a, January 25). Weaving a web of linguistic diversity. *Guardian Weekly.* Retrieved January 19, 2007, from http://www.guardian.co.uk/GWeekly/Story/0,3939,427939,00.html

Crystal, D. (2001b). *Language and the Internet.* Cambridge, UK: Cambridge University Press.

Cyr, D., & Trevor-Smith, H. (2004). Localization of Web design: An empirical comparison of German, Japanese, and United States Web site characteristics. *Journal of the American Society for Information Science and Technology, 55*(13), 1199-1208.

Default Language, The. (1999, May 15). *The Economist*, 67.

Emerson, T. (2000, March). Segmenting Chinese in Unicode. In *Proceedings of the 16th International Unicode Conference,* Amsterdam, The Netherlands. Retrieved January 19, 2007, from www.basistech.co.jp/knowledge-center/chinese/iuc-16-paper.pdf

Ethnologue. (2005). *Languages of the world.* Retrieved January 19, 2007, from http://www.ethnologue.com

EuroNews. (2005). *Many voices one vision.* Retrieved January 19, 2007, from http://www.euronews.net/feedback/en/1Channel.pdf

Global Reach. (2004). *Evolution of online populations.* Retrieved January 19, 2007, from http://global-reach.biz/globstats/evol.html

He,S. (2001). Interplay of language and culture in global e-commerce: a comparison of five companies' multilingual websites. In S. Tilley (Ed.), *Communicating in the New Millennium, Proceedings of the 19th Annual International Conference on Systems Documentation* (pp. 83-88). New York: The Association for Computing Machinery.

He, S. (2005). Multilingual Web sites in global electronic commerce. In M. Khosrow-Pour (Ed.), *Encyclopedia of e-commerce, e-government and mobile commerce* (pp. 823-827). Hershey, PA: Idea Group Inc.

He, S. (1999). A comparative study of three search engines for Chinese information retrieval on the Internet. In *Proceedings of the 20th National Online Meeting* (pp. 149-160). Medford, NJ: Information Today.

Hillier, M. (2003). The role of cultural context in multilingual Web site usability. *Electronic Commerce Research and Applications, 2*(1), 2-14.

HostLibrary. (2004). *Special report: Distribution analysis of multilingual domain names (IDNs).* Retrieved January 19, 2007, from http://www.hostlibrary.com/node/view/178

Hayashi, Y., et al. (1997). TITAN: A Cross-linguistic search engine for the WWW. In *Working Notes of AAAI Spring Symposium on Cross-Language Text and Speech Retrieval,* Stanford, CA (pp. 58-65).

Hopkins, R. (2000). *Multilingual Web sites: Benefits you can count on, headaches you can avoid.* Retrieved January 19, 2007, from http://www.weblations.com/eng/articles/art_2.htm

Internet World Stats. (2006). *Top ten languages used in the Web (number of Internet users by language).* Retrieved January 19, 2007, from http://www.internetworldstats.com/stats7.htm

Ishida R. (2006). *An introduction to multilingual Web addresses.* Retrieved January 19, 2007, from http://www.w3.org/International/articles/idn-and-iri/

Jenkins, M. (1997, July 8). The Web learns to speak in tongues. *Independent,* 4.

Kay, M. (1998). Multilinguality: Overview. In R. Cole et al. (Eds.), *Survey of the state of the art in human language technology* (pp. 229-265). Cambridge, UK: Cambridge University Press.

Khan, A. (2002). *Multi-linguality support for a maritime online portal.* Retrieved January 19, 2007, from http://www.sts.tu-harburg.de/people/pa.hupe/studenttheses/KhanSP.pdf

Knoppers, J. V. T. (1998). Global electronic commerce through localization and multilingualism. *Computer Standards and Interfaces, 20*(2), 101-109.

Language Facts. (2006). *Languages of the world.* Retrieved January 19, 2007, from http://www.ling.gu.se/projekt/sprakfrageladan/english/sprakfakta/eng-sprak-i-varlden.html

Lehman-Wilzig, S. (2001). Babbling our way to a new Babel: Erasing the language barriers. *The Futurist, 35*(3), 16-23.

Lin, Z. (1997). How to use CJK software to read Chinese, Japanese, and Korean on the Internet. *Computers in Libraries, 17*(11/12), 50-54.

Liu, N. (1997). Using the Internet to share information between China and the West. *Inspel, 32*(2), 103-109.

National Virtual Translation Center. (2006). *The range of world languages.* Retrieved January 19, 2007, from http://www.nvtc.gov/lotw/months/november/worldlanguages.htm

MacLeod, M. (2000). Language barriers. *Supply Management, 5*(14), 37-38.

Maynard, M., & Tian, Y. (2004). Between global and glocal: Content analysis of the Chinese Web sites of the 100 top global brands. *Public Relations Review, 30*(3), 285-291.

McKinsey, K. (2001). The mother of all tongues. *Far Eastern Economic Review, 164*(15), 38-39.

Mosquera, M. (2001). *Global e-commerce to hit $5 trillion in 2005.* Retrieved January 19, 2007, from http://www.techweb.com/wire/29105703

Payne, N. (2005). *10 reasons why you need a multilingual Web site.* Retrieved January 19, 2007, from http://www.workoninternet.com/article_4407.html

Pecchi, E., & Peters, C. (1997, May). Across languages and across cultures: Issues in multilinguality and digital libraries. *D-Lib Magazine.* Retrieved January 19, 2007, from http://www.dlib.org/dlib/may97/peters/05peters.html

Perrault, A. H., & Gregory, V. L. (2000). Think global, act local: The challenges of taking the Web site global. *International Journal of Special Libraries (INSPEL), 34*(3/4), 227-237.

Rieh, H.-Y., & Rieh, S.-Y. (2005). Web searching across languages: Preference and behavior of bilingual academic users in Korea. *Library & Information Science Research, 27*(2), 249-263.

Schneider, G. (2005). *Electronic commerce.* Boston: Course Technology.

Sullivan, D. (2004). *Major search engines and directories.* Retrieved from http://searchenginewatch.com/links/article.php/2156221

Seilheimer, S. (2004). Productive development of World Wide Web sites intended for international use. *International Journal of Information Management, 24,* 363-373.

Shen, S.-T., Woolley, M., & Prior, S. (2006). Towards culture-centered design. *Interacting with Computers, 19*(2), 1-33.

Singh, N., Zhao, H., & Hu, X. (2005). Analyzing the cultural content of Web sites: A cross-national comparison of China, India, Japan, and U.S. *International Marketing Review, 22*(2), 129-146.

Spethman, M. (2003). Web site globalization. *World Trade, 16*(11), 56-57.

St. Amant, K. (2005). E-mail call for chapters for the book, *Linguistic and Cultural Online Communication Issues in the Global Age.*

Stepanov, V. (2000, March 22). Multilingual capabilities of search engines. *Internet Librarian International 2000,* London. Retrieved January 19, 2007, from http://www.infotoday.com/ili2000/presentations/stepanov.doc

Starr, J. (2005). Design considerations for multilingual Web sites. *Information Technology and Libraries,* 107-116.

Tixier, M. (2005). Globalization and localization of contents: Evolution of major Internet sites across sectors of industry. *Thunderbird International Business Review, 47*(1), 15-48.

Tongues of the World. (2002, March 14). *The Economist.*

Tsikriktsis, N. (2002). Does culture influence Web site quality expectations? An empirical study. *Journal of Service Research, 5*(2), 101-112.

UNESCO WebWorld. (2006). *Multilingualism in cyberspace.* Retrieved January 19, 2007, from http://portal.unesco.org/ci/en/ev.php-URL_ID=16917&URL_DO=DO_TOPIC&URL_SECTION=201.html

Uszkoreit, H. (2006). *Language technology: A first overview.* Retrieved January 19, 2007, from http://www.dfki.de/~hansu/LT.pdf

Verdejo, et al. (2000). *Evaluating Wordnets in cross-language information retrieval: The ITEM search engine.* Retrieved January 19, 2007, from http://nlp.uned.es/pergamus/pubs/lrec00.pdf

Vogt, S. (2003). Going native to get global. *Information World Review, 191,* 18-19.

Wallraff, B. (2000). What global language? *The Atlantic Monthly, 286*(5), 52-66.

Wee, T. (2002, September 30). A historical background, current activities and issues of internationalized domain names. In *Proceedings of the JDNA-MINC Joint Meeting,* Tokyo, Japan. Retrieved January 19, 2007, from http://www.jdna.jp/event/020930-workshop/JDNA-speech1.pdf

Whatis.com. (2005). *Glocalization.* Retrieved January 19, 2007, from http://searchcio.techtarget.com/sDefinition/0,,sid19_gci826478,00.html

Wielansky, M. (2002). Internationalized domain names. *Journal of Adolescent & Adult Literacy, 45*(4), 302-308.

Wikipedia. (2006). *English language.* Retrieved January 19, 2007, from http://en.wikipedia.org/wiki/English_language

Wilson, C. (2005, April). Expanding the domain name system (DNS) with internationalised domain names (IDNs)—a victory for the internationalisation of the Internet or not? In *Proceedings of the 20th BILETA Conference: Over-Commoditised; Over-Centralised; Over-Observed: The New Digital Legal World?* Queen's University of Belfast, Ireland. Retrieved January 19, 2007, from http://www.bileta.ac.uk/Document%20Library/1/Expanding%20the%20Domain%20Name%20System%20(DNS)%20with%20Internationalised%20Domain%20Names%20(IDNs).pdf

Woldering, B. (2006). *Connecting with users: Europe and multi-linguality.* Retrieved January 19, 2007, from http://libraries.theeuropeanlibrary.org/press/200602_VALA_Woldering_Final.pdf

WorldLingo. (2006). *Instant Web site translator FAQ.* Retrieved January 19, 2007, from http://www.worldlingo.com/en/products/iwt_faq.html

Wu, P.-C. (2003). Complete folding of Han ideographs in internationalized domain names. *Computer Standards & Interfaces, 24*(1), 61-67.

Yunker, J. (2003). *Beyond borders: Web globalization strategies.* New Riders Press.

Chapter II
Linguistics of Computer-Mediated Communication: Approaching the Metaphor

Rosanna Tarsiero
Gionnethics, Italy

ABSTRACT

This chapter introduces the embodied metaphor as a means of studying the multifaceted relationship between computer-mediated communication (CMC) and culture. It argues that the embodied metaphor offers a more reliable framework, as opposed to both deterministic and dispositional perspectives, in that it provides a way to understand, explain, and frame the user's adaptive response to CMC. Furthermore, it also argues that culture has the potential to shape interactions by influencing the way metaphors are enacted. The author hopes that the aforementioned approach will provide a new understanding of the interplay between CMC and culture, informing future study design and bridging the existing gap between qualitative and quantitative research through action science.

INTRODUCTION

Since its inception, the study of language has been one of the most fascinating disciplines, combining neurophysiology, anthropology, social psychology, and sociology. In recent years, linguistics has been challenged by the advent of the "new media" whose sensory channel permutations are so high and whose sociological and psychological characteristics are so complicated and intertwined that they stress the very concept of "media discourse."

With the diffusion of the Internet for civil usage, people who had become familiar with using computers for computational, educational, and leisurely purposes also started engaging in human-to-human interactions *through* machines. Its remarkable simplicity, customization possibility, and flexibility quickly led e-mail to be the killer application of the Internet all over the world (ITU, 2001; Pew Internet and American Life Project , 2003). More recently with the birth of the Web 2.0 paradigm, the blending of old media into "new" hybrid spin-offs accelerated remarkably. Even text-written computer-mediated communication now has its sub-genres.

The spreading of the Internet and computer-mediated communication (CMC) also generated interest in computer science and social psychology researchers, who tried to make sense of it. However, in the attempt to provide an explanation, several technical, psychological, and social paradigms blossomed, yet none of them explicitly addressed the increasingly higher proportion of cross-cultural exchanges that had started happening after the Internet spread beyond where it was invented. Despite the diffusion of the Internet and CMC, the raising distributed location of their users, the increasing use of e-mail as a collaborative tool in distributed teams, and the amount of studies and information already collected, laymen and scholars alike neglect the role and weight culture has in shaping e-mail exchanges, and assume unrealistic stands that range from extreme determinism to extreme subjectivism.

The purpose of this chapter is to unfold the relationship between CMC and culture through language by exploring the metaphors behind CMC and how they are enacted in different cultural settings. In doing so, the author openly acknowledges and espouses Lakoff and Johnson's (2003) view of metaphors as cognitive schemes, and relativism as the interplay between subjectivism and objectivism.

In the "Background" section, the literature on CMC, on CMC linguistics, and on cross-cultural studies will spot some reasons of methodological concern. In the "Main Thrust of the Chapter" section, I will explore the metaphors behind CMC and look at them through the lenses of the literature in order to build a model of how culture affects spontaneously occurring computer-mediated exchanges. Finally, in the "Future Trends" section, I will discuss future scenarios and the limitation of the model, suggesting areas for possible refinement of the model and focus of future research.

BACKGROUND

Computer-mediated communication (CMC) and its linguistics were not specifically investigated with regard to how culture impacts them and their interplay. Therefore, this section is composed of three literature reviews: (a) works on CMC, (b) studies on CMC linguistics, and (c) cross-cultural research.

Computer-Mediated Communication

CMC has been extensively studied from psychological, sociological, and technological perspectives. Quantitative approaches outnumbered qualitative ones (Liu, 2002) and suffered with several methodological issues (Culnan & Markus, 1987; Garton & Wellman, 1995; Lamerichs & Molder, 2003; Walther, 1992, 2002). They focused on short-term interactions in artificially constituted groups whose members participated in the research project due to extrinsic rewards (such as higher course grade, extra-credit, money, etc.) or top-down imposition (such as work-related mandated task, workplace or coursework assignment completion, etc.), therefore the research design often became a confounding factor (Walther, 2002). In many of the aforementioned studies, participants were also asked to evaluate the effects of CMC within *close* categories (Lamerichs & Molder, 2003), pigeon-holing data range, participant experiences, and interpretative schemes in the name of a strong bias in favor of "objectivity."

Often, computer literacy went unaddressed/unassessed, even though it is known to adversely affect interaction outcome (Walther, 2002). Finally, in most quantitative works neither the culture of researchers nor participants was assessed and/or planned for in the design of the study. Despite all of these biases, the findings of these studies were surprisingly generalized to represent something "intrinsic to the brain" and CMC to be "clearly" inferior and devoid of advantages. Characteristically, quantitative research negative findings are attributed to negative characteristics of either the communication medium (*technological determinism*, epitomized by Daft & Lengel, 1984; Short, Williams, & Christie, 1976; Sproull & Kiesler, 1991), the task (*task determinism*), or the social group mechanisms (*sociological determinism*, Harasim, 1993; Postmes, Spears, & Lea, 1998; Postmes, Spears, Sakhel, & de Groot, 2001; Riva, 2002; Short et al., 1976; Spears & Lea, 1992; Suler, 1996; Turoff, 1978).

Qualitative studies focused on the hermeneutics of CMC. Most of them came to conclusions that drastically differ from their quantitative counterparts, and depicted online interactions as warm and meaningful (for a meta-analysis see Walther, Anderson, & Park, 1994). The growing body of case studies (Hiltz, 1985) and ethnographies (Baym, 1993, 1998, 2000; Hine, 2000; Turkle, 1995, 1997) usually does not suffer from heavy methodological limitations, and many scholars attribute research findings to participant preferences and choices. Nevertheless, they very rarely factor the participants' culture into the picture.

In tune with the most recent literature on online learning (Muleinburg & Berge, 2005; Mungania, 2003), Rice and Love (1987), Walther (1992, 1996), Herring (2001), and more recently Wood and Smith (2005) take a *dispositional* stand toward online exchange outcomes, according to which individuals adapt to a "lean" medium in order to build meaningful relationships. Again from online learning literature we know that three constructs are pivotal to engaging in online activities: (a) perceived control over one's behavior (*locus of control*, Rotter, 1966); (b) belief in one's capabilities to organize and execute the actions required to attain a goal (*self-efficacy*, Bandura, 1977); and (c) ability of having motivation from within (*intrinsic motivation*, Deci, 1975). These constructs are proven to be impacted by culture (Earley, Gibson, & Chen, 1999; Iyengar & Lepper, 1999; Santiago & Tarantino, 2002).

Despite the number of studies on CMC, neither the deterministic nor the dispositional stand fully account for how CMC determinants intertwine (Herring, 2001). Furthermore, neither addresses the impact of culture on CMC, while quantitative studies openly espouse determinism. However, if we postulate disposition to play a role in CMC production, no matter how small the role, the impact of culture on CMC can no longer be ignored.

Linguistics and Semiotic of CMC

Although the linguistic approach to e-mails provides a way to quantify spontaneous exchanges on an objective basis and can potentially bridge the gap between quantitative and qualitative findings (and their diverging interpretations), it has rarely been used (Lamerichs & Molder, 2003). Simplistically, social scientists and psychologists conceptualized e-mail as a form of written communication, which makes portability of findings into linguistics problematic (Herring, 2001). Even though linguistic scholars have repeatedly called e-mail a new genre (Ferrara, Brunner, & Whittemore, 1991) similar to what a transcript is to an oral exchange (Baron, 1998; Bordia, 1996; Lee, 1996; Herring, 1992, 2004), most studies involved caption and analysis of de-contextualized logs because the model of context is inadequate for e-mails (Herring, 1992, 2004; Jones, 2002).

In CMC, verbal cues are very strong. Communicators using a powerful language style are perceived as attractive, credible, and persuasive (Adkins & Brashers, 1995), and style is often more important than the actual exchange content

(Pennebaker, Mehl, & Niederhoffer, 2003). Individuals tend to progressively match each other's style and fine tune prepositions, articles, and other function words, regardless of the mutual liking or perceived quality of interaction (Niederhoffer & Pennebaker, 2002). Pronouns, words describing emotions, prepositions, and conjunctions have been associated with and can be used to predict a number of different outcomes (Newman, Pennebaker, Berry, & Richards, 2003). However, it is not known if these findings are also valid outside of North America.

CMC linguistics and discourse are unique because humans adapt to "lean" technologies by inventing paralinguistic expressions to reduce uncertainty and foster relationship formation (Herring, 2001; Walther, 1996; Walther & Tidwell, 1995), such as *unconventional orthography* to represent auditory information related to non-language sounds (MacKinnon, 1995), *quotes* from previous conversation to give the illusion of "taking turns" (Herring, 2001), *emoticons* to convey facial expressions (Herring, 2001; Rezabeck & Cochenour, 1994; Utz, 2000; Walther & D'Addario, 2001), and textual *representations of physical actions* to help contextualizing the conversation (Gumperz, 1982).

Counter-intuitively and despite the popularity of the "cues filtered out" approach, a number of *non-verbal cues* are available through CMC, such as *chronemics* (Hesse, Werner, & Altman, 1988; Liu & Ginther, 1999; Walther & Tidwell, 1995), which play a major role in impression formation (Liu, Ginther, & Zelhart, 2001; Walther, 1993), and relationship development (Hesse et al., 1988; Walther, 2002; Walther & Tidwell, 1995). Walther and Tidwell (1995) discovered that chronemics are contingent to e-mail orientation (i.e., task vs. social content). If task-oriented messages are sent at night, they are perceived as more dominant and "inherently" less social; the inverse is true for social messages. Also, quick replies to task-oriented messages and slow replies to social-oriented messages were rated more affectionate (Walther et al., 1994). This property can have a major impact on people living in different time zones.

Since the use and meaning of traditional verbal and nonverbal cues as well as paralanguage is modified by gender (Herring, 1992; Yates, 1993) and culture (Matsumoto, 2006), we can expect this to apply to e-mails as well.

Cross-Cultural Studies

There is no consensus on the definition of "culture," because its meaning varies with context (Chase, Macfadyen, Reeder, & Roche, 2002). Early attempts conceptualized it as a set of acquired dispositions uncritically mutuated from a homogeneous group (see Table 1 [Hofstede, 1983, 1991]; Table 2 [Hall, 1959, 1976]). More recently, Spencer-Oatey (2000) introduced an *interpretive* dimension according to which culture also influences interpretations of the meaning of others' behaviors.

The main objection of Hofstede's and Hall's frameworks is the passiveness of both the individual toward the group and the group toward collective brainwashing. Furthermore, such a group is also homogenous and often restricted to ethnicity or nationality. Since both Hofstede's and Hall's works are based on data collected in managerial settings decades ago, they might no longer be valid (i.e., cultural "shift"), be valid only in business-related settings, or a combination of the two. Representing culture-related concepts in binary pairs may also miss the importance of cultural nuances, at least with regard to some constructs (Abdelnour-Nocera, 2002). Finally, other authors question *en-bloc* the idea of culture as a group phenomenon, operationalizing it as social negotiation of meaning (Scollon & Wong-Scollon, 1995). However, these objections might, in turn, just be a cultural reaction of researchers from an individualistic society who experience difficulty in accepting the power of social group conditioning on the individual.

Table 1. Hoftstede's cultural dimensions

Indicator	Definition	Description
Power Distance Index (PDI)	Degree of status difference among people	High PDI: Inequalities of power/wealth are present within a society. Low PDI: Differences between citizen's power/wealth are undervalued.
Individualism (IDV)	Degree to which individual (vs. collective) achievement and interpersonal relationships are reinforced	High IDV: Individual is independent and placed above the group. Low IDV: The group and the individual are interdependent.
Uncertainty Avoidance Index (UAI)	Level of tolerance for uncertainty and ambiguity within the society	High UAI: Low tolerance for uncertainty and ambiguity. Low UAI: Less concern about ambiguity and uncertainty, and more tolerance for a variety of opinions.
Long-Term Orientation (LTO)	Degree to which long-term devotion to traditional, forward thinking values is embraced	High LTO: Values of long-term commitments and respect for tradition highly regarded. Low LTO: Long-term, traditional orientation and long-term traditions and commitments not very regarded.
Masculinity (MAS)	Degree of reinforcement of the traditional masculine work role model (achievement, control, power)	High MAS: High degree of gender differentiation. Low MAS: Low level of differentiation and discrimination between genders.

Table 2. Hall's classical patterns

Indicator	Definition	Description
Context	High vs. Low	High: Less written/formal information, strong boundaries, situational knowledge. Low: More written/formal information, rule-oriented, task-centered, explicit, and codified knowledge.
Time	Monochronic vs. Polychronic	Monochronic: Do one thing at a time, hard deadlines and plans, committed to the job, weak short-term ties, great respect for private property. Polychronic: Do many things at once, highly distractible, committed to people more than time, change plans often, strong long-term ties, borrow and lend things often and easily.
Space (proxemics)	High vs. Low Territoriality	High: All the spaces around the individual are big. Low: All spaces around the individual are small.

Most studies on "culture" spanned three main thematic areas: (a) verbalizations of situations, feelings, values, norms, and thought patterns (Maletzke, 1996); (b) culturally sanctioned behaviors (Hall, 1960, 1966); and (c) interpretations of situations, such as conflict and collaboration (Hofstede, 1983). Concepts like degree of context expressed through communication (Hall, 1976), time (Hall, 1976), tasks/relationships focus (Laurent, 1983; Trompenaars & Hampden-Turner, 1997), attitudes toward authority, adherence to "traditional" gender roles, group/individual focus (Hofstede, 1991), universalism/particularism pair (Trompenaars & Hampden-Turner, 1997), ways to deal with conflict, decision making, and collaboration (Kim & Bonk, 2002; Trompenaars & Hampden-Turner, 1997) still hold validity, even in limited contexts and even in CMC settings (Ess & Sudweeks, 2005).

Interestingly, in a study of 39 languages, the ones from less individualistic cultures allowed for pronouns to be dropped from sentences (Kashima & Kashima, 1998), while Hinkel (1995) discovered that the use of modal verbs between native-speaker undergraduates and ESL students differs, and such differences do not change with linguistic proficiency and exposure to other socio-cultural constructs. These results are grounds to think that different metaphors stand behind different languages and are expressed differentially according to cultures.

Though cross-cultural studies suffer with conceptual and methodological limitations, intercultural studies have leapfrogged the cultural issue by switching from not considering culture to speaking of "intercultural communication through the Internet" without deepening the concept, validity, reliability of "intercultural" dimension, and its measurements (Chase et al., 2002; Macfadyen, Roche, & Doff, 2004; Scollon & Wong-Scollon, 1995). For this reason, old intuitions can still be used to build more rigorous frameworks.

If the adaptation to CMC is even partially dispositional, culture can influence our motivation in e-mail interaction engagement. If language and paralanguage are differentially shaped by culture, culture will also shape CMC. If our knowledge derives from a given set of mental schemes (i.e., metaphors), culture will also modulate how we enact those metaphors.

MAIN FOCUS OF THE CHAPTER

Different cultures adopt different metaphors within their languages, each one denoting what is important in the *weltanschauung* of that specific culture. Because every culture symbolizes its world in a different way, if we want to understand how metaphors are used and what we can do "with" them, we are left to make sense of metaphors. In this section I will first explore which metaphors are used in computer-related language. I will then build a model that puts such metaphors in relation to literature from the previous section and, finally, define a list of e-mail indicators to be used when testing the impact of culture in e-mail exchanges.

Issues, Controversies, Problems

Lakoff and Johnson (2003) adopted an interesting perspective about the role and impact metaphors have in structuring and shaping our comprehension of phenomena around us. They claim that humans understand and experience all experiences through the use of previous physical and social knowledge. The connection between previous knowledge and present topic resides in the embodied *metaphor*, rooted in physical experience without which there would be no abstract reasoning. Since we build thoughts and actions through metaphors, by analyzing which metaphors humans use for a given topic, we can understand which cognitive schemes their minds use in representing, framing, and understanding the phenomenon, its consequences on the representation itself, and possible "actions" around that metaphor.

Possible claims against Lakoff and Johnson's theory focus on two concepts: (a) thought is complex and unconscious for the most part, and (b) abstract concepts are mostly metaphorical. Murphy (1996, 1997) explained technical (psycholinguistics) objections, while Gibbs summarized his stand in favor of the embodied metaphor concept (2003). From a behavioral perspective, it can be argued that we learn language from the environment, as a response to stimuli, and therefore it does not represent a complex unconscious scheme but just an adaptive reaction. Such reaction would however fail to explain why different metaphors are generated by the same person in a given situation. Alternatively, we could argue with Gumperz (1982) that metaphors are expressions of culture rather than manifestations of cognitive schemes. Such objection, however, would not explain why similar metaphors originate in different situations, environments, or cultures. Finally, a constructivist would object that objective meaning does not exist *per se,* but only as a negotiation process, therefore analyzing language from a metaphorical perspective de-contextualizes it.

The nouns commonly used when referring to the Internet refer to it as a "*space*" *with boundaries*, simultaneously away from where we are ("virtual") yet earthly present ("world"). Such space is partitioned into smaller ones ("chat *rooms*," "*channels*," "message *boards*") which mirror, in name and behavioral conventions, their "real" counterparts (greeting other participants when joining, introducing oneself, saying "bye" while leaving). Members already present in such space but unwilling to engage in a dialogue (*lurkers*) exist unobserved at its margins and posts, and messages are held together through "*threads*," which implicitly suggest: (a) the presence of discrete separators within public spaces, and (b) the virtual community to be a spatial object with a definite dimensionality brought about by the degree of willingness with which its members interact. Microsoft espoused this metaphor and gave it an adventurous gist through names such as "Explorer," "Windows," and "Outlook," inviting the user to "come out" into a broader space and "take a look."

Verbs describing Internet activities configure the "*Internet is an informal journey*" metaphor. Users "go to" one page from another, meanwhile they casually "chat" with whom they meet, superficially "browsing" information and "surfing" Web content.

The metaphors behind computers and their usage and communication in general go against the "negative" view of computers and CMC that emerged from the quantitative literature review. Schwartz (as cited in Lakoff, 2006) noted how the metaphor behind general machine functioning is "*machines are people*," in which the degree of functioning expresses their health while the computational power is seen as "thinking." Interestingly, the interaction between people and their computers also follows a human-to-human metaphor (Nass & Steuer, 1993). In the case of computer applications, they "go crazy," "freeze," and have to be "killed," "terminated," or "shut down." They do not just have a "life," but computer difficulties in handling the user's requests are seen as emotional trouble. This last trait conflicts with the metaphor of computers and their parts as working aids for their owners.

Espenson (as cited in Lakoff, 2006) noted how the metaphor behind communication is "*communication is linguistic communication*." Text-based or spoken communication features are considered more meaningful than nonverbal cues ("His eyes said yes." "His look spoke volumes." "I can read him like a book." "His gestures underlined what he was saying.").

Because some have advocated language as the source of cyberspace (Cicognani, 1998), and CMC has been represented in different cultures and languages through a relatively uniform set of metaphors, exploring these very same metaphors may teach us which tools to use when studying the correlations between culture and CMC.

Solutions and Recommendations

If the cognitive schemes applying to CMC result from the interplay of these metaphors, we can expect them to be critically enacted depending on the cultural framework of the interaction participants, that is, depending on the relative importance and acceptance each metaphor has in a given culture. Since both face-to-face (Kramsch & Anderson, 1999) and online communication patterns (Chase et al., 2002) do vary among cultures, variations in qualitative CMC characteristics might also exist.

The Internet is a space. Proxemics are not limited to physical ones (Riva & Galimberti, 1998) and hold validity also for avatars in virtual environments (Jeffrey & Mark, 1998). For these reasons, some behaviors enacted through e-mails and some specific characteristics of the e-mail can be perceived as an intrusion into the receiver's space and be accepted (or condoned) in situations that vary with cultures. I therefore posit that initiating an unsolicited exchange, the e-mail length, and the e-mail rate are *space domain indicators* (see Table 3). These indicators might characterize cultures in which a hard work ethic is paramount, and which either value privacy, time, and wide individual spaces (such as North America) *or* shyness, social order, and conformity (such as China and Japan).

The Internet is informal. Culture affects conversation shape and content, as well as degree of technology adoption (Buragga, 2002; Phillips, Calantone, & Lee, 1994). Novelty-seeking, innovative behaviors and authority and rule-probing/questioning through e-mails could be equally culturally sensitive. For these reasons, I posit the *formality domain indicators* (see Table 4) to be dependent on culturally sanctioned behaviors with regard to public expression of emotional content and to encompass: (a) e-mail orientation (social/task-related content), (b) number of emoticons per e-mail, (c) number of other paralinguistic cues (such as emoting, quoting, unconventional orthography, and textual representation of physical actions), and (d) degree of self-disclosure. Other such indicators reflect important culture-related attitudes toward (e) hierarchies and (f) traditions, while (g) e-mail linearity (defined as logical, low-context, step-by-step string of thoughts), and (h) e-mail length reflect the attitude toward reflexivity and extroversion, respectively. Such indicators might be particularly descriptive of cultures that value emotional self-disclosure, extroversion, multi-tasking, traditions, and hierarchies (such as the Mediterranean cultures).

The Internet is a journey. In monochronic cultures, time is compartmentalized, while in low-context ones, competition and order are highly

Table 3. Space domain indicators (virtual proxemics)

Indicator	Description
Initiation	Initiating an unsolicited exchange
Length	Length of an e-mail (word count)
Rate	Number of e-mails per day

Table 4. Formality domain indicators

Indicator	Description
Hierarchy	Attitudes toward hierarchies
Tradition	Attitudes toward traditions
Cues	Number of other paralinguistic cues (emoting)
Emoticons	Number of emoticons (per message or per person/message or overall)
Length	E-mail length (word count)
Linearity	E-mail linearity (step-by-step vs. loose content)
Orientation	Focused on tasks vs. relationships
SelfDisclosure	Degree of self-disclosure

valued. In monochronic low-context cultures, work is conceptualized as a "serious" activity during which no derailments and few distractions are allowed, while journeys are meaningful if they have a goal. I posit that *journey and working domain indicators* (see Table 5) are: (a) daily differences in e-mail orientation (either polychronic switching/multi-tasking between emotional and task-related e-mails, or monochronic string of work-related e-mails interspersed with occasional personal e-mails or remarks), (b) e-mail linearity (attitude toward context), (c) e-mail rate (number of tasks and kind of topics the sender has been engaged in daily), (d) e-mail length (extroversion and/or low-context culture), and (e) Walther and Tidwell's *chronemics* (to be analyzed together with e-mail orientation). This list of indicators could be useful in exploring polychronic low-context cultures, such as Australia.

Communication is linguistic communication. Low-context cultures rely more on written information, are more verbally explicit and less emotional, and ask for details whenever they do not possess all the information they desire to have (Gill, 1998). When looking at data from Europe on the differential use of text-based CMC, studies reveal higher e-mail usage in northern (low-context) countries and higher chat room usage in southern (high-context) nations (Greenspan, 2002). This can be explained by positing that in high-context cultures, e-mail is less frequent because asynchronous communication is perceived as *de facto* unable to convey "all the context" or not fast enough. For these reason I posit that (a) requests for clarifications, (b) e-mail orientation, and (c) e-mail linearity are *text domain indicators* (see Table 6). This indicator cluster might be an excellent fit to explore the low-context/high-context dimension.

Machines are people. Cultures that value long-term relationships and high-context cultures encourage people to behave traditionally and collaboratively toward the group (Kim & Bonk,

Table 5. Journey and working domain indicators

Indicator	Description
Chronemics	Walther's *chronemics*
Rate	E-mail rate
ΔLength	Daily differences in e-mail length (variations in e-mail word count)
ΔLinearity	Daily differences in e-mail linearity (variations in how the context is expressed)
ΔRate	Daily differences in e-mail rate (variations in e-mail daily number)
ΔOrientation	Daily differences in e-mail orientation (switching, multitasking, or any variation of subject of the task vs. relationship)

Table 6. Text domain indicators

Indicator	Description
Clarifications	Presence/absence of request for clarifications
Linearity	E-mail linearity (step-by-step vs. loose)
Orientation	Task vs. relationship

Table 7. People domain indicators

Indicator	Description
Collaboration	Attitudes toward collaboration
Hierarchy	Attitudes toward hierarchies
Tradition	Attitudes toward traditions
Critical Exchanges	Messages expressing critical viewpoints
Lexicon	Emotional e-mail lexicon
LongTerm	Existence of long-term exchanges with one or more in-groups

2002), and attend to long-term relationships. In individualistic cultures, abstract nouns and adjectives predominate, as opposed to interpersonal concrete verbs within collectivist cultures (Semin, Gorts, Nandram, & Semin-Goossens, 2002). Furthermore, collectivist cultures approach online decision making with different values (Rahmati, 2000; Thanasankit & Corbitt, 2000), while individualistic cultures have a higher rate of critical exchange (Reining & Mejias, 2004). For these reasons, I posit that (a) the attitude toward collaboration, (b) traditions, and (c) individuals, as well as (d) the emotional e-mail lexicon type (abstract nouns/adjectives vs. interpersonal concrete verbs), (e) the existence of long-term exchanges with one or more in-groups, and (f) the number of critical exchanges, are *people domain indicators* (see Table 7). These indicators can explore the individual/group dimension.

Looking at possible e-mail cultural aspects through the lenses of the five metaphors identified in the previous section allowed the author to identify 22 putative indicators to be put through validation in further studies.

FUTURE TRENDS

Possible Scenarios

Knowing how different cultures perceive e-mails and their characteristics will be a useful tool in various settings. For-profit and non-profit corporate managers alike could use these findings to align e-mail content, format, and policy to their organizational values, and informally manage enculturation, employee loyalty, and adherence to corporate culture. Human resource development managers and online educators could foster cross-cultural understandings and proper management of distant colleagues/partners, while also increasing knowledge sharing and collaboration whenever desired by tailoring their CMC to the audience's culture, motivation, and desired workplace or educational outcomes. Senior managers and supervisors could facilitate the social component of merging and outsourcing, by planning online hiring procedures and online job assignments, as well as disseminating electronic memos, news, and training manuals in a culturally sensitive way. Online trainers and educators could exploit online social learning by tailoring the format and content of e-mail exercises and interactions, which would allow them to differentially involve students in full respect with their cultural requirements, while still targeting desired learning objectives and outcomes. Professionals like health providers, academics, knowledge managers, and software developers would benefit worldwide from a blueprint for culturally sensitive interactions with peers for informal training, knowledge generation, and referral. Finally, job hunters and potential students could learn how to write personal electronic materials (i.e., electronic college applications, electronic curriculum vitae, cover e-mails) designed to appeal to like-minded individuals.

Limitations

The embodied metaphor's perspective main problem is that culture can influence CMC *prior* to the moment in which a metaphor is enacted or interpreted, since the attitude toward using technology is related to specific cultural aspects (Buragga, 2002; Gill, 1998; Hermeking, 2005; Van Belle & Stander, 2002). However, when analyzing spontaneously occurring e-mail exchanges, this objection seems less poignant because CMC users either belong to a technology-friendly culture or have already overcome their own cultural bias toward technology. The objection is still valid whenever e-mail usage is mandated, and is a serious pitfall of some research study designs if unaddressed. It remains to be assessed if spontaneous e-mail usage is an indicator of relative easiness in overcoming cultural biases in general.

Figure 1. Relationship among domain indicators

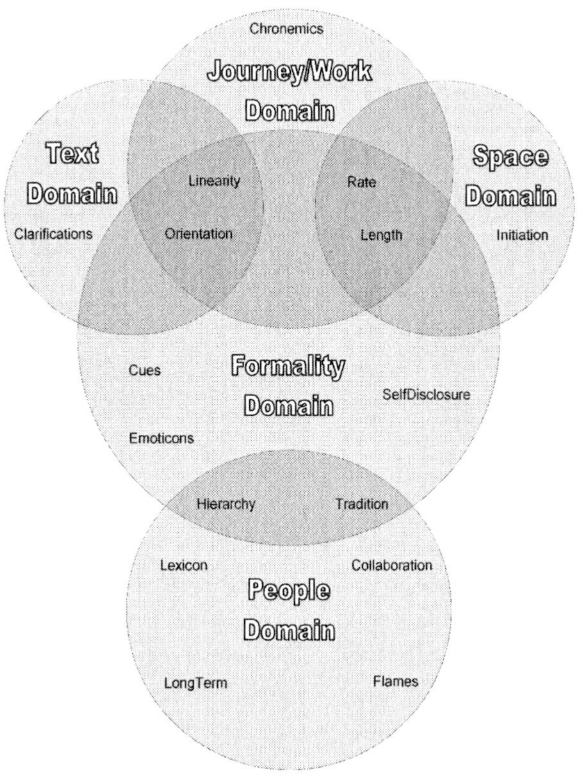

Table 8. List of indicators

Qualitative	Quantitative
Initiation	Length
Hierarchy	ΔLength
Tradition	Cues
Collaboration	Linearity
Clarifications	ΔLinearity
Orientation	Rate
ΔOrientation	ΔRate
SelfDisclosure	Emoticons
Critical Exchanges	Lexicon
LongTerm	Chronemics

The cross-cultural perspective *per se* is another possible limitation. Language on the Web "becomes" culture (Kramsch & Anderson, 1999), so CMC might benefit from an inter-cultural perspective instead (Bennet, 1993; Chase et al., 2002; Ma, 1996; Macfadyen et al., 2004). On the other hand, to the newcomer to the Internet and to individuals who have not yet criticized their culture, intercultural perspectives may paradoxically foster cultural insensitivity. The Internet itself may also generate *glocalization* in dealing with cultural aspects of CMC, especially in public fora where adherence to norms is strongly encouraged.

Although all five domains are developed (see Figure 1) and possess both unique and shared indicators, many indicators are not specific to a category, presenting significant overlapping.

That might reflect a poor metaphor choice as well as nuanced interconnections among metaphor dimensions, or frank interdependence. The best characterized domains are "People" (six indicators, four unique) and "Formality" (nine indicators, three unique). The latter appears to be the mainstay of the model, also allowing for use as indicator subset for gross model validation. The 22 indicators are evenly split between methodological categories (11 quantitative, 11 qualitative—see Table 8).

Testing of the Model

In order to build an objective theory that matches sample accuracy, it is mandatory to match the complexity of the data set with the one of the model. Among available methods, action science (see Argyris, Putnam, & Smith, 1985) appears to be the most fit because it explores events that: (a) occur spontaneously, (b) are composed of (or can be subdivided into) discrete quantifiable measurements, and (c) are not adequately explorable by an artificial reproduction of their core conditions.

The paucity of studies on CMC in "natural" settings and the impossibility to fake or eliminate "culture" point toward the use of observation of "spontaneous" interactions.

Previous studies on CMC belonged to either the positivistic model, where the researcher's epistemology and ontology are both objective, or the postmodern framework, in which the researcher's epistemology and ontology are both subjective. Because of the conditions in which both CMC and "culture" occur, the objective ontology and subjective epistemology of action science offer a fresh perspective. Action science also offers excellent guarantees about being the least biased method in these settings, since it does not make *a priori* hypotheses and does not embed interaction design into the study.

Finally, action science also shares its ontology with Lakoff and Johnson's (2003) model. The parallel between Argyris' cycle (observation, model building, testing, and action; Argyris et al., 1985) and metaphors as products of the interplay between objectivism and subjectivism is very strong. Metaphors can act as common ground between one cycle stage and the next one, or between cycles.

Researchers bring *into* the study their culturally susceptible biases, behaviors, and interpretations (Cray & Mallory, 1998), so a great deal of attention should be paid to study design criteria. Viable strategies are: (a) choosing an intercultural study design, and/or (b) employing multiple teams from different cultures observing, analyzing, and interpreting the same set of data so to either shrink the "cultural divide" or make it apparent.

Data selection has to be carefully done. Researchers should collect spontaneously occurring exchanges *before* enrolling participants-to-be, to prevent a change in their spontaneous motivation to participate into the exchange from affecting the linguistic of the data under study. Data can be gathered from public fora, Yahoogroups, Google groups, message boards, and listservs, as well as from any person of legal age and capacity to read, understand, and sign an informed consent privacy disclaimer describing type and scope of information data collection. Researchers should then contact and test participants via a standardized test, exploring as many independent cultural dimensions as possible, to determine which values participants do *individually* espouse. Researchers should code and analyze data before formal participant enrollment or testing, and in any case, without knowing the writer score while coding, scoring, or analyzing the message, so to avoid a source of interpretive bias. The obtained data could undergo a cluster analysis to see if taxonomy similar to the original model is produced.

Future Research

Even after successful testing of this framework, there will be further questions to address design and content of future research studies.

Among the contents to be addressed, a first big bulk of questions to explore relate to the difference between CMC in dyads and in groups, and the interplay between these differences and culture. These questions correspond to the study of possible intersections between the *people* and *text* domains (see Figure 1). How does the presence of more than two persons (group) change the dynamics of CMC exchanges? Are cultural expectations related to participation in group conversations? Are there other norms regarding public interactions that change with culture (for example, is the concept of "face" valid online as well)?

Another big set of questions to research lies in the interplay between constructs such as motivation, self-efficacy, and locus of control, with CMC via the cultural and cognitive metaphors applied to those constructs themselves. These questions would explore possible intersections between the *journey/working* and *people* domains (see Figure 1). Are motivation, self-efficacy, and locus of control linked to any specific cultural dimension? How does it impact on technology adaptation? How does it impact on the tendency to transcend one's culture?

Finally, a further set of questions explores the feedback that CMC may have on culture, which would refine the relationship between the *people* and *formality* domains (see Figure 1). Which kind of context is available in different cultures as far as e-mail exchanges are concerned? When, why, and how does the perception of cultural norms about e-mail change how different cultures "do e-mails"? What is the impact that signatures and pseudonyms have on CMC exchanges?

This chapter could provide for a set of CMC metrics that could quantify the impact culture has on CMC, thereby changing the current approach to CMC. Cultural inclusiveness, embedding a representation into the interaction, is not customizable enough for the needs of users, even if it claims to be so. It has to be pointed out how "cultural inclusiveness" is a concept germane to political correctness, which in turn is deeply rooted into the American culture and would therefore be yet another "symptom" of cultural imperialism. Intercultural sensitivity, on the other hand, recognizes and acknowledges different cultures in a framework in which one's culture is examined critically, while the other person is evaluated from within his/her culture (Bennett, 1993).

If this paradigm will be validated, it might really impact CMC research. The use of metaphors to make sense of CMC has never been attempted before and could be extended toward other forms of CMC such as IM. Findings from these studies could change the policies on the usage of CMC, with particular attention to workplace settings, and that in turn could change people's behavior and perceptions about CMC.

CONCLUSION

Despite the increasing use of asynchronous text-based CMC in cross-cultural settings, most of what we know about it comes from studies that did not take culture into any account. As a general rule, CMC research has either opted for "laboratory" conditions in the attempt to quantify measurable behavior, or used qualitative methods in the attempt to capture the complexity of field conditions, but results have been channeled into determinism of some sort. The approach this chapter illustrates is a promising one because it refuses to take "absolute" sides, yet addresses qualitative *and* quantitative issues in the attempt to match complexity of the problem with complexity of the method of study.

The way culture has been conceptualized and studied has shown to be outdated not merely because it dates back to the 1960s: methodological concerns arise toward the generalization of findings from special populations to the whole humankind. However, some of those concepts and findings can still be used to frame actual research. This chapter uses those findings as categories, in a way as "tags" to start from for classifying events and their properties, and then reframe them to explain everyday phenomena.

Building a model based on the metaphors that most frequently are associated with CMC and looking at those metaphors through frameworks accounting for cultural differences allows the author to find a list of qualitative and quantitative behavioral and linguistic metrics (see Table 8) potentially able to depict how culture impacts on CMC exchanges, going beyond the stands that have dominated the field so far. In fact, such metrics cut along disciplines and are related to one another in a way that also allows predictions about their change depending on starting conditions and changes that can occur in them.

This chapter mixes those overlapping disciplines that have most frequently studied CMC into a consistent framework that traces the path for future research. Action science is described as a way to conjugate metrics grounded in an interdisciplinary framework, while implementation details about future research suggest area of improvement.

REFERENCES

Abdelnour-Nocera, J. (2002). Context and culture in human computer interaction: 'Usable' does not mean 'senseful'. In F. Sudweeks & C. Ess (Eds.), *Cultural attitudes towards technology and communication 2002: Proceedings of the 3rd International Conference on Cultural Attitudes Towards Technology and Communication* (pp. 505-524). Murdoch, West Australia: School of Information Technology.

Adkins, M., & Brashers, D. E. (1995). The power of language in computer-mediated groups. *Management Communication Quarterly, 8*(3), 289-322.

Argyris, C., Putnam, R., & Smith, D. (1985). *Action science: Concepts, methods and skills for research and intervention.* San Francisco: Jossey-Bass.

Bandura, A. (1977). Self-efficacy: Toward a unifying theory of behavioral change. *Psychological Review, 84*(2), 191-215.

Baron, N.S. (1998). Letters by phone or speech by other means: The linguistics of e-mail. *Language & Communication, 18*(2), 133-170.

Baym, N. (1993). Interpreting soap operas and creating community: Inside a computer-mediated fan culture. *Journal of Folklore Research, 30*(2/3), 143-176.

Baym, N. (1998). The emergence of online community. In S. G. Jones (Ed.), *Cybersociety 2.0: Revisiting computer-mediated communication and community* (pp. 35-68). Thousand Oaks, CA: Sage.

Baym, N. (2000). *Tune, log on: Soaps, fandom, and online community.* Thousand Oaks, CA: Sage.

Bennett, M. (1993). Towards ethnorelativism: A development model of intercultural sensitivity. In M. Paige (Ed.), *Education for the intercultural experience* (pp. 21-71). Yarmouth, ME: Intercultural Press.

Bordia, P. (1996). Studying verbal interaction on the Internet: The case of rumor transmission research. *Behavior Research Methods, Instruments & Computers, 28*(2), 149-151.

Buragga, K. A. (2001). *An investigation of the relationship between national culture and the adoption of information technology.* Unpublished doctoral dissertation, George Mason University, USA.

Chase, M., Macfadyen, L., Reeder, K., & Roche, J. (2002). Intercultural challenges in networked learning: Hard technologies meet soft skills. *First Monday, 7*(8). Retrieved June 27, 2006, from http://firstmonday.org/issues/issue7_8/chase/index.html

Cicognani, A. (1998). On the linguistic nature of cyberspace and virtual communities. *Virtual Reality: Research, Development and Application, 3*(1), 16-24.

Cray, D., & Mallory, G.R. (1998). *Making sense of managing culture.* London: International Thomson.

Culnan, M. J., & Markus, M. L. (1987). Information technologies. In F. M. Jablin, L. L. Putnam, K. H. Roberts, & L. W. Porter (Eds.), *Handbook of organizational communication: An interdisciplinary perspective* (pp. 420-443). Newbury Park, CA: Sage.

Daft, R. L., & Lengel, R. H. (1984). Information richness: A new approach to managerial behavior and organization design. In B. M. Staw & L. L. Cummings (Eds.), *Research in organizational behavior* (Vol. 6, pp. 191-233). Greenwich, CT: JAI Press.

Deci, E.L. (1975). *Intrinsic motivation.* New York: Plenum.

Earley, P. C., Gibson, C. B., & Chen, C. C. (1999). "How did I do?" versus "How did we do?" Cultural contrasts of performance feedback use and self-efficacy. *Journal of Cross-Cultural Psychology, 30*(5), 594-619.

Ess, C., & Sudweeks, F. (2005). Culture and computer-mediated communication: Toward new understandings. *Journal of Computer-Mediated Communication, 11*(1). Retrieved February 15, 2006, from http://jcmc.indiana.edu/vol11/issue1/ess.html

Ferrara, K., Brunner, H., & Whittemore, G. (1991). Interactive written discourse as an emergent register. *Written Communication, 8*(1), 8-34.

Garton, L., & Wellman, B. (1995). Social impacts of electronic mail in organizations: A review of the research literature. In B. R. Burleson (Ed.), *Communication yearbook 18* (pp. 434-453). Thousand Oaks, CA: Sage.

Gibbs, R. W. (2003). Embodied experiences and linguistic meaning. *Brain and Language, 84,* 1-15.

Gill, S. P. (1998). The cultural interface: The role of self. In C. Ess & F. Sudweeks (Eds.), *Proceedings of Cultural Attitudes Towards Communication and Technology* (pp. 246-251). Murdoch, Australia: Murdoch University.

Greenspan, R. (2002). *Swedes, Danes and Dutch lead European usage.* Retrieved March 26, 2005, from http://www.clickz.com/stats/sectors/geographics/article.phpr/1107211

Gumperz, J. J. (1982). Contextualization conventions. In J.J. Gumperz (Ed.), *Discourse strategies* (pp. 130-152). Cambridge: Cambridge University Press.

Hall, E. T. (1976). *Beyond culture.* Garden City, NY: Doubleday.

Hall, E. T. (1966). *The hidden dimension.* Garden City, NY: Doubleday.

Hall, E. T. (1960). The language of space. *Landscape, 10,* 41-44.

Hall, E. T. (1959). *The silent language.* Garden City, NY: Doubleday.

Harasim, L. M. (1993) Networlds: Networks as social space. In L. M. Harasim, (Ed.), *Global networks: Computers and international communication* (pp.15-34). Cambridge, MA: MIT Press.

Hermeking, M. (2005). Culture and Internet consumption: Contributions from cross-cultural marketing and advertising research. *Journal of Computer-Mediated Communication, 11*(1). Retrieved February 19, 2006, from http://jcmc.indiana.edu/vol11/issue1/hermeking.html

Herring, S. C. (1992). *Gender and participation in computer-mediated linguistic discourse* (Doc. No. ED345552), Washington, DC: ERIC Clearinghouse on Languages and Linguistics.

Herring, S. C. (2001). Computer-mediated discourse. In D. Tannen, D. Schiffrin, & H. Hamilton (Eds.), *Handbook of discourse analysis* (pp. 612-634). Oxford: Blackwell.

Herring, S. C. (2004). Computer-mediated discourse analysis: An approach to researching online behavior. In S. A. Barab, R. Kling, & J. H. Gray (Eds.), *Designing for virtual communities in the service of learning* (pp. 338-376). New York: Cambridge University Press.

Hesse, B. W., Werner, C. M., & Altman, I. (1988). Temporal aspects of computer-mediated communication. *Computers in Human Behavior, 4*(2), 147-165.

Hiltz, S. R. (1985). *Online communities: A case study of the office of the future.* Norwood, NJ: Ablex Publishing.

Hine, C. (2000). *Virtual ethnography.* Thousand Oaks, CA: Sage.

Hinkel, E. (1995). The use of modal verbs as a reflection of cultural values. *TESOL Quarterly, 29*(2), 325-343.

Hofstede, G. (1991). *Cultures and organizations: Software of the mind.* London: McGraw-Hill.

Hofstede, G. (1983). Dimensions of national cultures in fifty countries and three regions. In J. Deregowski, S. Dzirawiec, & R. Annis (Eds.), *Explications in cross-cultural psychology* (pp. 389-407). Lisse, The Netherlands: Swets and Zeitlinger.

ITU. (2001, November 12-16). *Internet: Challenges and opportunities for Africa.* Retrieved February 19, 2006, from http://www.itu.int/AFRICA2001/forum/votingresults/inf8.html

Iyengar, S. S., & Lepper, M. (1999). Rethinking the value of choice: A cultural perspective on intrinsic motivation. *Journal of Personality and Social Psychology, 76*(3), 349-366.

Jeffrey, P., & Mark, G. (1998, March 16-17). Constructing social spaces in virtual environments: A study of navigation and interaction. In: K. Höök, A. Munro, & D. Benyon (Eds.), *Workshop on Personalised and Social Navigation in Information Space* (pp. 24-38). Stockholm: Swedish Institute of Computer Science (SICS).

Jones, R. H. (2002, March). *The problem of context in computer-mediated communication.* Paper presented at the Georgetown Roundtable on Language and Linguistics, Washington, DC.

Kashima, E. S., & Kashima, Y. (1998). Culture and language: The case of cultural dimensions and personal pronoun use. *Journal of Cross-Cultural Psychology, 29*(3), 461-486.

Kim, K.-J., & Bonk, C. J. (2002). Cross-cultural comparisons of online collaboration. *Journal of Computer-Mediated Communication, 8*(1). Retrieved March 16, 2006, from http://jcmc.indiana.edu/vol8/issue1/kimandbonk.html

Kramsch, C., & Andersen, R. W. (1999). Teaching text and context through multimedia. *Language Learning & Technology, 2*(2), 31-42.

Lakoff, G. (2006). *Communication is linguistic communication.* Retrieved March 26, 2006, from http://cogsci.berkeley.edu/lakoff/metaphors/Communication_Is_Linguistic_Communication.html

Lakoff, G. (2006). *Machines are people.* Retrieved March 26, 2006, from http://cogsci.berkeley.edu/lakoff/metaphors/Machines_Are_People.html

Lakoff, G., & Johnson, M. (2003). *Metaphors we live by* (rev. ed.). Chicago: Chicago University Press.

Lamerichs, J., & Te Molder, H. (2003). Computer-mediated communication: From a cognitive to a discursive model. *New Media & Society, 5*(4), 451-473.

Laurent, A. (1983). The cultural diversity of western conceptions of management. *International Studies of Management and Organization, 13*(1-2), 75-96.

Lee, J. Y. (1996). Charting the codes of cyberspace: A rhetoric of electronic mail. In L. Strate, R. Jacobson, & S. B. Gibson (Eds.), *Communication and cyberspace: Social interaction in an electronic environment* (pp. 275-296). Albany, NY: State University of New York Press.

Liu, Y. (2002). What does research say about the nature of computer-mediated communication: Task-oriented or social-emotion-oriented? *Electronic Journal of Sociology, 6*(1). Retrieved May 1, 2006, from http://www.sociology.org/content/vol006.001/liu.html

cultural diversity in business. London: Nicholas Brearley.

Turkle, S. (1995). *Life on the screen. Identity in the age of the Internet.* New York: Simon & Schuster.

Turkle, S. (1997). Constructions and reconstructions of self in virtual reality: Playing in the MUDs. In S. Kiesler (Ed.), *The culture of the Internet* (pp. 143-156). Mahwah, NJ: Lawrence Erlbaum.

Turoff, M. (1978). The EIES experience: Electronic information exchange system. *Bulletin of the American Society for Information Science, 4*(5), 9-10.

Utz, S. (2000). Social information processing in MUDs: The development of friendships in virtual worlds. *Journal of Online Behavior, 1*(1).

Van Belle, J.-P., & Stander, A. (2002). Gender differences in the perception and use of e-mail in two South African organisations. In F. Sudweeks & C. Ess (Eds.). *Proceedings of Cultural Attitudes Towards Technology and Communication 2002* (pp. 131-142). Murdoch, Australia: Murdoch University.

Walther, J. B. (1992). Interpersonal effects in computer-mediated interaction: A relational perspective. *Communication Research, 19*(1), 52-90.

Walther, J. B. (1993). Impression development in computer-mediated interaction. *Western Journal of Communication, 57,* 381-398.

Walther, J. B. (1994). Anticipated ongoing interaction versus channel effects on relational communication in computer-mediated interaction. *Human Communication Research, 20*(4), 473-501.

Walther, J. B. (1996). Computer-mediated communication: Impersonal, interpersonal and hyperpersonal interaction. *Communication Research, 23*(1), 3-43.

Walther, J. B. (2002). Time effects in computer-mediated groups: Past, present, and future. In P. Hinds & S. Kiesler (Eds.), *Distributed work* (pp. 235-257). Cambridge, MA: MIT Press.

Walther, J. B., Anderson, J. F., & Park, D. (1994). Interpersonal effects in computer-mediated interaction: A meta-analysis of social and anti-social communication. *Communication Research, 21*(4), 460-487.

Walther, J. B., & D'Addario, K. P. (2001). The impacts of emoticons on message interpretation in computer-mediated communication. *Social Science Computer Review, 19(3)*, 323-345.

Walther, J. B., & Tidwell, L. C. (1995). Nonverbal cues in computer-mediated communication, and the effect of chronemics on relational communication. *Journal of Organizational Computing, 5*(4), 355-378.

Wood, A. F., & Smith, M. J. (2005). *Online communication: Linking technology, identity and culture* (2nd ed.). Mahwah, NJ: Lawrence Erlbaum.

Yates, S. J. (1993). *The textuality of computer-mediated communication: Speech, writing and genre in CMC discourse.* Unpublished doctoral dissertation, Open University, UK.

Chapter III
Machine Translation as the Future of International Online Communication

Nicole St. Germaine-Madison
Texas Tech University, USA

ABSTRACT

American "ownership" of the World Wide Web is on the decline. In fact, two-thirds to three-fourths of Internet users are not native speakers of English. It thus does not require a leap of logic to imagine how many more non-U.S. users would buy from American e-commerce sites or even purchase American-manufactured products if they had access to translations of these e-commerce pages and technical documents in their native languages. The most oft-cited reason for not providing content in other languages, however, is the sheer cost involved of first hiring a translator to translate the original content and then keeping material updated. This chapter examines uses of machine translation as a mechanism for addressing these linguistic needs.

INTRODUCTION

The Internet used to be an American product. It was largely developed by researchers at MIT and various other universities as a means of connecting researchers and developers, and since its commercialization it has traditionally also been used as a vehicle for American information and American commerce. As a result, virtually all of the content was, and is, written in English (Internet Society, n.d.).

American "ownership" of the World Wide Web is on the decline. According to the Online Computer Library Center's annual Web Characterization Project, the most cited resource for World Wide Web statistics, while only 55% of sites of the World Wide Web were American in origin as of 2003, 72% of the total Web sites are in English. This statistic has remained stable since the first Web Characterization Project report in 1999.

Paradoxically, two-thirds to three-fourths of Internet users are not native speakers of English (Cronin 2001, p. 16). Nor do the majority of these non-native speakers necessarily prefer to read Web sites in English: 50% of Web users in Europe, 75% of users in Latin America, and 80% of users in Japan prefer to read Web sites in their native language over English (2001, p. 16). According to some estimates, the willingness of a target audience to read a Web site increases by 300% if the site is in their native language, even if they are competent in the other language the site is presented in (Forsyth, 2005) This disconnect represents a large gap where industry and e-commerce can step in to increase their markets. In 2003, the non-U.S. portion of the e-commerce market would account for 46% of a US$1.3 trillion market (2005, p. 16). It does not require a leap of logic to imagine how many more non-U.S. users would buy from American e-commerce sites or even purchase American-manufactured products if they had access to translations of these e-commerce pages and technical documents in their native languages.

Aside from e-commerce, other types of Web sites could also benefit from having a translation made available. Web sites representing international organizations and research applications would gain more members and more hits if these sites had available translations into other languages as well. Even the Society for Technical Communication (http://www.stc.org), an international organization, does not have content available in any language other than English.

The most oft-cited reason for not providing content in other languages is the sheer cost involved of first hiring a translator to translate the original content, and then keeping a subscription at a translation and localization firm in order to keep the site's information continually updated in all of the available languages. Most of these companies, such as SDL International (http://www.sdl.com) and Translation Services USA, hesitate to even promise a delivery time that is faster than two to three weeks *or more*, depending on the size of the Web site, the number of languages, and the commonality of the desired target languages. Each of these factors can greatly inflate the cost and time needed to complete the translation.

So what is a company to do? According to Cronin (2001, p. 112), "The best scenario for updating Web content is simultaneously updating all languages. This might not be possible except with automation." However, automated translation (or machine translation) is still not a reliable method for obtaining a completely accurate translation.

The main objectives of this chapter are:

1. To explain the need for rapid, inexpensive translation for online content
2. To present different case studies for how companies are approaching the production of different language sites
3. To explore the differences of translating for the Web vs. translating for print, as well as to discuss the implications for translation
4. To explain machine translation: a brief history, how it works, and its reliability
5. To explore current uses of machine translation for the Web
6. To make specific recommendations for optimizing the use of machine translation for the Web

THE NEED FOR RAPID, INEXPENSIVE TRANSLATION FOR THE WEB

In 2005, it was estimated that around three-quarters of the approximately one billion people who were conducting business or communicating online were living outside of the U.S., and most of these people did not speak English (Schenker, n.d.). Taken with the staggering size of the profits and opportunities lost by monolingual Web sites, this fact has provided the impetus needed for many sites to move beyond their former reliance on English as a vehicle to capture world markets or world audiences.

In an effort to garner international attention and sales, many companies are moving toward translating and localizing content for their various markets or perceived audiences. According to Cronin (2001, p. 14):

Most professional Web sites contain continuously updated and revised content, sometimes referred to as streaming content. Most of today's professional Web sites are updated frequently, are provided in multiple languages, and offer a high degree of personalization. The main challenges in maintaining multilingual (or global) Web sites is internationalizing the site architecture, balancing global/translated versus local content, automating translation workflows, and keeping multilingual content in sync with the source language.

As products and information are rapidly updated, the Web site (or Web sites if there are versions for different countries) must be similarly updated in a timely manner in order to have maximum impact on the world market.

Further, more information that used to be solely available in print is being made available online. O'Hagan and Ashworth state:

Product and service information has been one of the traditional markets for translation and is now being increasingly made available online. In some cases, such information might only be available online since this saves the physical printing and distribution cost of the paper-based documentation. This in turn means that translations of product brochures, manuals or PR literature are now also published in the electronic mode. (2002, p. 9)

Technical communication is becoming an increasingly online enterprise. One of the basic tenets of technical communication has always been to make sure that the audience has access to the documentation (Markel, 2003). As more of that audience has access to the Internet, the benefits of online publication of documentation, such as the ability to cut printing costs and easily update information, have become attractive reasons to switch to online publication.

Since technical documentation has been increasing its presence online, translations of this documentation have also increased their presence online. Here the benefit of fast and easy updating of documentation has become a problem. Each time a site or document is updated, it must also be translated. Human translators can rarely keep up with the workload that these sites require, and if they could keep up, the work would be prohibitively expensive. Estimates for translation run between 30 and 60 cents or more per word, making the cost of translating a large corporate Web site run easily into the thousands for just one language (Horton, 1993). Many companies are turned off by such high estimates, yet they may not be aware of any other alternatives to using a human translator (O'Hagan & Ashworth, 2002).

STRATEGIES FOR RAPID DISSEMINATION OF INFORMATION: TRANSLATION AND LOCALIZATION

Examples of companies that are increasing their global presence and providing translated Web sites for each of its major markets can be found all over the Web. However, strategies for coordinating content and managing translations can differ widely. The following two case studies present two different approaches for producing localized and translated content for online purposes.

Case Study: Avon—Producing Different Content from the Ground Up

As of 2005, Avon Products has built and maintained a separate Web site for each of its 44 markets (available from the global Avon page at http://www.avoncompany.com/world/). In an effort to provide optimal localization and translation for each market, Avon had each market office produce its own version of the Web site. This seemed like a good idea during the planning phase. However, Avon found that keeping all of the sites coordinated with similar information was an impossible task. Some sites, such as the Venezuelan Avon site, updated frequently and employed professional layout and design. Others were five years behind the U.S. site in terms of content and quality (Orr, 2005).

The end result was a very inconsistent Web experience for international users, with each individual site providing a very distinct message according to the site designer's level of expertise and availability of resources. Avon eventually resorted to sending out U.S. experts to coach designers of different sites in order to achieve a similar message and look, but tailored to the audience for each market (Orr, 2005).

As a result of such experiences, many companies are first producing the U.S. content, and then changing what must be localized in order to address audience needs for each cultural market, and last, translating the content. Keeping the content similar allows for faster and more consistent updating of all of the content, whereas starting from the ground up (such as was the case with Avon) not only makes the updating slower and less consistent, but it also can create a cluster of Web sites with wildly different themes and messages (Orr, 2005).

Case Study: Caterpillar—Using Plain English and Machine Translation Technology

The Caterpillar Company has used a different strategy for producing multiple-language documentation, both in print and for online use on its site, http://www.cat.com. The Caterpillar company has developed "Caterpillar Technical English" (or CTE), which uses a lengthy list of hundreds of stock words and phrases that are approved for use in order to simplify translation. Each of these stock words or phrases gets an approved translation from a human editor, and once that has been accomplished, these translated phrases are stored in the translation memory (TM) to be used again and again in a machine translation program. Using this system allows the machine translation program to automatically recognize each of the words of phrases stored in the translation memory and match it with the specific translation that was approved by the human translator, therefore avoiding errors in translation that could be introduced by the program. This strategy is particularly useful for highly technical terms (Cronin, 2001, p. 114).

Caterpillar's efforts have been recognized widely throughout the industry as perhaps the best way to produce documentation for multiple sites. The way Caterpillar approaches translation reaps the benefit of a quick, low-cost translation through machine translation technology, yet takes steps to ensure that the translation is an accurate one by obtaining and storing translation of difficult or technical terms or phrases from a human translator (Cronin, 2001).

TRANSLATING FOR THE WEB vs. FOR PRINT

One of the reasons why developing different language versions of Web sites is so difficult and there are so many strategies for translation and localization is because there are several differences between translating for the Web and translating for print. These differences usually make using a human translator more difficult, more time consuming, and more expensive than employing a human translator for print documentation. According to O'Hagan and Ashworth (2002), translating for the Web differs from translating for print in the following ways:

1. The readership of the text is unspecified and can mean an extremely wide range of native-speaker populations.
2. The text will be read on screen rather than on paper, at least in the first instance of use.
3. The text may be read in any order, and therefore in different contexts, depending on which hyperlink the reader may follow.
4. The text is subject to much more frequent change than is paper-based text.
5. The text may need to be adapted to the target market readers, involving content changes.
6. The text may contain multimedia components, such as audio and extensive graphics and icons, whose cultural appropriateness may need to be considered against the target-culture norms.

The readership of the text is unspecified and can mean an extremely wide range of native-speaker populations. In regard to the readership, the wide variety of potential audiences has often been cited as an excuse for not translating or localizing Web sites. Since the Web is, in theory, available to anyone with a computer and Internet access, the potential audience grows from a few thousand people to over one billion people. Obviously, no one Web site can take into account the needs and preferences of this many people. This is where market research comes in. If the company decides that a significant amount of its audience comes from a limited range of native-speaker populations, then it can concentrate its efforts on translating and localizing for these specific audiences.

If the company did not want to limit itself to even a few audience possibilities based on research, the beauty of machine translation is that even the least accurate of commercial machine translation programs can instantly and inexpensively provide at the minimum a gist translation (meaning not 100% accurate, but enough to get the message across) to even more language groups.

The text will be read on screen rather than on paper, at least in the first instance of use. Translating online text involves an understanding of online media. For example, navigation, alt tags, and other markers that are specific to online usage must also be translated, which means that the translator must have a certain amount of technological savvy to locate and accurately translate these markers (O'Hagan & Ashworth, 2002).

Further, people read differently on screen than they do online; studies have shown that users tend not to read all of the text (especially extended blocks of text), and they are hesitant to read anything that they cannot immediately perceive the value of (O'Hagan & Ashworth, 2002). The fact that people read 20-30% more slowly online, and that reading online is more tiring, also contributes to this hesitancy to read the entire text (Dillon, McKnight, & Richardson, 1998).

The text may be read in any order, and therefore in different contexts, depending on which hyperlink the reader may follow. Text coherence is a major consideration in translation. However, since the audience can choose any number of orders to read the text or ways in which to follow links, it is extremely difficult to make sure that the content for each link makes sense in light of the content of the previous links. Because of this lack of imposed order, translators have far less control over the overall coherence of the text, and the process of checking for coherence can be much more time consuming, as the coherence of the overall site is often checked link by link rather than in consideration of the previous information (O'Hagan & Ashworth, 2002).

The text is subject to much more frequent change than is paper-based text. This is perhaps the issue that leads to the biggest issue with using human translators for the Web: increased cost and bigger time delays. As previously mentioned, human translators rarely have the time or resources to translate changes anywhere near real-time, most often taking two to three weeks, and they are also much more expensive than using machine translation technology. Further, when the content changes are for multiple language versions of a site online, this may mean that several translators need to be employed to translate each language version for every update made. Large sites often update quite frequently, many times a day, making near-real-time updating impossible (O'Hagan & Ashworth, 2002).

The text may need to be adapted to the target market readers, involving content changes. Just like in print, content changes must be completed before a translator gets the content for translation. For each language version, significant content changes may need to be made in consideration of the cultural content and rhetorical preferences of the audience (O'Hagan & Ashworth, 2002). For example, many corporate Web sites such as Coca-Cola (www.coca-cola.com) localize content for each country page before the pages are translated. Translators usually are not responsible for such content changes, but the localization effort can mean additional delays before the site can be translated.

The text may contain multimedia components, such as audio and extensive graphics and icons, whose cultural appropriateness may need to be considered against the target-culture norms. The flexibility of online content can be both a blessing and a curse when it comes to localization. As the amount of content on a page increases, the amount of localization efforts must rise accordingly. When video and other multimedia components are added, if the goal is to make each language site contain the same types of content, this may mean adding subtitles or even producing different multimedia content localized for each site. Further, the appropriate method for translation and localization must be chosen, and these methods may be individual for each country. For example, a video clip produced in the U.S. may be able to be translated into Spanish for a Mexican audience, but if the film is inappropriate for this audience, a different film might have to be produced or alternate content found to substitute for the film. All of these decisions take time and further slow the process of getting the site updated and translated (O'Hagan & Ashworth, 2002).

MACHINE TRANSLATION: HOW DOES IT WORK?

Because using human translators can be more time consuming and expensive than using a machine translation program, many companies such as Caterpillar are exploring ways to use machine translation technology to produce fast, inexpensive translations.

Translation software has been used by the government since the 1950s on mainframe systems. Once the Pentium chip was invented and desktop computers became a reality, translation software became available to anyone who wanted it (Budiansky, 1998, p. 81).

These systems have improved somewhat, in that now machine translation is capable of looking at adjacent words and arriving at some sort of contextually adequate (at times) rendering of the word. Before it used to be very literal—the computer would look up a word and replace it with an equivalent in the other language without looking at context at all.

However, machine translation systems are still far from perfect. The problem with machine translation is that it is an "AI (artificial intelligence) complete" problem, meaning that in order to have a perfectly functioning machine translation system, the system must be able to "read" and process information similarly to how a human reads and processes information. The debate is whether computers have a mind in the sense that humans have a mind, as adherents to "strong AI" believe, or if the goal of achieving this type of understanding in computers is outside our reach, as is the argument of "weak AI" (Searle, 1980).

At the crux of this issue is the fact that at this point, computers cannot understand the information that they process, and as a result, they cannot understand the information that is being translated (see discussion of the Chinese room experiment in Searle, 1980). This type of understanding is a key element for grasping the context of the sentence being translated. Without understanding the context, an appropriate translation cannot be made. For example, the computer cannot understand the text to know that in the case of the English sentence "He beat the system," the verb "beat" refers to winning rather than a physical beating.

Although AI researchers are continually trying to find ways to make computers understand the information they are given or make them simulate understanding, as of yet, computers have not been able to do this. The best that even our most modern machine translation systems can do is use the information that has been programmed in by a person (Trujillo, 1999).

At first, machine translation programmers set out to do just that; they took thousands of syntactic and linguistic rules and programmed them into the system (called "parsing"). However, the problem of context was not solved by programming in language rules alone. What the programmers found was that language and its relationship to context was more complex than they had previously anticipated; the hundreds of programmed-in rules became thousands as more mistakes were uncovered, then tens of thousands once the rules from one language were sorted and "poured" into the rules for another language. The result was literally millions of potential mistakes, and that is just between two different languages (Budiansky, 1998). Modern machine translation systems often boast of the capability to translate to and from more than 20 languages—using a simple parsing system, the potential for error is nearly endless.

When programmers began to realize that they could not possibly rely on a parsing machine translation system, they began to explore other ways to use machine translation which might better simulate the understanding of context. One solution was to feed in thousands of pages of text in two different languages that had been translated by a human translator. Using these millions of sample translated sentences, the machine then statistically determined the probable word order and context of the sentence, and selected the most probable translation. The developers of these systems hoped that the context problem could be solved by simple mathematics. For example, if the words "book" and "criminal" had a high probability of occurring in a sentence where "book" meant "to process a criminal" (rather than the noun "book"), the machine translation system should deduce the context of the sentence from the probability of that association. The goal was to "force" the machine translation system to consider the context of the sentence, even if the system could not understand it.

Many supporters of the parsing method of machine translation scoffed, feeling that machine translation could not possibly work without the nuances and subtleties of linguistics included into the programming. Surprisingly, statistically based machine translation systems have an accuracy rating of around 75%, only 10-15% lower than most parsing-based MT programs (Budiansky, 1998).

While statistically based machine translation systems worked reasonably well, they did not solve the context issue as hoped, nor did they solve the problems with errors that the parsing systems made. Language is just too flexible and full of meaning, hidden or otherwise, to completely eliminate contextual errors based on statistics or parsing alone. However, one important development that did come out of these statistically based systems was translation memory, a tool which stores approved translations of certain words or phrases. As translation memory was used in the case of Caterpillar, a human translator could program in some of the more technical or difficult words or phrases to translate, thus leaving the machine translation system to translate the more common terms. This development has helped make machine translation much more exact, especially when the text to be translated is highly technical in nature. While translation memory is a work-around for machine translation and does not help the program translate more effectively, it does allow for a more accurate translation overall, and one which needs less attention from a human translator than the traditional method.

Today, most commercial machine translation systems do not exclusively use one method or the other. Variations of these two methods (parsing and statistical) now number in the hundreds as research has progressed. Instead, most machine translation systems use two or more of these methods as filters. For example, one method which is highly accurate in regard to word-for word translation might be employed first, then a method which is more accurate for proper syntax could be used next, and so on, until a translation that is as accurate as possible is achieved. Using these methods as filters helps eliminate many of the mistakes that can result from using one method alone. Further, combining translation methods in this manner is said to improve translation quality by an average of 10% (Ren & Shi, 2002, p. 361). Some of these methods are up to 93% accurate, as high as commercial (non-classified) machine translation is reputed to achieve (2002).

Obviously, 93% accurate is still far from perfect, and not as accurate as a human translation. However, for the purposes of rapid and inexpensive translation, 93% can provide more than the mere echo or "gist" of the original document. Until AI researchers develop a way to make computers understand the context of the sentences in the way that humans are do, 93% is likely as good as the MT technology will get.

CURRENT AND FUTURE USES OF MACHINE TRANSLATION FOR THE WEB

Many users are familiar with popular translation engines for the Web such as Altavista's Babel Fish (http://babelfish.altavista.com/). When many users think of online machine translation, they think of free services such as this one. Often, experience with free online translation services causes many to think poorly about machine translation. Entire articles, such as Steven Budiansky's (1998) "Lost in Translation," have been written to satirize the translation results of such machine translation tools. However, before one makes a judgment about machine translation, one must look at the true purpose of such free services.

Babel Fish, for example, is powered by SYSTRAN, the industry leader in machine translation. However, the intention of Babel Fish and other free translation sites is not so much to get you to use their services, but rather to purchase the more robust versions produced by those same companies (Reinhardt, 2002). In other words, you may find yourself simultaneously impressed by the ability of Babel Fish to translate your site, yet dismayed by the inaccuracies it introduces. Therefore, a visit to Babel Fish might prompt you to visit the real powerhouse behind the free translation: SYSTRAN.

SYSTRAN has recently been making inroads into this emerging online market. Its Web site (http://www.SYSTRANsoft.com) states that, "SYSTRAN offers the widest variety of language pairs of any true machine translation product, excluding simple word-lookup products. Translation systems are available for 28 mono-directional language pairs." SYSTRANnet is SYSTRAN's commercially available tool for translating Web pages, and it is downloadable from the SYSTRAN Web site and offers a bidirectional language pair for $30.

Autodesk, a U.S.-based software company and the developer of AutoCad, was the first company to utilize SYSTANnet on its commercial Web site, starting in 2001. Autodesk, like many other companies, had special difficulties with its Web sites in terms of translation. According to SYSTRAN:

With more than 60% of its business conducted outside the United States, Autodesk is a case study for the challenges of information dissemination and management in a multilingual environment. One of Autodesk's most pressing challenges is supporting customers across many languages in a cost-efficient manner. Autodesk provides customer support through a database of more than 10,000 articles that are accessible from its Web site. The articles, which average 1,000 words in length, are written in English only, are highly technical, and are specific to design issues for various industries.

With SYSTRANnet, Autodesk made 10,000 of its technical and support documents available in over 20 language pairs, and for less than half the cost that employing a human translator for a one-time (not constantly updated) translation would cost (Schenker, 2005). The results, according to Mirko Plitt, process analyst in Autodesk's Worldwide Localization department, have been astonishing. On SYSTRAN's Web site, he states:

Innovative customization approach was the only answer to our international customers' need for a multilingual product support knowledge base: translations produced by general-purpose machine translation systems are of little use to our non-English-speaking clients, and a translation workflow involving human intervention was not a realistic option. The specific machine translation solution developed by SYSTRAN maximizes the benefit our customers get from the Product Support Web site and further increases the quality of service provided by Autodesk.

Autodesk is an interesting example of the possibilities of machine translation for online uses, not only with SYSTRAN, but with other machine translation companies as well. Many companies with international markets may not be aware that that there is a happy medium between a Babel Fish "gist" translation and a human translator, or that other companies are using such technology successfully. The success of SYSTRANnet with Autodesk demonstrates that even companies with highly technical documents can successfully utilize machine translation technology.

OPTIMIZING MACHINE TRANSLATION FOR THE WEB

Of course, certain strategies can be employed to help increase the probability that the translation on the Web site will be an accurate and useful one. First, consider employing what some companies refer to as a style sheet to use certain words and phrases consistently. These phrases can then be checked by a human translator for accuracy and added to the translation memory for use again and again. Employing a human translator to translate a few stock words and phrases one time costs pennies compared to using one to translate an entire site. This strategy is especially useful with complex or highly technical information because it takes the guesswork out of the most difficult to translate words and phrases in the document. Many companies, such as Caterpillar, have used this strategy successfully for years, both when employing a human and a machine translator (Cronin, 2002).

Next, do not make documents more difficult to translate than they need to be. Whenever possible, use simple (or "plain") English, avoiding colloquialisms or other phrases that are difficult to translate even under the best circumstances. Writing for translation should ideally begin during the planning process, rather than after the document is already written (Cronin, 2002).

Last, because different cultures prefer different rhetorical styles and even different types of content, localize the content and wording before the text goes to the translator. As in the Coca-Cola Web site example, a successful international site must have culturally appropriate content as well as accurate translations. In order to save embarrassment and costly mistakes, usability testing or checking the content with a subject-matter expert is a useful strategy to make sure that the content is culturally appropriate. Finding these mistakes early can keep the content from having to be updated and re-translated multiple times.

CONCLUSION

Today, companies and organizations have several options and strategies available for utilizing machine translation technology for their Web sites. If only a gist translation is needed, Altavista and other sites that offer free translation utilities offer downloadable utilities that can be linked to through plugging in a single line of code into your Web site's code. If a more robust translation is needed, downloads such as SYSTRANnet and other machine translation utilities meant strictly for online use are available and still relatively inexpensive compared to a human translator. Other functions such as translation memory can add further accuracy for machine translations.

No matter what strategy for utilizing machine translation is chosen, it is certain that more companies will become aware of the potential of using machine translation for their Web sites. As the Internet becomes more global and less insular in design and use, a newer, more diverse crop of users has been emerging—users who do not necessarily accept the fact that they have to read online content in English alone. Taking advantage of the available resources, such as machine translation, can only increase the Web site's exposure and build sales in international markets.

REFERENCES

Budiansky, S. (1998). Lost in translation. *The Atlantic Monthly,* 81-82.

Cronin, M. (2001). *Translation and globalization.* New York: Routledge.

Dillon, A., McKnight, C., & Richardson, J. (1998). Reading from paper versus reading from screens. *The Computer Journal, 31,* 457-464.

Forsyth, S. (2005). *Online translation—the future??* Retrieved from http://www.translation-services-usa.com/articles/online-translation.shtml

Horton, W. (1993). The almost universal language: Graphics for international documents. *Technical Communication, 43,* 682-693.

Internet Society. (n.d.). *A brief history of the Internet.* Retrieved from http://www.isoc.org/internet/history/brief.shtml

Markel, M. (2003). *Technical communication* (7th ed.). Cranbury, NJ: Bedford/St. Martin's.

O'Hagan, M., & Ashworth, D. (2002). *Translation-mediated communication in a digital world.* Buffalo, NY: Multilingual Matters.

OCLO. (n.d.). *Web characterization: Country and language statistics.* Retrieved from http://www.oclo.org /projects/archive/wep/stats/intnl.htm

Orr, A. (2005). *Lost in translation.* Retrieved from http://www.cmomagazine.com/read/050105/lost_translation.html

Reinhardt, A. (2002). It's all Greek to these sites. *Business Week, 3792,* 18.

Ren, F., & Shi, H. (2002). A multi-engine translation approach to machine translation. *International Journal of Information Technology & Decision Making, 1,* 349-361.

Schenker, J. (n.d.) *The gist of translation.* Retrieved from http://time.com/time/digital/printout/0,9869,166805,00.html

Searle, J. (1980). Minds, brains, and programs. *Behavioral and Brain Sciences, 3,* 417-457.

Trujillo, A. (1999). *Translation engines: Techniques for machine translation.* New York: Springer-Verlag.

Chapter IV
Classifying Web Users: A Cultural Value-Based Approach

Wei-Na Lee
University of Texas at Austin, USA

Sejung Marina Choi
University of Texas at Austin, USA

ABSTRACT

In today's global environment, a myriad of communication mechanisms enable cultures around the world to interact with one another and form complex interrelationships. The goal of this chapter is to illustrate an individual-based approach to understanding cultural similarities and differences in the borderless world. Within the context of Web communication, a typology of individual cultural value orientations is proposed. This conceptualization emphasizes the need for making distinctions first at the individual level, before group-level comparisons are meaningful, in order to grasp the complexity of today's global culture. The empirical study reported here further demonstrates the usefulness of this approach by successfully identifying 16 groups among American Web users as postulated in the proposed typology. Future research should follow the implications provided in this chapter in order to broaden our thinking about the role of culture in a world of global communication.

INTRODUCTION

As the adoption of media technology such as the Internet rapidly spreads around the world, communication across cultures increases. Individuals from diverse cultural groups interact with each other regardless of physical distances. On the one hand, such increased communication between cultures might facilitate cultural convergence on the global scale (Kincaid, 1988; Rogers & Kincaid, 1981). On the other hand, online technology's capability to offer individualized communication might further fragment the global culture as people with similar values, outlooks, and interests across the world pursue their personal agendas via the decentralized electronic media (Choi & Danowski, 2002).

Culture has been a focal issue in global communication. More specifically, cultural similarities and differences have been considered the key to understanding cross-cultural human interactions. Extensive research to date has provided ample evidence of differences between cultures in terms of communication styles and messages. Implicit to this line of research is the assumption that members of a culture are likely to exhibit a pattern of social perception and behavior common within the culture, but different from that of another culture. Given this paradigm of conceptualizing culture, most cross-cultural comparisons are made at the national or cultural level, that is, between nations or cultures, while overlooking potential variations among individuals within a culture.

In today's fast-changing media environment, people are exposed to various cultures through a multitude of channels and formats. While still adhering to the dominant values of the culture in which they belong, people these days rely on multiple frames of cultural reference simultaneously to construct their individual cultural orientations. For these reasons, it would be too simplistic to assume that everyone in the same culture displays the same pattern of thinking and behavior. In fact, individuals' cultural orientations within the same culture could vary widely (Campbell, 2000). Therefore, a thoughtful investigation of today's technology-mediated global culture needs to start from exploring fundamental cultural value orientations at the individual level.

Foremost among the major dimensions of cultural orientations are individualism and collectivism. Generally considered as polar opposites of each other, individualism emphasizes the concept of self, whereas collectivism focuses on other-directedness. Departing from this dichotomous view, recent research has suggested a more in-depth conceptualization of individualism and collectivism where, depending on whether equality (horizontal) or hierarchy (vertical) is underscored, the following four types of orientations can be identified: (1) *horizontal individualism (uniqueness)*, where one can be unique and independent while still maintain status equality with others; (2) *vertical individualism (achievement)*, where one strives to be the best and enjoys privileges that come with it; (3) *horizontal collectivism (cooperativeness)*, where interdependence and equality in status are valued; and (4) *vertical collectivism (dutifulness)*, where people submit to the social hierarchy ascribed by their in-groups (Triandis, 1995, 2001; Triandis & Suh, 2002). Initial empirical evidence supports the viability of this four-way typology in detecting differences across national cultures (Nelson & Shavitt, 2002) and individual differences within a single culture (Lee & Choi, 2005).

Understanding similarities and differences in cultural orientations is the key for successful global online communication. Since the Web has emerged as an ideal medium for tailored communication for people across the world, it is imperative to obtain a baseline understanding of cultural values held by those who are users of the Web. As cultures increasingly interconnect on the Web and national borders gradually vanish, these insights will help prepare us for a future world community that is likely to be dominated by technology-mediated communication. At this juncture, research on cultural values in global communication should focus on the individual, not the nation or culture. Therefore, based on the aforementioned four-way typology, the goal of this chapter is to propose and empirically demonstrate a comprehensive classification framework for assessing cultural orientations at the individual level.

To accomplish the goal, this chapter first explains individualism and collectivism as dimensions of culture, and reviews relevant research developments in this area. Then, a thorough explication of an in-depth typology that encompasses sufficient complexity to reflect cultural differences among individuals is provided. Next, this chapter reports results from an empirical study that classified and compared Web users in the U.S., using the proposed individual level typology. This chapter concludes with implications of the findings and directions for future research.

THE CONCEPTUALIZATION OF INDIVIDUALISM AND COLLECTIVISM

The constructs of individualism and collectivism have been widely regarded as instrumental in helping explain differences between cultures. This can be seen from the vast amount of literature employing these ideas to account for differences in communication patterns and content, business practice, and preferences for communication styles and persuasive message appeals (Cho, Kwan, Gentry, Jun, & Kropp, 1999; de Mooij, 1998; Hall, 1984; Han & Shavitt, 1994; Hofstede, 1983, 1984; Miracle, Chang, & Taylor, 1992). The following sections provide a review of the defining characteristics of individualism and collectivism, and the conceptual advancements in this area.

Individualism and Collectivism

From an analysis of survey data collected from more than 50 countries around the world, Hofstede (1980) identified individualism and collectivism as one of the several fundamental dimensions of culture. He further demonstrated how these constructs can be characterized in people's social perception and behavior. In individualistic cultures, typified as autonomous and independent, people's personal goals are usually valued over the goals of their in-groups. Hence, people's behaviors are primarily based on their own attitudes rather than the norms of their in-groups. In contrast, people in collectivistic cultures are interdependent within their in-groups, and the goals of their in-groups take priority over their own. Collectivists generally behave according to the norms of their in-groups, and place much emphasis on group harmony and hierarchy. In short, collectivists tend to do what is expected of them whereas individualists tend to do what they find personally fulfilling (Triandis, 1995).

Hofstede's (1980, 1984) one-dimensional conceptualization places individualism and collectivism at the opposite ends of a continuum. A majority of the focus in research since then has been placed on explaining national differences using these constructs (Gudykunst & Ting-Toomey, 1988). In applications, comparisons are made by defining nations as residing at one or the other of those two extremes, or between them. Given today's global environment and frequent border-crossing of people and ideas, however, the notion of a homogeneous population within a culture and the nation-based view of those constructs and comparisons may no longer hold true (Singlis & Brown, 1995). Put simply, not every person in an individualistic culture is an individualist. Nor does it mean that people in a collectivistic culture are all collectivists.

Research evidence over the years further suggests that the dichotomous conceptualization of individualism and collectivism in a bipolar manner is limited. There is a great need to expand the conceptualization from uni- to multidimensional in order to capture the complexity of cultural orientations and offer a comprehensive understanding (Singlis, Triandis, Bhawuk, & Gelfand, 1995; Triandis & Gelfand, 1998).

Horizontal / Vertical Individualism / Collectivism

Expanding on the above view, Triandis (1995, 2001) suggested that there are, in fact, different types of individualism and collectivism. Upon careful examination, for example, Korean collectivism is not entirely the same as the collectivism among the Japanese. The individualism in France is different from American individualism. Among the many dimensions that can further distinguish individualism and collectivism is the horizontal-vertical aspect of social relationships. In essence, both individualism and collectivism may be *horizontal* (emphasizing equality) or *vertical* (emphasizing hierarchy). Underneath the horizontal orientation is the assumption that people see themselves as being essentially similar to others in their social relationships. On the contrary, the vertical dimension highlights hierarchy as the key to social relationships where people perceive themselves as being different from others. Adhering largely to Shintoism, the Japanese are highly egalitarian while exhibiting a very strong sense of cooperation. In contrast, social hierarchy is the guiding principle for attitude and behavior in Korea. Korean people have a tendency to value family and group hierarchy, and sacrifice their personal goals for group goals. Research has further shown that some individualistic cultures such as France and Sweden place more emphasis on equality by focusing on "doing one's own things," whereas other individualistic cultures such as the U.S. tend to emphasize hierarchy by embracing superiority.

From this conceptualization, four distinct types of cultures can be identified: (1) *horizontal individualism (HI-uniqueness)*, where people strive to do their own thing and be unique; (2) *vertical individualism (VI-achievement)*, where people strive to be distinguished and the best in competition with others; (3) *horizontal collectivism (HC-cooperativeness)*, where people merge their selves with their in-group and underscore interdependence, harmony, and common goals with others; and (4) *vertical collectivism (VC-dutifulness)*, where people submit to the authorities of the in-group and are willing to sacrifice themselves for their in-group (Triandis, 2001; Triandis & Suh, 2002). Although all individualistic people share the tendency of being independent and giving more priority to personal goals over group goals, in HI, people have little interest in acquiring high status, unlike those in VI. Likewise, in HC, people respect group goals, but do not simply give in to authorities, much different from those in VC.

While this four-way typology was suggested to help identify distinct prototypes of cultures, it also shows promise for an in-depth understanding of within-culture variations (Lee & Choi, 2005). At the individual level, however, the four types are often orthogonal and not mutually exclusive. In other words, people can and may exhibit a number of combinations of these cultural orientations. Given today's global environment where cultures are increasingly fused via a variety of media technology, it is important to recognize the growing need for a more in-depth typology that captures the intricacy of cultural values.

Idiocentrism and Allocentrism

Shifting the focus from the group to the individual, the terms "idiocentrism" and "allocentrism" have been put forth to refer to personal individualism and collectivism, respectively (Yamaguchi, Kuhlman, & Sugimori, 1995). From this conceptual standpoint, a person's cultural orientation is not automatically equated with his or her cultural or national membership. At the individual level of analysis, people's cultural values may no longer coincide with the culture in which they belong. That is, there are idiocentrics (those who possess individualistic characteristics) in collectivistic cultures and allocentrics (those who show collectivistic orientations) in individualistic cultures (Triandis, McCusker, & Hui, 1990). In other words, both idiocentrics and allocentrics can and should be found in any given culture. However, the proportions of the two groups might vary given the dominant values at the national or cultural level (Triandis & Suh, 2002).

Based on the horizontal/vertical individualism/collectivism conceptualization and the need for individual level of understanding, a more thorough classification of cultural orientations is outlined in the following section.

CLASSIFYING INDIVIDUAL CULTURAL ORIENTATIONS

Research evidence to date has demonstrated the utility of Triandis' typology. Studies using the four prototypical patterns (HI, VI, HC, VC) to classify cultures generally assume that one particular type among the four prevails as the cultural orientation. In theory, however, people can have a mixture of attributes as defined by these four types simultaneously. Therefore, the original typology might not effectively describe all possible combinations of the various types of cultural orientations that people exhibit. A more comprehensive multidimensional classification, which identifies those pure types documented in the original conceptualization plus the hybrid types, should serve as a useful starting point to capture differences in individual cultural orientations.

Triandis' distinction between horizontal and vertical individualism and collectivism is primarily developed for group-level comparisons. As an extension, the proposed individual-level typology classifies people based on the extent to which they exhibit patterns of attributes along *horizontal and vertical idiocentrism and allocentrism.* Since a person's orientation can be assessed as either high or low on the four different dimensions, this expanded classification, presented in Table 1, identifies 16 possible individual cultural orientation groups. Below is a description of the major groups under this framework.

Horizontal Idiocentrics and Vertical Idiocentrics

Horizontal and vertical idiocentrics are those individuals who exhibit idiocentric tendencies with either a horizontal or vertical orientation.

Table 1. Individual cultural orientation classification

Group	Name	HI	VI	HA	VA
1	Horizontal Idiocentrics	***High***	Low	Low	Low
2	Vertical Idiocentrics	Low	***High***	Low	Low
3	Horizontal Allocentrics	Low	Low	***High***	Low
4	Vertical Allocentrics	Low	Low	Low	***High***
5	All-Around Idiocentrics	***High***	***High***	Low	Low
6	All-Around Allocentrics	Low	Low	***High***	***High***
7	The Horizontals	***High***	Low	***High***	Low
8	The Verticals	Low	***High***	Low	***High***
9	Bi-Idiocentric Allocentrics	Low	***High***	***High***	Low
10	Bi-Idiocentric Allocentrics	***High***	Low	Low	***High***
11	Tri-Idiocentric Allocentrics	Low	***High***	***High***	***High***
12	Tri-Idiocentric Allocentrics	***High***	Low	***High***	***High***
13	Tri-Idiocentric Allocentrics	***High***	***High***	Low	***High***
14	Tri-Idiocentric Allocentrics	***High***	***High***	***High***	Low
15	Null	*High*	*High*	*High*	*High*
16	Null	Low	Low	Low	Low

They score high on horizontal or vertical idiocentrism and low on all of the other dimensions. As idiocentrics, these people value the self and independence. However, *horizontal idiocentrics* focus on self-development and self-reliance, whereas *vertical idiocentrics* are competitive and strive to be better than the rest.

Horizontal Allocentrics and Vertical Allocentrics

People who score high on horizontal or vertical allocentrism and low across all of the other dimensions are referred to as horizontal or vertical allocentrics. *Horizontal allocentrics* work toward group harmony and regard relationships among individuals in the group to be more or less equal. In contrast, *vertical allocentrics* put group goals above individual desires and obey social hierarchy closely.

All-Around Idiocentrics and All-Around Allocentrics

Although distinctions have been made between an orientation toward equality (horizontal) vs. hierarchy (vertical), people could still display both simultaneously. Those who are classified as high on both horizontal and vertical dimensions of idiocentrism, but low on both dimensions under allocentrism, may be regarded as *all-around idiocentrics*. With a focus on the self over groups, their social relations encompass tendencies toward hierarchy as well as equality. In other words, all-around idiocentrics keenly compete with others for status and recognition, as well as strive to be themselves and independent of others. Likewise, those who score high on both horizontal and vertical dimensions of allocentrism, but low on both dimensions of idiocentrism, are referred to as *all-around allocentrics*. Since their social relationships evolve around their in-groups both horizontally and vertically, all-around allocentrics strive for group harmony and hierarchy at the same time. For both all-around idiocentrics and allocentrics, the dominance of either the horizontal or vertical dimension could be a function of other factors.

The Horizontals and the Verticals

People who are predominantly horizontal or vertical in their orientation could behave with a mixture of both idiocentric and allocentric tendencies. They are referred to as *the horizontals* or *the verticals*. The horizontals are classified as high on the horizontal dimension under both idiocentrism and allocentrism, but low on both vertical dimensions. The opposite is true for the verticals. It appears that the horizontals and the verticals value equality and hierarchy, respectively, above and beyond their self or group orientation. With their deep-rooted view of hierarchy-centered social relationships, for example, the verticals respect the authorities in their in-group while endeavor to achieve high social status.

Bi-Idiocentric Allocentrics and Tri-Idiocentric Allocentrics

Bi-idiocentric allocentrics represent a portion of the population who simultaneously display attributes from two diametrically different dimensions. This group includes those who are high on vertical idiocentrism and horizontal allocentrism but low on horizontal idiocentrism and vertical allocentrism. In addition, people who score high on horizontal idiocentrism and vertical allocentrism but low on vertical idiocentrism and horizontal allocentrism also qualify as bi-idiocentric allocentrics. In a similar vein, *tri-idiocentric allocentrics* are people who are high on three out of the four dimensions and low on the remaining one. The level of complexity is significantly higher in this tri-idiocentric allocentric group and, consequently, it could be difficult to disentangle their attitudinal and behavioral differences based solely on cultural orientations.

The Null Group

Those who are equally high or low across all four dimensions are classified as the null group. Theoretically, it is plausible that some individuals simply do not exhibit a strong inclination toward any particular orientation. Likewise, some people might display an equally strong tendency toward all four orientations. Consequently, it might be a rather challenging task to comprehend these null groups' cultural orientations due to their even predisposition along different dimensions. In this situation and other similarly complex situations outlined above, the proposed classification should be used in tandem with other cultural constructs in order to shed light on the multifaceted nature of individual cultural orientations.

CULTURAL ORIENTATIONS AND GLOBAL COMMUNICATION ON THE WEB

With the rapid diffusion of technology, the Web is quickly becoming a significant part of people's daily lives. From information gathering to entertainment, from shopping to personal communication, the Web is omnipresent. As the penetration of the Web reaches the general population, the online population in the U.S. becomes demographically diverse (Schlosser, Shavitt, & Kanfer, 1999). Research evidence shows that this trend is occurring worldwide as well. Web users in different parts of the world become similar to each other in terms of their demographic characteristics and general Web use patterns (Chen, Boase, & Wellman, 2002). As a result, people within the U.S. are less similar to each other than they are to others outside the U.S. In other words, there could be more differences within a country than there are between countries.

The Web is a distinctive medium very much characterized by its ability to offer audience-controlled exposure, selectivity, and interactivity (Wolin, Korgaonkar, & Lund, 2002). Due to the unique nature of the Web, personal characteristics of Web users need to be taken into consideration in order to understand their social behaviors. Among the various demographic and psychological characteristics, Web users' cultural orientations should be of prime importance in this endeavor.

In communications research, individualism and collectivism have served as a useful means to compare similarities and differences in styles and content across cultures (de Mooij, 1998; Hofstede, 1980, 1983). In addition, individualism and collectivism appear to be related to the distinction between low vs. high context. Generally speaking, low-context communication, which is usually straightforward, explicit, and direct, is common in individualistic cultures, whereas communication in collectivistic cultures is highly context dependent, typified by abstract, implicit, and indirect messages (Hall, 1976, 1984).

These constructs of cultural values offer viable means for examining differences in computer-mediated communication between cultures (e.g., Callahan, 2005; Wurtz, 2005). However, as the Web continues to facilitate intercultural flows of ideas and communication between people from all over the world in the forms of message boards, e-mails, newsgroups, and so forth, the need for a thorough investigation of cultural values at the individual level becomes more pertinent (Hewling, 2005). As Scollon and Wong-Scollon (2001) state, "Cultures do not talk to each other; individuals do" (p. 138). Differences in individual cultural orientations might be the most critical factor in understanding Web users from diverse cultures. In this light, the individual-level typology proposed in this chapter should serve as a useful tool for further scrutinizing cultural orientations and classifying Web users.

THE STUDY

A Web survey with a sample drawn from an online panel of consumers in the U.S. was undertaken in October 2003 to provide empirical evidence for the proposed individual-level typology. The method and results of the study are reported in this section.

Method

Sample. Participants were recruited from an established online consumer panel consisting of Web users with diverse demographic characteristics. Those Web users participate in Web-based studies at regular intervals over a period of time for various rewards. The demographics of the online panel track well with the latest online population trend figures—consisting of predominantly female, young, non-Hispanic white, higher education and household income Web users (Pew Internet & American Life Project, 2005). Of the original 1,101 surveys completed, a total of 1,033 were included in the sample after eliminating incomplete surveys.

More than half of the respondents were female (58.5%). The average age of the respondents was 40 years old with a range from 23 to 80 years. Caucasians constituted the majority of the study participants (79.3%), followed by Hispanics (8.2%) and Asians (5.6%). Over half of the respondents were married (54.5%), while one-third (30.5%) were single. The respondents were relatively well-educated, with the vast majority of them having a college degree or higher (98.9%). In terms of economic status, 70.8% of the respondents were employed full time, and about half of the respondents reported an annual household income of $50,000 or higher. Table 2 provides a description of the sample characteristics.

Data collection procedure. An invitation e-mail announcement was sent to members of the online panel. The invitation e-mail contained a brief description of the study and directed panel members to the study site. All those who participated in the study were eligible to win a drawing of a $150 cash prize.

Measures. The questionnaire consisted of four main sections. In the first part of the questionnaire, respondents were asked to indicate how much time on an average weekday they used television, radio, newspapers, magazines, and the Web. Next, respondents' cultural orientations were gauged via the four-way typology measures developed by Triandis (1995). As the scale, originally developed for group-level measurements, was deemed equally suitable for individual-level assessments (Lee & Choi, 2005), the measures were adopted without modifications. Each of the four dimensions was measured via a four-item, seven-point, Likert-type scale: (1) horizontal idiocentrism—HI, (2) vertical idiocentrism—VI, (3) horizontal allocentrism—HA, and (4) vertical allocentrism—VA. Measures tapping the cultural orientations included statements such as "I'd rather depend on myself than others" (HI), "It is important that I do my job better than others" (VI), "If a coworker gets a prize, I would feel proud" (HA), and "Parents and children must stay together as much as possible" (VA).

The third section of the survey examined respondents' experience with the Web. An eight-item, seven-point, Likert-type scale asked for respondents' beliefs about their Web skills and their perceived degree of challenges in their experience using the Web (Novak, Hoffman, & Yung, 2000). Respondents' demographic characteristics such as gender, age, employment, annual household income, ethnicity, highest education level attained, and marital status were obtained toward the end of the survey. The specific items for the major constructs and their respective reliability coefficients are shown in Table 3.

Table 2. Sample characteristics

Characteristic		Frequency	Percent*
Gender	Male	422	58.5%
	Female	604	40.9%
Age	20-29	260	25.2%
	30-39	341	33.0%
	40-49	151	14.6%
	50-59	167	16.2%
	Over 60	104	10.1%
Employment	Full-time	727	70.4%
	Part-time	113	10.9%
	Unemployed	187	18.1%
Ethnicity	Caucasian	814	78.8%
	Hispanic-American	84	8.1%
	Asian-American	57	5.5%
	African-American	14	1.4%
	Native American	3	0.3%
	Multiracial	20	1.9%
	International	22	2.1%
	Other	13	1.3%
Marital Status	Single	313	30.3%
	Married	560	54.2%
	Divorced	59	5.7%
	Living with someone	75	7.3%
	Separated	6	0.6%
	Widowed	11	1.1%
	Other	3	0.3%
Education	Vocational/technical school (2 yrs)	1	0.1%
	Some college	10	1.0%
	College graduate (4 yrs)	579	56.1%
	Master's degree	283	27.4%
	Doctoral degree	46	4.5%
	Professional degree (MD, JD, etc.)	112	10.8%
Household Income	Under $10,000	24	2.3%
	$10,000-$19,999	25	2.4%
	$20,000-$29,999	51	4.9%
	$30,000-$39,999	105	10.2%
	$40,000-$49,999	120	11.6%
	$50,000-$74,999	173	16.7%
	$75,000-$99,999	159	15.4%
	Over $100,000	309	29.9%
	Other	44	4.3%

*Note: * The base of the percentages was the total sample size of 1033.*

Table 3. Specific items for the key measures

Cultural Dimensions
Horizontal Idiocentrism (α = .64)
I'd rather depend on myself than others.
I rely on myself most of the time; I rarely rely on others.
I often do "my own thing."
My personal identity, independent of others, is very important to me.
Vertical Idiocentrism (α = .66)
It is important that I do my job better than others.
Winning is everything.
Competition is the law of nature.
When another person does better than I do, I get tense and aroused.
Horizontal Allocentrism (α = .69)
If a coworker gets a prize, I would feel proud.
The well-being of my coworkers is important to me.
To me, pleasure is spending time with others.*
I feel good when I cooperate with others.
Vertical Allocentrism (α = .64)
Parents and children must stay together as much as possible.
It is my duty to take care of my family, even when I have to sacrifice what I want.
Family members should stick together, no matter what sacrifices are required.
It is important to me that I respect the decisions made by my groups.*

Web Experience
Web Skills (α = .87)
I am extremely skilled at using the Web.
I consider myself knowledgeable about good search techniques on the Web.
I know somewhat less about using the Web than most users. (R)
I know how to find what I am looking for on the Web.
Web Challenges (α = .66)
Using the Web does not challenge me. (R)
Using the Web challenges me to perform to the best of my ability.
Using the Web provides a good test of my skills.
I find that using the Web stretches my capabilities to my limits.

Note: All of the items are measured on a seven-point scale.

** Item was dropped from creating an index score due to the low factor loading.*

Results

Horizontal/vertical idiocentrism/allocentrism scale validation for individual-level differentiations. Because the group-level measures of horizontal and vertical individualism and collectivism have been previously validated for individual level of analysis (Lee & Choi, 2005), detailed discussion on the horizontal/vertical idiocentrism/allocentrism scale validation is omitted here. In brief, all of the items significantly loaded on the constructs that they were intended to tap into, although two items with a factor loading below 0.4 were dropped from further analysis. Table 4 reports the factor loadings of the indicators for each latent variable and the goodness-of-fit indices of the measurement model. Items for each construct were averaged to form an index score. The descriptive statistics of the four cultural dimensions are shown in Table 5, and the correlations among the constructs are reported in Table 6.

Horizontal/vertical idiocentrism/allocentrism classification. Given the ethnic diversity in the American culture, differences in cultural orienta-

Table 4. Factor loadings of indicators

Factors	Indicators	Unstd.	Std.
Horizontal Idiocentrism	I'd rather depend on myself than others.	1.00	0.72
	I rely on myself most of the time; I rarely rely on others.	1.30	0.72
	I often do "my own thing."	0.74	0.40
	My personal identity, independent of others, is very important to me.	0.54	0.40
Vertical Idiocentrism	It is important that I do my job better than others.	1.00	0.59
	Winning is everything.	1.42	0.68
	Competition is the law of nature.	1.04	0.52
	When another person does better than I do, I get tense and aroused.	1.07	0.53
Horizontal Allocentrism	If a coworker gets a prize, I would feel proud.	1.00	0.65
	The well-being of my coworkers is important to me.	1.03	0.76
	I feel good when I cooperate with others.	0.70	0.58
Vertical Allocentrism	Parents and children must stay together as much as possible.	1.00	0.66
	It is my duty to take care of my family, even when I have to sacrifice what I want.	0.68	0.57
	Family members should stick together, no matter what sacrifices are required.	0.99	0.62

Note: All coefficients are significant ($p < .001$).

Goodness of fit statistics: χ^2 (71) = 561.07, $p < .001$, GFI = .93, AGFI = .89, CFI = .83, RMSEA = .08

Table 5. Descriptive statistics of cultural dimensions

Variables	M	SD
Horizontal Idiocentrism	5.39	0.92
Vertical Idiocentrism	4.03	1.05
Horizontal Allocentrism	5.72	0.84
Vertical Allocentrism	5.32	1.03

Note: All items were measured on a seven-point scale (N=1008).

Table 6. Covariance and correlation matrix of the cultural dimensions

	HI	VI	HC	VC
Horizontal Idiocentrism	1.00	0.33**	0.01	0.03
Vertical Idiocentrism	0.52	1.00	-0.19**	0.09*
Horizontal Allocentrism	0.01	-0.32	1.00	0.29**
Vertical Allocentrism	0.04	0.13	0.39	1.00

*Note: * $p < .01$, ** $p < .001$*

Correlations are in the lower triangle and covariances in the upper triangle.

tions among different ethnic groups were examined first. While significant differences between respondents with different ethnic backgrounds were indeed observed, the relatively small sizes of several ethnic groups did not allow for meaningful comparisons across groups in further investigation. Furthermore, an ethnically homogeneous group of Web users was deemed appropriate for the purpose of assessing the utility of the individual-level typology for detecting within-culture variations. For further classification, therefore, the sample included a single majority ethnic group, 814 Caucasian respondents, after eliminating other small ethnic groups. The Web users in the sample were first divided into high vs. low groups on each of the four cultural elements using a median split. The median scores were 5.5 (HI), 4.0 (VI), 5.7 (HA), and 5.3 (VA) respectively. Using the proposed typology, these Web users were then classified into one of the 16 cultural groups depending on their locations along the four cultural dimensions.

Table 7 reports the number of respondents classified as belonging in each of the groups. In support of the viability of the individual-level typology for detecting within-culture variations, all 16 types of cultural orientations were represented and successfully identified among members of the single culture tested here. Surprisingly, among the 16 groups, *all-around allocentrics* appeared to be the largest cultural group with 102 people. The next major groups identified included two *null* and two *tri-idiocentric allocentric* groups, with the number of members ranging from 68 to 78. While 62 *all-around idiocentrics* comprised the sixth largest group, two other types of allocentrics followed as the next largest groups, with 56 people and 48 people classified as *vertical allocentrics* and *horizontal allocentrics*, respectively. Furthermore, fewer people were categorized as *horizontal idiocentrics* (39) or *vertical idiocentrics* (29) than *the verticals* (46) or *the horizontals* (43). Taken together, these results show that, contrary to the literature in which the U.S. is constantly characterized as a predominantly individualistic culture at the national culture level, collectivistic tendencies were commonly observed in the cultural orientations among American Web users at the individual level.

Of note was that there appeared to be a large number of people who exhibited both idiocentric and allocentric orientations, belonging in *bi-idiocentric allocentrics* or *tri-idiocentric allocentrics*. Consistent with previous theoretical discussion on the potential influence of the global culture in the Web environment, the Web users examined in this study indeed fused a combination of cultural values across the vertical/horizontal and the idiocentric/allocentric dimensions to guide their social perception and behavior. In

Table 7. Number of people in cultural orientation groups

Group	Name	HI	VI	HA	VA	No. of Respondents
1	Horizontal Idiocentrics	*High*	Low	Low	Low	39
2	Vertical Idiocentrics	Low	*High*	Low	Low	29
3	Horizontal Allocentrics	Low	Low	*High*	Low	48
4	Vertical Allocentrics	Low	Low	Low	*High*	56
5	All-Around Idiocentrics	*High*	*High*	Low	Low	62
6	All-Around Allocentrics	Low	Low	*High*	*High*	102
7	The Horizontals	*High*	Low	*High*	Low	43
8	The Verticals	Low	*High*	Low	*High*	46
9	Bi-Idiocentric Allocentrics	Low	*High*	*High*	Low	10
10	Bi-Idiocentric Allocentrics	*High*	Low	Low	*High*	37
11	Tri-Idiocentric Allocentrics	Low	*High*	*High*	*High*	30
12	Tri-Idiocentric Allocentrics	*High*	Low	*High*	*High*	72
13	Tri-Idiocentric Allocentrics	*High*	*High*	Low	*High*	69
14	Tri-Idiocentric Allocentrics	*High*	*High*	*High*	Low	25
15	Null	*High*	*High*	*High*	*High*	68
16	Null	Low	Low	Low	Low	78

Note: (N=814 Caucasians)

summary, within a seemingly homogenous cultural group, the diversity of individuals' cultural orientations was observed using the individual-level typology.

Characteristics across cultural groups. In the next phase of the analysis, media use, Web experience, and demographic characteristics of the people classified into the different cultural groups were examined. When asked to indicate the amount of time spent on an average weekday watching television, reading magazines and newspapers, listening to the radio, and using the Web, a great majority of the respondents reported using all of the media on a daily basis. The media use patterns appeared to be relatively consistent across the groups, and no significant differences between the groups were observed. Not surprisingly, the Web was the most popular media, with about 19% of the respondents using it for more than five hours a day, followed by television, radio, newspapers, and magazines. Table 8 reports respondents' time spent with each of the media. Additionally, respondents' perceived skills and challenges pertaining to the Web were assessed. Respondents' self-judged Web skills were quite high with an average rating of 5.85 on a seven-point scale, whereas they perceived the Web as not very challenging, displaying a mean score of 2.90.

As summarized in Table 9, people from the different cultural orientation groups showed similar demographic characteristics as well. Across almost all of the demographic categories, *all-around allocentrics* were identified as the major type of cultural orientation. This is similar to the overall rankings of the groups. One exception was that single people were mostly *all-around idiocentrics,* whereas the two most common

Table 8. Time spent with media on an average weekday

	Television	Radio	Newspapers	Magazines	The Web
Do not use on a daily basis	38 (4.7%)	43 (5.3%)	227 (27.9%)	205 (23.8%)	1 (0.6%)
30 min.	68 (8.4%)	209 (25.7%)	320 (39.3%)	415 (54.7%)	84 (12.2%)
1 hour	87 (10.7%)	167 (20.5%)	153 (18.8%)	109 (8.7%)	120 (15.7%)
1 hr. 30 min.	68 (8.4%)	76 (9.3%)	33 (4.1%)	14 (1.2%)	78 (7.6%)
2 hours	155 (19.0%)	95 (11.7%)	33 (4.1%)	40 (6.4%)	145 (16.3%)
2 hr 30 min.	47 (5.8%)	31 (3.8%)	7 (0.9%)	0 (0.0%)	31 (2.3%)
3 hours	117 (14.4%)	36 (4.4%)	18 (2.2%)	16 (2.3%)	87 (11.6%)
3 hr 30 min.	21 (2.6%)	6 (0.7%)	0 (0.0%)	1 (0.0%)	11 (0.6%)
4 hours	82 (10.1%)	31 (3.8%)	7 (0.9%)	2 (0.6%)	65 (7.6%)
4 hr 30 min.	11 (1.4%)	5 (0.6%)	1 (0.1%)	1 (0.0%)	8 (1.7%)
5 hours	26 (3.2%)	23 (2.8%)	4 (0.5%)	0 (0.0%)	23 (2.3%)
More than 5 hr.	90 (11.1%)	87 (10.7%)	6 (0.7%)	5 (1.2%)	155 (20.3%)

Note: The base of the percentages was the sample size of 814 Caucasians.

cultural orientations among married people were *all-around allocentrics* and *vertical allocentrics*. Of additional interest was that among those who worked full time, *all-around idiocentrics* were identified as the second frequent type, following *all-around allocentrics*.

SUMMARY

Understanding culture is central to global communication because it serves as a meaningful platform that helps articulate communication needs and delivery for people around the world. As the penetration of the Web increases and technology-mediated communication proliferates, cultural gaps between countries and regions are often said to become narrower. Yet, multiplicity, as opposed to uniformity, of cultural values is still commonly observed. The decentralized and individualized nature of today's media technology might have resulted in further fragmentation of people's cultural orientations, since reinforcements from others with similar views and preferences could be just a click away. Perhaps cultural convergence takes place between like-minded people across nations, whereas divergence of cultural values might be witnessed among people within a country. A

Table 9. Demographic characteristics of cultural orientation groups

No.	Group Name	Gender		Employment			Marital Status		
		Female	*Male*	*Full-time*	*Part-time*	*Unemployed*	*Single*	*Married*	*Others*
1	Horizontal Idiocentrics	24	15	25	3	10	14	14	11
2	Vertical Idiocentrics	13	15	27	1	1	10	15	4
3	Horizontal Allocentrics	31	17	31	9	8	5	34	9
4	Vertical Allocentrics	33	23	38	**10**	8	11	**42**	1
5	All-Around Idiocentrics	41	21	**52**	5	3	**29**	21	12
6	All-Around Allocentrics	**64**	**36**	**61**	**13**	**28**	11	**76**	**15**
7	The Horizontals	30	13	32	3	8	11	21	11
8	The Verticals	21	25	35	4	7	8	37	1
9	Bi-Idiocentric Allocentrics	5	5	6	2	2	3	7	0
10	Bi-Idiocentric Allocentrics	23	14	22	5	10	8	26	3
11	Tri-Idiocentric Allocentrics	12	18	22	3	5	3	20	7
12	Tri-Idiocentric Allocentrics	**48**	24	48	8	**16**	19	36	**17**
13	Tri-Idiocentric Allocentrics	24	**45**	50	4	**14**	**26**	36	6
14	Tri-Idiocentric Allocentrics	20	5	19	4	2	8	10	7
15	Null	37	**30**	46	**10**	12	19	36	12
16	Null	**53**	25	**59**	9	9	**23**	**39**	**16**
	Total	479	331	573	93	143	208	470	133

Note: The three highest numbers per each group are highlighted in bold type.

thoughtful investigation is warranted to unravel the dynamic role of communication technology in culture change.

In today's global environment, geographical perimeters are blurred and people are exposed to many different cultures through various means. Global trends, growing communication between cultures, and shifting frames of cultural reference make the scrutiny of individual cultural orientations a pressing issue. Based on recent conceptual developments of individualism and collectivism, an in-depth typology classifying individual cultural orientations is proposed in this chapter. Altogether, 16 individual groups are delineated. This theoretical classification was tested with a sample of American Web users where all 16 groups were identified. While appearing homogeneous on the surface with the same national membership and ethnicity, as well as similar income and education levels, Web users in the study showed a wide assortment of cultural orientations as postulated by the classification typology. This empirical evidence provides the much-needed impetus for future research using a cultural value-based approach to understanding users of technology-mediated communication around the globe.

Three general theoretical issues are evidenced from the proposed classification and empirical findings. First, the simple fact that the individual-level typology was successfully implemented suggests that, in today's highly interconnected world, it is necessary to start assessing cultural level constructs at the individual level. Other cultural value dimensions such as power distance, uncertainty avoidance, and long-term orientation might also benefit from this approach.

Second, given today's global environment, people with complex cultural orientations are not only common but should be expected. This can be seen from the identification and prevalence of the horizontals, the verticals, and the bi and tri-idiocentric allocentrics in the study. These groups were theoretically derived and empirically verified. This observation further illustrates the importance of an individual-level approach to understanding the role of culture in global communication. Finally, because of the complex mixing of cultural value orientations in some groups, it might be necessary that the proposed classification be further explored in conjunction with other constructs such as communication context dependency and situational variations, to name but a few. Among the multitude of cultural values characterizing the groups, for example, a certain dimension might become particularly pertinent depending on the context. This endeavor will help propel valuable progress toward a comprehensive framework of culture and global communication.

Certainly, the individual cultural value orientation approach does not negate the need for group-level comparisons. In fact, group-level comparisons are of practical concern because operationally they provide effective means for the differentiation and implementation of cross-country communication, both in terms of content and format. After applying the proposed classification to individuals, comparisons of dissimilar groups within and similar groups across countries or cultures are possible. The traditional national or cultural borders are no longer an issue of concern in this conceptualization. Future research should follow the implications provided in this chapter in order to broaden our thinking about the role of culture in a world of borderless communication.

REFERENCES

Callahan, E. (2005). Cultural similarities and differences in the design of university Web sites. *Journal of Computer-Mediated Communication, 11*(1). Retrieved February 15, 2006, from http://jcmc.indiana.edu/vol11/issue1/callahan.html

Campbell, A. (2000). Cultural identity as a social construct. *Intercultural Education, 11*(1), 31-39.

Chen, W., Boase, J., & Wellman, B. (2002). The global villagers: Comparing Internet users and uses around the world. In B. Wellman & C. Haythornthwaite (Eds.), *The Internet in everyday life* (pp. 74-113). Malden, MA: Blackwell.

Cho, B., Kwan, U., Gentry, J. W., Jun, S., & Kropp, F. (1999). Cultural values reflected in theme and execution: A comparative study of U.S. and Korean television commercials. *Journal of Advertising, 28*(4), 59-73.

Choi, J. H., & Danowski, J. (2002). Making a global community on the Net—global village or global metropolis?: A network analysis of Usenet newsgroups. *Journal of Computer-Mediated Communication, 7*(3). Retrieved December 9, 2005, from http://jcmc.indiana.edu/vol7/issue3/choi.html#theories

de Mooij, M. (1998). *Global marketing and advertising: Understanding cultural paradoxes.* Thousand Oaks, CA: Sage.

Gudykunst, W. B., & Ting-Toomey, S. (1988). *Culture and interpersonal communication.* Newbury Park, CA: Sage.

Hall, E. T. (1976). *Beyond culture.* Garden City, NY: Doubleday & Company.

Hall, E. T. (1984). *The dance of life.* Garden City, NY: Doubleday/Anchor.

Han, S.-P., & Shavitt, S. (1994). Persuasion and culture: Advertising appeals in individualistic and collectivistic cultures. *Journal of Experimental Social Psychology, 30,* 326-350.

Hewling, A. (2005). Culture in the online class: Using message analysis to look beyond nationality-based frames of reference. *Journal of Computer-Mediated Communication, 11*(1). Retrieved December 15, 2005, from http://jcmc.indiana.edu/vol11/issue1/hewling.html

Hofstede, G. (1980). *Culture's consequences.* Beverly Hills, CA: Sage.

Hofstede, G. (1983). Dimensions of national cultures in fifty countries and three regions. In J. Deregowski, S. Dzirawiec, & R. Annis (Eds.), *Explications in cross-cultural psychology* (pp. 389-407). Lisse, The Netherlands: Swets and Zeitlinger.

Hofstede, G. (1984). *Culture's consequences: International differences in work-related values.* Beverly Hills, CA: Sage.

Kincaid, D. L. (1988). The convergence theory and intercultural communication. In Y. Y. Kim & W. B. Gudykunst (Eds.), *Theories in intercultural communication* (pp. 280-298). Newbury Park, CA: Sage.

Lee, W., & Choi, S. M. (2005). The role of horizontal and vertical individualism and collectivism in online consumers' response toward persuasive communication on the Web. *Journal of Computer-Mediated Communication, 11*(1). Retrieved December 15, 2005, from http://jcmc.indiana.edu/vol11/issue1/wnlee.html

Miracle, G. E., Chang, K. Y., & Taylor, C. R. (1992). Culture and advertising executives: A comparison of selected characteristics of Korean and U.S. television commercials. *International Marketing Review, 9*(4), 5-17.

Nelson, M. R., & Shavitt, S. (2002). Horizontal and vertical individualism and achievement values: A multimethod examination of Denmark and the United States. *Journal of Cross-Cultural Psychology, 33*(5), 439-458.

Novak, T. P., Hoffman, D. L., & Yung, Y.-F. (2000). Measuring the customer experience in online environments: A structural modeling approach. *Marketing Science, 19*(1), 22-42.

Pew Internet & American Life Project. (2005). *Who's online: Demographics of Internet users* (February-March 2005 tracking survey). Retrieved July 29, 2005, from http://www.pewinternet.org/trends/User_Demo_05.18.05.htm

Rogers, E. M., & Kincaid, D. L. (1981). *Communication networks: Toward a new paradigm for research.* New York: The Free Press.

Schlosser, A. E., Shavitt, S., & Kanfer, A. (1999). Survey of Internet users' attitudes toward Internet advertising. *Journal of Interactive Marketing, 13*(3), 34-54.

Scollon, R., & Wong-Scollon, S. (2001). *Intercultural communication* (2nd ed.). Oxford: Blackwell.

Singlis, T. M., & Brown, W. J. (1995). Culture, self, and collectivist communication: Linking culture to individual behavior. *Human Communication Research, 21,* 354-389.

Singlis, T. M., Triandis, H. C., Bhawuk, D., & Gelfand, M. J. (1995). Horizontal and vertical dimensions of individualism and collectivism: A theoretical and measurement refinement. *Cross-Cultural Research, 29,* 240-275.

Triandis, H. C. (1995). *Individualism and collectivism.* Boulder, CO: Westview.

Triandis, H. C. (2001). Individualism-collectivism and personality. *Journal of Personality, 69,* 907-924.

Triandis, H. C., & Gelfand, M. J. (1998). Converging measurements of horizontal and vertical individualism and collectivism. *Journal of Personality and Social Psychology, 74,* 118-128.

Triandis, H. C., & Suh, E. M. (2002). Cultural influences on personality. *Annual Review of Psychology, 53,* 133-160.

Triandis, H. C., McCusker, C., & Hui, C. H. (1990). Multimethod probes of individualism and collectivism. *Journal of Personality and Social Psychology, 59*(5), 1006-1020.

Wolin, L. D., Korgaonkar, P., & Lund, D. (2002). Beliefs, attitudes and behavior towards Web advertising. *International Journal of Advertising, 21*(1), 87-113.

Wurtz, E. (2005). A cross-cultural analysis of Websites from high-context cultures and low-context cultures. *Journal of Computer-Mediated Communication, 11*(1). Retrieved December 15, 2005, from http://jcmc.indiana.edu/vol11/issue1/wuertz.html

Yamaguchi, S., Kuhlman, D. M., & Sugimori, S. (1995). Personality correlates of allocentric tendencies in individualist and collectivist cultures. *Journal of Cross-Cultural Psychology, 26,* 658-672.

Chapter V
Intercultural Computer-Mediated Communication Between Chinese and U.S. College Students

Yun Xia
Rider University, USA

ABSTRACT

This study incorporates the effects of culture and computer-mediated communication (CMC) in the investigation of Chinese college students' use of English in communication with U. S. college students. Ethnography of communication was used as the method to uncover four language patterns in Chinese college students' use of English: others-oriented talk, mentor-mentee talk with limited relationship development, adoption of others' talk, and icebreaker talk. Chinese cultural values showed a strong effect on Chinese college students' use of English in CMC. Chinese cultural values included Chinese significance of personal relationships, Chinese collectivism, Chinese understanding of authority, and Chinese use of CMC. The characteristics of CMC showed minimal effect as either a constraint or a way of encouragement in the four language patterns.

INTRODUCTION

Although English is a global language in intercultural computer-mediated communication (CMC), patterns, rules, and premises of English use vary across different cultures. Decades ago, Hymes (1968) proposed assumptions regarding speech and language in a cultural group. First, speech and language of a group constitutes a system. Second, speech and language vary cross-culturally in function. Third, speech and language activity of a community should be the primary object of attention. Thus, the study of particular use of English language in a cultural group in CMC should be the focus that can uncover the resources of different communication contexts. The interpretation of meanings of those resources should provide guidelines for intercultural CMC in return. In this chapter, ethnography of communication was used as the research method to describe, analyze, and interpret Chinese college students' English language activities and Chinese cultural resources regarding language activities that constructed a distinctive intercultural communication between Chinese and U.S. college students.

BACKGROUND

CMC and Intercultural Communication

In the literature of intercultural computer-mediated communication, there are two major perspectives. On the one hand, the focus on communicative effects of CMC argues that CMC with its unique characteristics can facilitate and contribute to intercultural communication. On the other hand, the evaluation of cultural factors in communication leads to the conclusion that communicative styles in CMC as well as attitudes and perceptions toward CMC are subject to cultural effects. However, few studies in the literature have incorporated both CMC characteristics and cultural effects in the examination of intercultural CMC. The current study adopts both perspectives in the evaluation of Chinese cultural effects and CMC's effects on Chinese college students' English communication with U.S. college students.

CMC includes different formats: e-mail, chatgroups, virtual worlds, and the World Wide Web. Across different formats, CMC is primarily in the form of written language. Compared with face-to-face communication, the written communication in CMC is "leaner" because the social context cues in face-to-face communication are missing (Kiesler & Sproull, 1992; Walther & Parks, 2002). The missing social context cues include people's race, gender, social class, and accent (Sproull & Kiesler, 1991). The direct communicative effect of this characteristic is CMC's potential for fostering flagrant and hostile language known as "flaming" (O'Sullivan & Flanagin, 2003). CMC is also ranked as low in media richness with less immediate feedback and limited personalization (Daft & Lengel, 1986; Trevino, Lengel, & Daft, 1987). Media with low richness communicate low social presence. In CMC, users may pay less attention to others' presence. As a result, CMC makes communication more task oriented (Walther & Burgoon, 1992). Empirical studies (Jonassen & Kwon, 2001; Mazur, 2004) report that during group problem-solving activities, CMC participants produce fewer, but more task-oriented, messages than do face-to-face communication participants.

In intercultural communication, these characteristics of CMC encourage more participation from people who may normally feel intimidated in face-to-face interaction. As Warschauer and De Florio-Hansen (2003) stated, people in CMC do not know whether others are a dog, Black or White, male or female, or rich or poor. Thus, people who are less outspoken in face-to-face situations may contribute more in CMC with anonymity or less exposure (Simons, 1998; Tella & Mononen-Aaltonen, 1998). In some cultures, people take a passive rather than an active role in interaction. With low social presence, CMC's task-oriented communication characteristic offers people an opportunity to make more contributions (Warschauer, 2000; Warschauer & De Florio-Hansen, 2003). CMC's absence of nonverbal cues such as frowning and hesitating also makes interactions less intimidating and thus allows people in cultures that are less dominant to play a more active role in intercultural communication (Warschauer, 2000; Warschauer & De Florio-Hansen, 2003).

On the other hand, CMC as well as other media is subject to cultural influences (Leonardi, 2003; Chase, Macfadyen, Reeder, & Röche, 2002). Studies show that the differences between individualistic and collectivistic cultures affect the levels of satisfaction with e-mail (Chung, 1992;

Scollon & Scollon, 1995; Simons, 1998). With social relationships constructed in language, an individualistic culture allows people to be more satisfied with task-oriented, short, and straightforward e-mail that has minimal context cues. With special forms of discourse that carefully preserve the boundaries between those who are inside members of the group and all others who are not members of the group, a collectivistic culture may require more context information for a successful communication in e-mail (Chen, 1998; Gunawardena et al., 2001).

The differences between high- and low-context culture also affect the use of CMC as well as other communication technologies (Chen, 1998; Chung, 1992; Leonardi, 2003; Xia, 2006). A high-context culture has most of the information in communication implicit (Hall, 1976). The meanings are to be inferred from the context or internalized in communicators themselves. A low-context culture has information explicit in the transmission of verbal messages. In a low-context culture, people tend to use verbal expressions explicitly, value self-expression and eloquent speech, and express their opinions directly. Studies (Chen, 1998; Chung, 1992) suggest that high-context cultures will be less satisfied with relationship building and feelings expressed in CMC. Low-context cultures, on the other hand, are likely to find CMC useful for expressing specifics in a short and straightforward style.

The differences between monochronic and polychronic culture also impact intercultural communication in CMC (O'Dowd, 2001). Monochronic cultures, such as the U.S. and northern European countries, are task oriented with emphasis on scheduling activities according to deadlines. Polychronic cultures, such as countries from the Mediterranean and the Middle East, prefer to combine activities and consider deadlines as objectives rather than as anything certain (Taylor, 2000). Communication between the two cultures can break down if people in a polychronic culture do not answer e-mails on time or do not work on a timetable for a synchronous CMC interaction (O'Dowd, 2001). Meanwhile, people from a monochronic culture may appear to be demanding to polychronic people when they try to get things done as fast as possible (O'Dowd, 2001).

Language Analysis in Intercultural CMC

Because CMC is primarily text based, the analysis of language use becomes critical to understand the nature of CMC and the relationship between language and social practice in a culture (Herring, 2001). However, the literature is particularly sparse on how particular language behaviors function in intercultural CMC. Only a few studies have explored unique cultural experience through language interaction in CMC. O'Dowd (2003) investigated intercultural experience in CMC and identified key characteristics of e-mail exchange that helped develop intercultural communicative competence. Choi and Danowski (2002) studied power-play negotiations of dominant and minority cultures in CMC through the analysis of structural patterns of online communication flow. In a study of how linguistic interaction patterns changed over time among a geographically and ethnically diverse group of young people, Cassell and Tversky (2005) uncovered the ways in which people from different cultural backgrounds constituted a community, spoke in a collective voice, and formed a unique culture. Through the analysis of written reports and interview notes, Ma (1996) analyzed the recurring themes in CMC between East Asian and U.S. students. After proposing and testing five propositions about intercultural CMC, Ma found that both East Asian and U.S. students were more direct in CMC than in face-to-face communication.

In the study of culture through language analysis in CMC, Herring and her colleagues played a significant role. According to Herring (1996), individual electronic messages are internally organized text, and texts' distinctive schematic organization can be uncovered through language analysis. After studying one female-dominated and one male-dominated listserv, Herring found that messages posted by women were more interactive and contained more information while men tended to express their views critically. With the discovery of distinctive schematic text organization, Panyametheekul and Herring (2003) studied language use and culture in turn-taking of a Thai chat room. They found that Thai females had more frequent participation than males. This finding broke cultural expectations about the roles of women in Thai society and the role of women online.

From the intersection of the reviewed bodies of literature, the following are three research questions about Chinese college students' use of English in intercultural CMC with the U.S. college students. With the three research questions, the study attempts to examine the effect of Chinese culture and characteristics of CMC on Chinese students' use of English in CMC.

RQ1: *What are the distinctive language patterns in Chinese college students' use of English in intercultural CMC with U.S college students?*

RQ2: *How does Chinese culture affect Chinese college students' use of English in intercultural CMC with U.S. college students?*

RQ3: *How do the characteristics of CMC affect Chinese college students' use of English in intercultural CMC with U.S. college students?*

MAIN FOCUS OF THE CHAPTER

Method

In the study of this chapter, ethnography of communication was used as the research method. Ethnography of communication fills the gap between ethnography and linguistics. There is a widespread relationship between a language and a culture. The actual uses and patterns of use of language in different cultural groups need ethnographic description. The focus of ethnography of communication is on the way that communication in a cultural group is patterned and organized as systems of communicative events. An ethnographer of communication is "a naturalist, who watches, listens, and records communicative conduct, in its natural setting" (Philipsen, 1992, p. 7). With non-participant observation of intercultural CMC scripts and in-depth interviews in ethnography of communication, the researcher is able to describe, analyze, and interpret Chinese college students' language activities that construct distinctive communication patterns.

The international exchange program in a private university of the U.S. was used as the research field. In this exchange program, a group of Chinese students come to the university and study business administration for a year each spring. In the fall before they come to the U.S., the university recruits U.S. students as Chinese students' mentors. Still in China, each Chinese student communicates with his or her mentor through e-mail and other formats of CMC about culture, classes, campus housing, food, and weather. In this study, a group of 12 Chinese students was solicited as the participants. The participants were contacted and asked to keep their CMC scripts. After the participants arrived in the U.S., CMC scripts were collected and in-depth interviews were conducted. The researcher audiotaped all interviews, and all tapes were transcribed. All interviews were conducted in English.

In the analysis, elements of Carbaugh's study (1989) about different cultural terms for talk were used. The messages from the terms for talk could be about communication, sociality, and personhood. Those elements enabled the researcher to start with the metacommunicative vocabulary in the interview transcripts that might indicate the way the participants chose to communicate with their U.S mentors and about how they related to the U.S. mentors. With the metacommunicative vocabulary as the guideline, the researcher read and reread the transcripts of all recorded tapes and collected CMC scripts. During reading, patterns in the participants' language behaviors were identified. After that, the transcripts and scripts were reassembled and reorganized according to the patterns. Examples and stories in Chinese participants' communication were used to explain how their language behaviors were affected by Chinese culture and CMC characteristics.

Findings

In the analysis, four patterns that constituted Chinese participants' English language use in CMC with their U.S. mentors emerged: others-oriented talk, mentor-mentee talk with limited relationship development, adoption of others' talk, and icebreaker talk. For each of the four patterns, exemplars were used to illustrate Chinese participants' use of English language. Some uses were experienced by individuals, but others were more universal.

Others-Oriented Talk

In Chinese participants' communication, there was a clear pattern that showed much consideration of others. Some participants termed this pattern as "others-oriented talk." The "others-oriented talk" was illustrated by Chinese participants' extreme politeness, consideration of others as the priority, and talk about others in communication.

Although Chinese participants were more direct in CMC than in face-to-face communication, Chinese participants showed extreme politeness in CMC. The politeness markers, such as "please," "thank you very much," and "I appreciate," appeared in Chinese participants' e-mails and texts of instant messaging all the time. For example, one Chinese participant asked about the weather in the U.S.: "If it is not too much trouble, would you please tell me whether it is very cold in winter or very hot in summer there." Sometimes, the U.S. students were even annoyed by the extreme politeness. One Chinese participant spent two paragraphs explaining why she did not e-mail back for two weeks. Her U.S. mentor started the response: "It is OK. You are not ignoring me. It is a waste of time explaining too much. Everyone is busy. I understand. Let's talk about some serious stuff." In the interview, Chinese participants explained that being polite showed their consideration of others because they were seeking help from their U.S. mentors.

Others-oriented talk was also about how Chinese participants talked about others rather than themselves. The talk about others could be a perfect example of Ma's (1996) finding about East Asians' indirect online communication that directed around another person other than himself or herself. Most Chinese participants loved to talk about their families, friends, and classmates. One Chinese participant mentioned in e-mail that she would bring a laptop to access the Internet. She could keep close contact with her parents and friends. With a computer camera, she could even see her parents as if she was around them everyday. On the other hand, most U.S students engaged in individual talk that was featured by personal plans and personal interests. One U.S. mentor talked about her trip to Puerto Rico and the Virgin Islands in the summer. She also talked about her love of dancing: "Dancing is my life. I'm actually the vice president of the dance team at the university and choreographer. So I like to go out to clubs and also we perform also at Basketball games and things like that."

Others-oriented talk was directly related to the consideration of others as the priority in communication. Chinese participants tended to yield convenience to others and incorporate others' feelings in CMC. One participant showed her consideration of others in the e-mail: "As your holiday break is coming, you must be extremely busy with your courses these days. I will be embarrassed to bother you with my questions. I will wait until you have all the tests pass." After several weeks of e-mail communication, some of the participants started to talk about using synchronous CMC, such as MSN or AOL instant messaging systems. The major problem of using synchronous CMC was the time difference. Chinese participants always took the first step to offer the adjustment. One participant said in the e-mail: "Well, since you are busy with class work, I can stay late until 11 o'clock or midnight. It will be your 10 or 11 o'clock in the morning. It is easier for you."

In the interview, Chinese participants used Chinese culture about the significance of others in a person's life to explain others-oriented talk. In one participant's words, others were an integral part of personal life: "I don't even realize I talk about others a lot in the e-mails to my mentor. Well, for me, without others I will be like a leaf without branches and trunks. I will dry and die." In Chinese culture, the significance of others is closely related to Chinese values about smooth and harmonious personal relationships (Yang, 1994). Personal relationships are often encountered when seeking to establish friendships or even business partnerships in China. The way to establish personal relationships is to develop networks of mutual dependence. Personal relationships depend on the creation of obligation and indebtedness to each other. Thus, all participants thought that their questions and queries might add "extra burden" for their U.S. mentors. They all felt "indebted" to others whenever they sent questions to their mentors.

As for the effect of CMC characteristics on others-oriented talk, Chinese participants regarded e-mail as well as other formats of CMC as a constraint for their indirect way of communication. In one participant's words, e-mail was "direct and formal" communication. E-mail did not work well as a way of communication for Chinese indirect communication. Since Chinese participants still engaged in an indirect way of communication, the characteristics of CMC as a constraint made Chinese participants uncomfortable. Interestingly enough, one participant talked about his changed attitude toward e-mail after coming to the U.S.: "I feel much more comfortable to use e-mail here. I think I am adopting the U.S. direct, straightforward way of communication. E-mail offers me the way to engage in direct communication."

Mentor-Mentee Talk with Limited Relationship Development

Chinese participants used the term "mentor-mentee talk" to summarize the intercultural CMC between them and their U.S. mentors. Chinese participants always remembered their role as a mentee and the U.S. mentors' role as a mentor. The role as a mentee defined what a Chinese participant could talk about and in what way a Chinese participant could talk.

First, mentor-mentee talk was illustrated by the topics that Chinese participants chose in CMC. In summary, there were six categories of topics that Chinese participants used: weather, schoolwork, textbooks, family, clothes, and campus housing. These topics were "safe for a mentee." In the interview, Chinese participants explained that they never met their mentors and all they could talk about were the "general stuff in life they need to know." Second, mentor-mentee talk was also about the way that Chinese participants chose to communicate with their mentors. Chinese participants' extreme politeness also emerged in mentor-mentee talk. For example, one Chinese participant expressed his appreciation for the answer about campus housing: "It is very kind of you to take such trouble to give me such a detailed explanation." In the end of his e-mail, he expressed his appreciation again: "Please accept my gratefulness. Thank you very much." In the interview, Chinese participants explained that they felt honored and grateful to have the U.S. students as their mentors. They should show great respect for their mentors in communication.

On the other hand, the U.S. students never treated Chinese participants as their mentees. They tried to make friends with Chinese participants. Thus, the U.S. students engaged in a different way of talk. Chinese participants termed their U.S. mentors' talk as "friendship talk." In their first e-mail, the U.S. mentors introduced themselves and always concluded with their willingness to be friends of Chinese participants: "*Please feel free to contact me with any questions or concerns you may have about coming to the U.S. When you arrive in January, I will be ready and waiting to give you a tour of campus and help you get acquainted with the U.S. Think of me as a friend, you can approach me with anything.*"

When one Chinese participant expressed the worry that she might bother her mentor too much, the U.S. mentor e-mailed back: "You are not bothering me. Do not worry. We are friends."

In the interview, Chinese participants talked about the effect of Chinese culture on their mentor-mentee talk. Some Chinese participants mentioned that the meaning of "mentor" in Chinese was close to "teacher" or "consultant." In Chinese culture, both teacher and consultant have authority in their specialized field. As mentees, Chinese participants thought that they should respect their mentors' authority in explaining life specifics in a U.S. college. In addition, Chinese participants felt that they were not confident with their English language skills. The U.S. students automatically became authorities in language. Chinese participants could not treat their U.S. mentors as their friends. Being equal as friends, they did not show enough respect for the U.S. mentors.

In terms of the effect of CMC on mentor-mentee talk, Chinese participants had a limited relationship development that was encouraged by the characteristics of CMC. The limited relationship development between Chinese participants and their U.S. mentors was reflected in the way addressing each other evolved from formal to casual and no addressing at all. At the beginning, both Chinese participants and their U.S. mentors addressed each other very formally. One U.S. mentor initiated the contact in the first e-mail with "Dear Christina Yi Zhang" and concluded the e-mail with "Yours, Laurie Smith." The Chinese participant also answered with "Dear Laurie Smith" and concluded with the term "sincerely" and her full name. After the second week, the U.S. mentor started to address her Chinese mentee with "Hey, Chris" and ended with only "L." After the fifth week, the U.S. mentor used "Hey, hey" and sometimes started e-mails without addressing. The Chinese participant also switched to "hi" as the start and concluded with "c" in the end. In the interview, the Chinese participant acknowledged that she used formal addressing because she did not know her mentor. After more communication, she "accumulated enough information" in CMC and felt a little bit closer to her mentor. She could use a casual way of addressing.

However, the relationship development in CMC was limited by Chinese participants' mentor-mentee talk. In one Chinese participant's words, the relationship was "on the surface," activated by the need to be as casual and comfortable as possible in communication. With mentor-mentee talk, the relationship could never be developed to "friendship" because of the unequal status. One Chinese participant explained about the limited relationship development: "Do we feel closer to each other after dozens of e-mails? Yes, but I still don't feel he is my friend. I am asking too much and giving too little. It is not equal. We can't be friends."

Adoption of Others' Talk

In intercultural CMC, Chinese participants adopted their U.S. mentors' talk when they learned the unique online discourse, native English expressions, and U.S. culture. The adoption was encouraged by Chinese participants' need to learn native language, and Chinese culture underlining solidarity and integration with others. Text-based CMC provided the context that made the adoption possible.

Chinese participants were often confused by the unique English discourse in CMC. After learning the discourse, Chinese participants adopted them in return. In one Chinese participant's e-mail, she asked about the meaning of "ill talk to ya soon." In the response, the U.S. mentor explained: "it is ... kind of like a simple way to say goodbye online ... keep in touch ... ill speak with you in the near future ... things like that ... and 'ya' just is slang for 'you'. 'ill' is a abbreviation for 'I'll.' In online talk, you use lower case letters as much as possible." Interestingly enough, in the next e-mail, this Chinese participant concluded the e-mail with "ill take [talk] to ya soon." All participants mentioned that they learned to use "live English" through CMC. One participant said that he learned to conclude e-mails with "If you have questions, feel free to ask me." Whenever he answered his mentor's questions, he always concluded with this sentence.

In the communication, Chinese participants also learned to adapt to the U.S. culture. When Chinese participants asked about the weather, they had to use degrees. There was confusion between metric and English system. Chinese participants used Celsius and the U.S. mentors used Fahrenheit. One U.S. mentor asked for the conversion to Fahrenheit when a Chinese participant used Celsius: "What is –2 Celsius? Can you tell me what's in Fahrenheit? Is it cold like –2 Fahrenheit?" The Chinese participant answered back: " –2 Celsius is like 30 degrees Fahrenheit. It is cold. But is not like deadly cold. I think 0 Celsius is about 32 Fahrenheit and 10 Celsius is about 50 Fahrenheit. I will use Fahrenheit from now on. It is easy for me to do the conversion. I got to be ready for the system anyway."

In interviews, Chinese participants explained their adoption of some English expressions as helping them to learn native English. Such adoption happens in interacting between people of a second language and native speakers. The adoption can be explained by behavioral adjustments people make during communication. It matches communication accommodation that results in favorable social attractiveness, communication effectiveness, and cooperativeness (Giles, Coupland, & Coupland, 1991). On the other hand, Chinese participants also used Chinese collectivism culture to explain their adoption of the U.S. mentors' talk. As an essential theme in the literature of Chinese use of CMC, Chinese culture has been used widely to explain Chinese communication behaviors (Chen, 1998; Chung, 1992; Giese, 2003; Weber, 2002). Chinese collectivism culture emphasizes solidarity and integration with others, and prioritizes the needs of the group over the needs of the self. Being collectivistic, Chinese people emphasize in-group belonging, personal interdependence, and social harmony with others (Cheng, 1987; Lu, 1997). In one participant's words, their adoption of the U.S. mentors' talk led to similarity instead of dissimilarity that was "good for both sides."

In addition, CMC characteristics allowed Chinese participants to adopt others' talk. With text-based communication, CMC did not have "intimidating factors," such as body movements, vocal tones, and facial expressions. With only text in communication, Chinese participants would be free to ask questions about language, learn U.S. culture, and improve their English. One participant said: "In a face-to-face interaction, I would be embarrassed to ask my mentor to explain the meaning of a English word. But in e-mail, I don't have to care about how my mentor thinks." In addition, text-based CMC provided the context for Chinese participants to read, reread, and reflect on the U.S. mentors' talk before their adoption. Some Chinese participants admitted that in an oral communication, their mentors' talk would slip away and they would have nothing to learn in the adoption.

Icebreaker Talk

Chinese participants were always careful about what they should talk about in CMC with the U.S. mentors. They tended to stay with "general topics" that were safe for both sides. Some participants termed this talk as "icebreaker talk." Chinese participants defined "icebreaker talk" as "talk about the surface," "talk as contact," and "talk without specifics." In an icebreaker talk, Chinese participants picked up general topics and never went into specifics. For example, one Chinese participant asked about textbooks: "Would you like to tell me where I buy textbooks? Are they expensive?" In the interview, the participant explained that she did not want to ask for specifics, such as textbooks in her major, exact prices for textbooks, required textbooks, and optional textbooks. She did not expect specifics over CMC. She said that the question functioned more as a way to initiate and maintain the communication with her mentor.

Chinese participants attributed their icebreaker talk to different cultural understandings of CMC. In the U.S. culture, e-mail as well as other formats of CMC become one part of routine communication practice. However, in China, e-mail and other formats of CMC are not popular as ways of communication. Guo (2005), along with other colleagues in the Chinese Academy of Social Sciences, conducted a survey tracking Internet use in five major Chinese cities. According to the survey, over two-thirds of people never or seldom use e-mail, instant messaging, online discussion, and chat groups as ways of communication. Close to 40% of the survey participants check e-mail once or less than once a week. Most survey participants (about two-thirds) use CMC on the Internet for news, especially entertainment-related news and online games. Not surprisingly, Chinese participants in this study mentioned that they would prefer face-to-face interaction to CMC as a way of communication. One Chinese participant offered an example about a typical communication event in China: "If you have questions or problems, you need meet people. Sit down with a couple of tea. Take it easy. Negotiate back and forth." All participants thought that different formats of CMC functioned as a good way to initiate a contact, but face-to-face interaction was the best for further communication.

Chinese participants also talked about the effect of CMC characteristics on their icebreaker talk. All participants agreed that e-mail or instant messaging functioned well at the beginning between them and their mentors. In one participant's words, e-mail served as an icebreaker. Without context information, it was easy for Chinese participants to begin the communication with their U.S. mentors. After the beginning, e-mail turned out to be a constraint between them and their mentors. Most participants wished that they could have face-to-face interaction after a few weeks of e-mail exchange.

FUTURE TRENDS

Summary of the Study

In this study, there were four language patterns in CMC between Chinese participants and their U.S. mentors: others-oriented talk, mentor-mentee talk with limited relationship development, adoption of others' talk, and icebreaker talk. The four language patterns were affected by both cultural and CMC effects. First, in others-oriented talk, Chinese participants showed extreme politeness, considered others as the priority, and talked about others in communication. Chinese culture about the significance of personal relationships formulated others-oriented talk. CMC constrained others-oriented talk with its directness and written formality. Second, in mentor-mentee talk, Chinese participants talked according to their role as mentees. Chinese participants respected their mentors because of the Chinese meaning of "mentor" as "teacher" or "consultant." CMC only allowed a limited relationship development between Chinese participants and their U.S. mentors. Third, in the adoption of others' talk, Chinese participants learned the unique online discourse, native English expressions, and the U.S. culture. Both Chinese participants' need to learn native language and Chinese collectivism culture encouraged Chinese participants to accommodate to others' talk. With its text-based communication, CMC allowed Chinese participants' learning and adoption of their U.S. mentors' talk. Fourth, Chinese participants regarded CMC as a good icebreaker in intercultural communication. The lack of context cues made CMC appropriate for icebreaker talk and at the same time constrained further communication.

Discussion of Implications of Research for Current Situations

In each of the four language patterns of Chinese participants' communication, cultural values and effects play a significant role. For others-oriented talk, it is Chinese culture about the significance of personal relationships. For mentor-mentee talk with limited relationship development, it is Chinese understanding of the term "mentor" as "teacher" or "consultant." For adoption of others' talk, it is Chinese participants' need to learn the native language and Chinese collectivism culture that encourage communication accommodation. For icebreaker talk, it is CMC's unpopularity as a way of communication in China and Chinese preference for face-to-face communication.

Chinese culture that is under the strong Confucian influence becomes essential to explain Chinese use of CMC on the Internet. When Bockover (2003) evaluated the impact of Confucian values, she argued that traditional Chinese values are still strong in Chinese use of CMC. Critical to Confucian values is the concept of a person that is always an essential part of a larger social group, and "personal agency" is always socially defined (Bockover, 2003, p. 164). Thus, In China, the main moral goal has been harmonious interdependence in personal relationships. Bockover argued that the Internet with its autonomy, independence, and free trading of information could be at odds with Chinese cultural values. In the study of the dynamics and political function of the transnational Chinese cultural sphere on the Internet, Yang (2003) argued that cultural repertoire with Chinese values and traditions regulates communications and actions of Chinese participants from mainland China, Hong Kong, Taiwan, and North America. In the study of Chinese policy making and implementation on the Internet, Weber (2002) defined Chinese culture as "diffuseness culture" that emphasizes relationship closeness and harmony in social structure, as opposed to Western individualistic "specifity culture." These cultural differences lead to interpretation problems of the Internet policies in China. Adding to the literature, the findings of the study offer empirical evidence for the impact of culture on intercultural CMC.

On the other hand, the characteristics of CMC function as either a constraint or a way of encouragement in this study. The supplemental effect of CMC characteristics contradicts the findings about CMC's determining effect on intercultural CMC (Ma, 1996; Tella & Mononen-Aaltonen, 1998). For example, among Ma's (1996) widely cited five propositions about intercultural CMC, two propositions are directly derived from CMC's determining effect. First, East Asians tend to show greater self-disclosure in CMC conversations than in face-to-face conversations. Second, status difference is unnoticeable in intercultural CMC. In the current study, Chinese participants' icebreaker talk about "safe" and "general" topics is contradictory to the proposition about more self-disclosure in CMC. Chinese participants' talk from the mentor-mentee relationship does not match the proposition about unnoticeable status difference in intercultural CMC. One explanation is the situational use of CMC that varies from culture to culture, context to context, and people to people (Warschauer, 2000). Another explanation may be about Ma's study sample that consists of Chinese students attending U.S. universities and Taiwan universities. Under the U.S. cultural impact, Chinese students attending U.S. universities may not incorporate much Chinese culture in intercultural CMC. Even Chinese students attending Taiwan universities may not incorporate much Chinese culture in CMC because of strong globalization in Taiwan. In the current study, Chinese participants have never communicated with native speakers in such a scope before. These Chinese participants showed much more cultural effect on their language patterns in CMC.

Chinese culture underlines the patterns of communication that Chinese students have with their U.S. mentors. The patterns can be explained by Philipsen's (1992) theory of two codes in speaking: a code of honor and a code of dignity. Chinese participants communicate with their U.S. mentors from a code of honor. In a code of honor, "the individual is conceived a role or character, a persona" (Philipsen, 1992, p. 109). In the communication with the U.S. mentors, Chinese participants kept the roles of mentees and perceived their U.S. counterparts in the roles of mentors. In a code of honor, society along with others exists before an individual. In this study, Chinese participants had others-oriented talk and adopted others' talk for similarity and integration with others. On the virtual stage of intercultural CMC, Chinese participants used the talk in CMC as an icebreaker contact instead of actual communication. In a code of honor, Chinese participants viewed their talk as fulfilling their role as a mentee in the social order with their U.S. mentors.

On the other hand, the U.S. mentors communicate with Chinese participants from a code of dignity. In U.S. culture, a code of dignity is "concerned with the person qua person, as someone who is made up of unique feelings, ideas, and attitudes" (Philipsen, 1992, p. 113). In a code of dignity, communication functions to express individual intent and will. The U.S. mentors engaged in individual talk that featured personal hobbies, personal interests, and personal plans. The U.S. mentors also engaged in friendship talk, treating Chinese participants as equal individuals in intercultural CMC.

Implications of Research for Future Situations

The findings in the study call attention to social constructivism in future studies of communication technologies. In literature about communication technologies, social constructivism advocates that people in a particular culture shape social behavioral patterns regarding technology use (Fischer, 1992; Fulk, 1993; Fulk & Boyd, 1991). Through negotiations and struggles among people, the meanings in people's use of technologies are internally constructed and redefined in changing cultural contexts (Barley, 1990; Leonardi, 2003; Xia, 2006). The application of social constructivism ideas calls for more analysis of technology use for a particular culture in future studies. The findings about the determining cultural effect on Chinese participants' use of English in CMC with their U.S. mentors offer an example of cultural analysis of technology fit from a social constructivism perspective. In the current study, the four language patterns of Chinese participants' communication call for the evaluation of a particular culture as a part of social construction process.

Suggestions for Future Applications of the Findings

It is interesting to consider the U.S. students' responses in discussing this chapter's future applications of to intercultural CMC. Some U.S. students felt strange with Chinese participants' talk about families, friends, and classmates. Some U.S. students felt annoyed by Chinese participants' paragraphs of apology. Some U.S. students did not understand Chinese participants' "touch and run" talk. These reactions can be explained by the discrepancy in self-perceptions between the East Asian and U.S. students. Even if the East Asians thought of themselves as direct in CMC, the U.S. students still found them extremely polite, reserved, and indirect (Ma, 1996). How to change this misunderstanding should be the focus of future studies.

The findings in the current study uncover the importance of understanding cultural resources regarding language activities in intercultural CMC. The assumed relation between one language and one culture explains significant meanings in Chinese participants' communication with the U.S. students. Although Chinese participants use English in CMC, the rules and meanings regarding language come from their native culture. As Hymes (1968) proposed in the assumption about speaking in a cultural group, Chinese participants' use of English language in CMC consists of a unique system, and their actual language activities should be the focus of study. Knowledge of cultural resources about language activities is the key to mutual understanding. The findings in the current study about Chinese culture in Chinese participants' others-oriented talk, mentor-mentee talk, adoption of others' talk, and icebreaker talk can help their U.S partners as well as people from other cultures understand diverse cultural resources in the future intercultural communication over CMC.

Suggestions for Future Studies

With the fast development and diffusion of CMC technology as well as cultural globalization, interaction in CMC across different cultures is becoming more and more common. English as a global language is the key for intercultural CMC. As Cassell and Tverksy (2005) reviewed, studies of the particular use of online English in different cultures are sparse. The future research of intercultural CMC should focus more on the particular rules, patterns, and meanings of English language use in different cultures and uncover cultural resources regarding English language activities in CMC.

In the study, there are 12 Chinese participants. Although small samples are common for in-depth qualitative studies, the generalization of the findings to other Chinese students is unknown. The relation between cultural resources and language use in CMC may not be generalized to other groups of people. Further examination with large samples is needed for other groups. Discussing different language patterns in CMC individually helps to describe themes, patterns, and ideas clearly. However, it does not represent the interconnections among them. Apparently, Chinese participants' others-oriented talk can affect their adoption of others' talk. More research is needed to unpack these interconnections among different language patterns in Chinese use of English in CMC.

CONCLUSION

The study incorporated both cultural effects and CMC characteristics in the investigation of Chinese college students' use of English in communication with U. S. college students. Ethnography of communication was used as the method to uncover four language patterns in Chinese college students' use of English: others-oriented talk, mentor-mentee talk with limited relationship development, adoption of others' talk, and icebreaker talk. Chinese cultural values showed strong effect on Chinese college students' use of English in CMC. Chinese cultural values included Chinese significance of personal relationships, Chinese collectivism, Chinese understanding of authority, and Chinese use of CMC. The characteristics of CMC showed minimal or supplemental effect as either a constraint or encouragement in the four language patterns.

REFERENCES

Barley, S. R. (1990). The alignment of technology and structure through roles and networks. *Administrative Science Quarterly, 35,* 61-103.

Bockover, M. (2003). Confucian values and the Internet: A potential conflict. *Journal of Chinese Philosophy, 30*(2), 159-175.

Carbaugh, D. (1989). Fifty terms for talk: A cross-cultural study. *International and Intercultural Communication Annual, 13,* 93-120.

Cassell, J., & Tversky, D. (2005). The language of online intercultural community formation. *Journal of Computer-Mediated Communication, 10*(2). Retrieved March 10, 2006, from http://jcmc.indiana.edu/vol10/issue2/cassell.html

Chen, G. M. (1998). Intercultural communication via e-mail debate. *The Edge: The E-Journal of Intercultural Relations, 1*(4). Retrieved March 10, 2006, from http://www.interculturalrelations.com/v1i4Fall1998/f98chen.htm

Cheng, L. L. (1987). *Assessing Asian language performance: Guidelines for evaluating limited-English-proficient students.* Rockville, MD: Aspen.

Choi, J. H., & Danowski, J. (2002). Making a global community on the Net—global village or global metropolis?: A network analysis of Usenet newsgroups. *Journal of Computer-Mediated Communication, 7*(3). Retrieved March 1, 2006, from http://jcmc.indiana.edu/vol7/issue3/choi.html

Chung, J. (1992, November). *Electronic mail usage in low-context and high-context cultures.* Paper presented at the Annual Meeting of the Speech Communication Association, Chicago.

Chase, M., Macfadyen, L. P., Reeder, K., & Röche, J. (2002). Intercultural challenges in networked learning: Hard technologies meet soft skills. *First Monday, 7*(8). Retrieved July 13, 2006, from http://firstmonday.org/issues/issue7_8/chase/index.html

Daft, R. L., & Lengel, R. H. (1986). Organizational information requirements, media richness and structural design. *Management Science, 32,* 554-571.

Fischer, C. S. (1992). *America calling: A social history of the telephone to 1940.* Berkeley, CA: University of California Press.

Fulk, J. (1993). Social construction of communication technology. *Academy of Management Journal, 36,* 921-950.

Fulk, J., & Boyd, B. (1991). Emerging theories of communication in organizations. *Journal of Management, 17,* 407-447.

Giese, K. (2003). Construction and performance of virtual identity in Chinese Internet. In K. C. Ho, R. Kluver, & K. C. C. Yang (Eds.), *Asia.com: Asia encounters the Internet* (pp. 193-211). London: Routledge.

Giles, H., Coupland, N., & Coupland, J. (1991). Accommodation theory: Communication, context, and consequences. In H. Giles, J. Coupland, & N. Coupland (Eds.), *Contexts of accommodation developments in applied sociolinguistics* (pp. 1-69). Cambridge: Cambridge University Press.

Gunawardena, C. N., Nolla, A. C., Wilson, P. L., Lopez-Islas, J. R., Ramírez-Angel, N., & Megchun-Alpízar, R. M. (2001). A cross-cultural study of group process and development in online conferences. *Distance Education, 22*(1), 85-121.

Guo, L. (2005). *Surveying Internet usage and impact in five Chinese cities.* Beijing, China: Research Center for Social Development, Chinese Academy of Social Sciences.

Hall, E. T. (1976). *Beyond culture.* Garden City, NY: Anchor.

Herring, S. C. (1996). Two variants of an electronic message schema. In S. C. Herring (Ed.), *Computer-mediated communication: Linguistic, social and cross-cultural perspective* (pp. 81-109). Philadelphia: John Benjamins North America.

Herring, S. C. (2001). Computer-mediated discourse. In D. Tannen, D. Schiffrin, & H. Hamilton (Eds.), *Handbook of discourse analysis* (pp. 612-634). Oxford: Blackwell.

Hymes, D. (1968). The ethnography of speaking. In J. Fishman (Ed.), *Readings in the sociology of language* (pp. 99-138). The Hague: Mouton.

Jonassen, D. H., & Kwon, H. (2001). Communication patterns in computer mediated versus face-to-face group problem solving. *Educational Technology Research and Development, 49*(1), 1042-1629.

Kiesler, S., & Sproull, L. S. (1992). Group decision making and communication technology. *Organizational Behavior and Human Decision Processes, 52,* 96-123.

Leonardi, P. M. (2003). Problematizing "new media": Culturally based perceptions of cell phones, computers, and the Internet among United States Latinos. *Critical Studies In Media Communication, 20,* 160-179.

Lu, S. (1997). Culture and compliance gaining in the classroom: A preliminary investigation of Chinese college teachers' use of behavior alteration techniques. *Communication Education, 46,* 10-28.

Ma, R. (1996). Computer-mediated conversations as a new dimension of intercultural communication between East Asian and North American college students. In S. C. Herring (Ed.), *Computer-mediated communication: Linguistic, social and cross-cultural perspectives* (pp. 173-186). Philadelphia: John Benjamins North America.

Mazur, J. (2004). Conversation analysis for educational technologists: Theoretical and methodological issues for researching the structures, processes and meaning of online talk. In D. H. Jonassen (Ed.), *Handbook for research in educational communications and technology* (pp. 1073-1099). Mahwah, NJ: Lawrence Erlbaum.

O'Dowd, R. (2001). In search of a truly global network: The opportunities and challenges of online intercultural communication. *CALL-EJ Online, 3*(1). Retrieved March 10, 2006, from http://www.clec.ritsumei.ac.jp/english/callejonline/6-1/o_dowd.html

O'Dowd, R. (2003). Understanding the "other side": Intercultural learning in a Spanish-English e-mail exchange. *Language Learning & Technology, 7*(2), 118-144.

O'Sullivan, P., & Flanagin, A. (2003). Reconceptualizing 'flaming' and other problematic messages. *New Media & Society, 5*(1), 69-94.

Panyametheekul, S., & Herring, S. C. (2003). Gender and turn allocation in a Thai chat room. *Journal of Computer-Mediated Communication, 9*(1). Retrieved March 10, 2006, from http://jcmc.indiana.edu/vol9/issue1/panya_herring.html

Philipsen, G. (1992). *Speaking culturally: Explorations in social communication.* Albany: State University of New York Press.

Scollon, R., & Scollon, S. (1995). *Intercultural communication.* Oxford: Blackwell.

Simons, G. (1998). Meeting the intercultural challenges of virtual work. *Language and Intercultural Learning, 16*(1), 13-15.

Sproull, L. S., & Kiesler, S. (1991). *Connections: New ways of working in the networked organization.* Cambridge: MIT Press.

Taylor, N. (2000, March). *Cross-cultural issues in business.* Paper presented at the meeting of TESOL Spain 2000. Madrid, Spain.

Tella, S., & Mononen-Aaltonen, M. (1998). *Developing dialogic communication culture in media education: Integrating dialogism and technology.* Helsinki: Media Education Publications.

Trevino, L. K., Lengel, R. H., & Daft, R. L. (1987). Media symbolism, media richness, and media choice in organizations: A symbolic interactionist perspective. *Communication Research, 14,* 553-574.

Walther, J. B., & Burgoon, J. K. (1992). Relational communication in computer-mediated interaction. *Human Communication Research, 19*(1), 50-88.

Walther, J. B., & Parks, M. R. (2002). Cues filtered out, cues filtered in: Computer-mediated communication and relationships. In M. L. Knapp & J. A. Daly (Eds.), *Handbook of interpersonal communication* (pp. 529-563). Thousand Oaks, CA: Sage.

Warschauer, M. (2000). Language, identity, and the Internet. In B. Kolko, L. Nakamura, & G. Rodman (Eds.), *Race in cyberspace* (pp. 151-170). New York: Routledge.

Warschauer, M., & De Florio-Hansen, I. (2003). Multilingualism, identity, and the Internet. In A. Hu & I. De Florio-Hansen (Eds.), *Multiple identity and multilingualism* (pp. 155-179). Tübingen: Stauffenburg.

Weber, I. (2002). Communicating styles: Balancing specifity and diffuseness in developing China's Internet regulations. *Journal of Intercultural Studies, 23*(3), 307-321.

Xia, Y. (2006). Cultural values, communication styles, and use of mobile communication in China. *China Media Research, 2*(2), 64-74.

Yang, G. (2003). The Internet and the rise of a transnational Chinese cultural sphere. *Media, Culture and Society, 25*(4), 469-490.

Yang, M. M. (1994). *Gifts, favors, and banquets: The art of social relationships in China.* Ithaca, NY: Cornell University Press.

Section II
Global Virtual Communities and International Online Collaboration

Chapter VI
Culture and Communication Challenges in Virtual Workspaces

Bolanle A. Olaniran
Texas Tech University, USA

ABSTRACT

The issue of globalization is one that contemporary organizations are constantly trying to unravel. However, globalization has brought about or reinvents the concept of workspaces such that work is no longer confined to a traditional space in a specific geographical setting whereby individuals are restricted to face-to-face meetings and interactions. Rather, globalization has created the idea of virtual workspaces wherein collaborations and communication interactions occur with the aid of information and communication technologies in order to allow collaboration among workers across different geographical boundaries. Virtual workspaces, however, also bring into the foreground the issue of culture and its effect on behaviors and communication interactions. This discussion explores the role of culture and its implications for virtual workspaces. At the same time, an attempt is made to address adaptation from traditional to virtual workspaces. The objective of researching and reporting on the transition of traditional to virtual workspaces is to offer organizations and their members or other practitioners a way to address differences in cultures while fostering effective collaboration in virtual workspaces which facilitates the globalization process.

INTRODUCTION

Virtual teams represent individuals who are gathered together to perform a particular task in an environment where the communication occurs through a host of information communication technology systems, often referred to as computer-mediated communication (CMC) media. The essence of a virtual team can be described in two formats: (1) those occurring in collocated groups whereby individuals and participants are in close physical proximity, and (2) global virtual teams in which participants or members in the group interact from geographically dispersed locations and are not in the same physical proximity. It is essential to distinguish the two types of virtual teams because there is a critical variable "culture" that holds different implications for task performance and outcomes. For instance, individuals in collocated virtual teams may not have to worry about culture as much as those interacting in the global virtual team do. Thus the primary emphasis in this chapter is on the latter (i.e., global virtual team) and the role of culture in the context of global virtual interaction. This is especially important given that, in today's contemporary business environment, the norm is that business is no longer confined to national geographic borders; rather, much organizational work transcends national boundaries, and as a result, the tasks occur in global or cross-cultural virtual contexts. Notwithstanding, there is a significant oversight in the research as to the role of culture in global virtual environments. Most research emphasizes the technological effects or characteristics of the media in global virtual teams (GVTs) to the detriment of how culture influences the ensuing interactions among GVT members (Anawati & Craig, 2006; Olaniran, 2004). At the same time culture has been identified as an important variable in project management (e.g., Eggington, 1993), as well as posing a significant challenge for cross-cultural virtual teams (Dube & Pare, 2001; Olaniran, 2004, 2006). Therefore, the goal of this chapter is to explore culture, identify some challenges that culture poses for communication interaction in GVT, and attempt to offer suggestions for addressing the problems. First however, it is necessary to define culture and explore why it is important to devote attention to the idea of culture in GVT.

According to Keesing (1974) not everyone shares precisely the same theory of cultural code, and not all individuals share the same ideals about their culture; consequently, culture is viewed not just as a mere collection of symbols fitted together by an analyst, but instead as a system of knowledge, shaped and constrained by the way the human brain acquires, organizes, and processes information and creates an "internal model of reality" (Keesing, 1974, p. 89). However, as individuals learn to maneuver within their cultures, they develop their individual implicit theories of communication that overlap with the implicit personal theories of communication that other members of the same culture use to guide communication and interaction (Gudykunst & Kim, 2003; Olsen, 1978). In essence, it is argued that:

> *Our cultures influence our behavior directly through the norms and rules we use to guide our behavior when we interact with others. Our cultures also indirectly affect our communication through individual characteristics we learn when we are socialized into our cultures.* (Gudykunst & Kim, 2003, p. 16)

Therefore, if culture is germane to how individuals view themselves as a part of a unique social collective (i.e., identity) and eventual communication interaction, then it is contended that culture in essence influences any organizational activities and project management, be it face-to-face or virtual settings. This view has been echoed by Hall (1959) who argued that "culture is communication and communication is culture" (p. 159). Similarly, Hofstede, contends that culture is learned rather than inherited because it is derived from "one's social environment" and as such it is described as "the collective programming of the mind" that differentiates members of one human group from the other (Hofstede, 1980). Another well-known definition of culture was offered by Kroeber and Kluckholn (1952) by relating culture to behavior. They suggest that culture represents patterns of ideas and values that shape peoples' behavior.

Culture, however, is often neglected in the virtual workspace literature in spite of its significance. Perhaps one of the key reasons for the oversight is the fact that the globalization process in which standardization is often the norm is believed to be moving the world and its inhabitants toward the notion of universal culture (Olaniran, 2001a). Specifically, an argument has been made that with increased immigration and globalization, a new identity or "third" identity representing shifting hybridization of pre-existing national or historic cultural patterns is developing (Ess & Sudweeks, 2005; McSweeney, 2002). On the contrary, a shift or whatever change in cultural patterns has been found to be restricted to economic changes at best, whereas cultural norms relating to relational patterns have been practically non-existent or remain unaltered (Smith, 2002). Others have agued that just because globalization has increased the influx of people through immigration from different cultures does not imply that all cultures are steering toward universal ideals (Olaniran, 2001a, 2004; Smith, 2002). Gimenez (2002) concurs and specifies that "while some corporate differences are disappearing with globalization, local cultures and local meanings are still diverse. De Mooij (2003) also suggests that globalization is not converging cultures, but rather diverging them. Thus, attention must be given to the role of national and local cultures that influence communication interactions taking place in the virtual team collaborations, and general virtual workspaces in the process of meaning sharing.

In order to understand the importance of culture in the global virtual workspaces, it is important to examine the variations in different national cultures. Thus, Hofstede's dimension of cultural variability (Hofstede, 1980, 1996, 2001) offer a useful platform. There are four dimensions of cultural variability, which consists of power distance, uncertainty avoidance, individualism, and masculinity (Hofstede, 1980). Power distance (PD) is defined as "the extent to which the less powerful members of institutions and organizations accept that power is distributed unequally" (Hofstede & Bond, 1984, p. 418). Uncertainty avoidance (UA) is "the extent to which people feel threatened by ambiguous situations and have created beliefs and institutions that try to avoid these" (Hofstede & Bond, 1984, p. 419). Individualism-collectivism (IDCV) plays on the fact that in individualistic cultures, "people are supposed to look after themselves and their family only," while in collectivistic cultures, "people belong to in-groups or collectivities which are supposed to look after them in exchange for loyalty" (Hofstede & Bond, 1984, p. 419). Masculinity-femininity (MSFM) refers to countries "in which dominant values in society are success, money and things," while femininity refers to countries "in which dominant values are caring for others and quality of life" (Hofstede & Bond, 1984, pp. 419-420).

Notwithstanding, Hofstede's dimension has faced some criticism recently, regarding whether his claim generalizes to national culture, given that his data was collected from IBM with representatives from 50 countries (Ess, 2002; McSweeney, 2002). As a matter of fact, a special issue of the *Journal of Computer-Mediated Communication* was devoted to the dimension of cultural variability (Ess & Sudweeks, 2005). In spite of the criticism, it was found that certain aspects of the dimensions hold. For example, Hermeking (2005) finds a strong correlation between individualism dimension and Internet usage, and a strong negative correlation between high uncertainty avoidance and Internet usage. Another study finds consistency with Hermeking's support of Hofstede's dimension of individualism and uncertainty avoidance when looking at Internet usage (Barnett & Sung, 2005). These findings are significant, in the sense that most of the research findings in the special issues were devoted to the use of communication technology media (i.e., Internet) that are essential for making communication in global virtual workspaces a reality. Furthermore, past research used these four dimensions to operationalize cultural differences and their effects on uncertainty reduction in intercultural communication encounters (Gudykunst,

Chua, & Gray, 1987; Olaniran, 1996; Roach & Olaniran, 2001). Also, Smith (2002) responds to McSweeney's (2002) criticism of Hofstede's dimensions, indicating that the bases upon which McSweeney attempts to discard the utility of Hofstede's dimensions was in itself erroneous, because McSweeney only selectively chose two countries (i.e., cultures) whose profiles did not fit the dimensions of cultural variability while ignoring several countries that do (Smith, 2002).

For the purpose of this chapter, it needs to be stressed that the dimension of individualism-collectivism and the power distance aspects of Hofstede's model are particularly useful when exploring communication interactions among GVT members. One reason is based on the fact that individualism and power distance dimensions directly influence how one communicates or interacts with others (Olaniran 2004). The other is that both dimensions focuses on relationship and relational development, which are believed to be less subject to change in spite of any hybridization that may be taking place in any given culture (Smith, 2002). For instance, regardless of economic changes that might have moved individuals from collectivistic toward individualistic tendency (i.e., material wealth and accumulation), a large number of those individuals still maintain their identity and hold on to norms guiding different relationships (i.e., co-workers, superior-subordinates interactions, marriages, and families) (Smith, 2002).

CULTURAL DIMENSIONS AND VIRTUAL WORKSPACES

The focus of global virtual workspaces is less on the technology than the interaction that occurs in the process of accomplishing organizational and personal goals. Heaton (1998) agrees with this notion when she claims that the focus of computer-supported cooperative work (CSCW) is less on working with computer technology than on working with "each other" through computers. This is not to belittle the roles that technology plays in the process of virtual workspaces. After all, virtual interactions cannot occur in the absence of the technologies that makes interaction possible. However, the focus on participants' interactions in virtual workspace allow one to understand how social context or culture influences peoples' use of technology and perhaps the selection of technology to use. For instance, different cultural perspectives underlie the way workers and researchers organize themselves and their work in CSCW in Japan when compared with Scandinavia. In particular, the idea of democracy (individualistic tendency), which is valued among Scandinavians, influences the way they frame their CSCW choices in relation to how they view participation, building up resources, helping the weak cut across boundaries, hierarchy, and others, which also affects how they work together as equal members in teams (Heaton, 1998). On the other hand, the same priority on democracy was not shared by the Japanese, whose work ethic is primarily driven by the notion of authority (i.e., high-power distant culture) and thus evidenced in their CSCW interactions. Japanese culture invokes "justification" for decisions to stress contextual awareness and nonverbal communication in their CSCW systems. As a result, the most important thing to Japanese people in group interaction is face-to-face communication to create the atmosphere or *kuuki* (Heaton, 1998). Consequently, the biggest challenge facing Japanese people deals with how kuuki can be established in virtual interaction or transmitted across distance.

Both individualistic (e.g., Scandinavia and USA) and collectivistic (Japanese) cultures have communication styles that they prefer as alluded to above. However, the communication style preferences are either in low context or high context. According to Hall (1976), low-context communication is one in which a majority of the information is in the explicit code. In contrast, a high-context communication style has most of the information in the physical context or internalized in the person, while very little is in the explicit code. Thus, high-context communication preference puts emphasis on the need to understand the implicit meaning embedded in a message. The challenge for global virtual teams and interactions is individuals' discerning which communication style is being used in order to get at the true meaning of messages.

Scholars have attempted to answer the question of task-media fit when they focused on the idea of media richness—which was an attempt to determine or rank communication technology media based on the degree to which they are able to convey multiple social cues (Daft & Lengel, 1986; Daft, Lengel, & Trevino, 1987). In essence, face-to-face (FTF) medium is considered as the richest medium, while videoconferencing was considered next to FTF given the tendency to support immediate feedback, nonverbal cues, personalization, and language variety (Hiltz, Johnson, & Turoff, 1986; Connolly, Jessup, & Valacich, 1990). Information communication technologies as a whole are considered to be less friendly, emotional, or task oriented (Garton & Wellman, 1995; Rice & Love, 1987). By implication, the media richness theory focuses its attention on the physical characteristics of communication media as a strong determinant of outcomes in interactions. Another theoretical perspective, the social influence perspective, argues quite the contrary that the social contexts in which technology media are used contribute to the communication outcomes by influencing users' perceptions (e.g., Desanctis, & Poole, 1994; Fulk, Schmitz, & Steinfield, 1990; Olaniran, 1995, 1996; 2001a; Walther, 1995). Notwithstanding the views of the two different camps, it has been found that both technology media and the social factors are important and should be taken into consideration when using technology in virtual interaction (Olaniran, 1995, 2004; Yoo & Alavi, 2001).

Different media are able to convey different levels of cues through bandwidth. The implication speaks to the underlying factor in selection of and preference for a particular communication media in global virtual workspaces. The social influence perspective did not directly address culture, but its emphasis on social context indirectly speaks to its influence on preference, selection, and eventual use of a given technology media. Olaniran (2006) argues that cultural categories (high or low context and power distance) have implications for implicit and explicit communication tendencies and the general propensity to use technology in virtual interactions. Consequently, cultural factors tend to influence how individuals use or view communication technologies, and interpretations drawn from messages through them. As alluded to earlier, it might be difficult to get people to use certain computer-mediated technology systems in power distant cultures where status dictates every aspect of interpersonal communication (Devereaux & Johansen, 1994). For example, a study revealed that Korean and Japanese employees do not use e-mail to communicate with their superiors because of the perception that e-mail may be conceived by supervisors to be rude; as a result, individuals opt for alternative communication media deemed to be more respectful (Lee, 2002). On the contrary, Americans and other Western cultures do not share the same perception of respect and do not perceive the use of e-mail to be rude. Similarly, Kawachi (1999) found Japanese people, who represent a collectivistic and high-power distant culture, were less willing to use the Internet for collaboration activities (i.e., virtual communication) and that the Internet is primarily used by Japanese people to search and print information, translation, and entertainment.

Not only does the underlying culture influence the choice of technology media, it also adds certain complexity to virtual collaborative projects in terms of the respective tasks and the need to negotiate and build relationships among participants. It has been noted that one of the key problems global virtual teams face in building meaningful relationships is the challenge with "trust," which is compounded by cultural differences (Olaniran, 2004). Specifically, trust in global virtual teams is considered fragile and temporal (Holton, 2001; Jarvenpaa & Leidner, 1999). Furthermore, the perceived lack of multiple "cue-carrying capacity" of communication technology media is not conducive to facilitating trust, rather it is believed to hinder it and consequently lead users to question their suitability for virtual teamwork and group processes that require a certain degree of cohesion through personal and socio-emotional factors. For example, the Japanese always insist on having face-to-face interactions prior to virtual interaction in an attempt to establish the atmosphere that is conducive to relational building (Heaton, 1998). Also, study indicated that in African cultures, where significant emphasis is put on relationships, it was found that when e-mail was used for interactions, people habitually follow up the e-mail with communication in a more traditional medium (e.g., telephone and face-to-face) to determine whether the message was received (McConnell, 1998), and if the message had the desired effect (Olaniran, 2001a, 2006).

Another obstacle to relational development in virtual workspaces is the temporary nature of the project and the lack of common history among participants. The lack of history makes it difficult for members to draw on common experience in their attribution process or message interpretation (Crampton, 2002; Olaniran, 2001a, 2004). Furthermore, when group members are geographically dispersed, they are often less concerned about social information such that they fail to seek adequate social and contextual information to support their interpretation or attribution (Anawati & Craig, 2006; Crampton, 2002; Olaniran, 1994, 2004; Walther, 2002).

Some studies of communication media emphasize the lack of nonverbal cues in these technologies, leading researchers to conclude that by nature these media are "impersonal." Some of the studies offer that increased time is needed for participants to "orient" themselves and become familiar with other members, to the extent that differences between traditional face-to-face and technology-mediated communication should converge with increased time (Olaniran, 1994; Walther, 1994). On the other hand, when a communication technology medium offers non-verbal cues and immediate feedback as in videoconferencing, the geographical dispersion of the members makes time zone differences a hindrance such that videoconferencing is not an option in some cases (Olaniran, 2004, 2006). Notwithstanding, research has found that even in asynchronous computer-mediated interaction, members offer messages that are socially warm and relationally positive from the beginning of the interaction (Walther & Burgoon, 1992; Walther, 1997). The findings in essence contradict the idea or notion that communication technology media (especially, asynchronous) are lacking in social cues, and thus are not capable of conveying personal and social messages as originally claimed. Furthermore, it has been found that time alone has neither shown clear-cut differences, nor accounts for the differences between the computer-mediated and face-to-face medium (Olaniran, 2001b; Potter & Balthazard, 2002; Walther, 2002). However, with global virtual workspaces compounded by culture, it would seem that time by itself is not a sufficient criteria to develop cross-cultural communication competence. For example, Vroman and Kovacich (2002) found that the idea of "deadlines" is liberally interpreted in virtual teams such that it creates conflicts among participants.

An alternate explanation for the social warmth in communication technology was found to be "anticipation of future interaction," which is based on the premise that when individuals using online communication technologies are working in a virtual environment intend to meet at some point in the future, they tend to engage in communication activities where they actively seek and provide social information about one another (Olaniran, 1994, Walther, 1994). In other words, anticipation of the future encounter or meeting opportunities allows group members to behave and interact in a manner that is norm compliant and resulting in attempts to focus on the social dimension of communication through communication of socially warm messages. The anticipation effect on communication interaction with technologies is not confined to time (i.e., short- or long-term duration) (Olaniran, 2001b, 2004; Walther, 1994).

There are other problems in virtual workspaces. Olaniran (2004) identified a few that include fragmented communication, confusion during teleconferences with people on different pages of documents, failure to return phone calls or respond to inquiries, and key group members left off e-mail distribution lists. As a whole, managers or team leaders are faced with the difficulty of transferring group culture subtleties across distance such as protocol for determining task performance.

From a cultural standpoint, silence is another key challenge in virtual workspaces. Silence can be interpreted differently by individualistic and collectivistic cultures which would ultimately alter perceptions and attributions. For example, silence as a reaction to a request or business proposal would be perceived negatively among Americans, Europeans, and Arab business persons, while it is cherished among listening or collectivistic cultures (e.g., in East Asia and West Africa), where silence is considered part of social interaction and not necessarily a "failure to communicate" (Lewis, 2000; Olaniran, 2004). More importantly, silence is intensified in an asynchronous mode of virtual interaction, as participants from a listening culture either choose not to use the medium, or wait to collect information requested before responding to a particular request as a show of respect to the message sender. Asynchronous media presents a time lag between messages, and when unknown, it can result in misunderstanding where other participants attribute non-communication to "lack of manners," thereby negatively affecting relationships (Crampton, 2001; Olaniran, 2004; Pauleen & Yoong, 2001).

SOLUTIONS / RECOMMENDATIONS

Virtual workspaces and teams differ from traditional workspaces in many ways. As indicated above, virtual teams face challenges from geographical distance, time difference, cultural differences—all of which affect the selection of communication media and the development of interpersonal relationship that fosters team or group cohesion. Notwithstanding, it is possible to have effective virtual interaction by overcoming these challenges. For example, Waldir Arevoedo (a research analyst at Gartner Inc.) suggests the use of different technology media to allow for different cues such as video, text, and audio in an attempt to overcome the barriers of time, distance, and culture (Solomon, 2001). This recommendation is not without its own problem; for instance it was mentioned earlier that geographical distance along with different time zones may limit the degree to which synchronous communication media can be used in virtual interactions. For example, it was found that virtual workers primarily use e-mail and telephone, and that videoconferencing and computer conferencing were rarely used, if used at all (Sivunen & Valo, 2006). However, when multiple communication media are possible, it may help overcome certain problems. For instance instant feedback is lacking in e-mail or bulletin board, whereas IM or teleconferencing offer instant feedback and, as a result, the ability to facilitate immediacy among participants.

The selection of multiple communication channels by itself is not going to overcome problems attributable to culture outright. As a result, it is suggested that organizations carefully choose their team leaders. This is important because the team leaders face the challenge and responsibility of selecting the technology to be used in virtual workspaces and for managing teams effectively (Olaniran, 2004; Sivunen & Valo, 2006). While selection of communication technology is at times based on team leaders' preferences, it is important that the decision takes into consideration team members' preferences for certain media, along with the ease of access to the media. Such a decision would require awareness of virtual team members' cultures and implies that media usage must be taken into account for the process to be effective. The foundation for using team members' preference in media selection is germane to accommodating communication partners in a way that enhances effective communication. Furthermore, the approach is to support the social influence theory of media choice (Carlson & Davis, 1998; Fulk et al., 1990; Sivunen & Valo, 2006).

The need to adapt communication behavior in virtual interaction is pressing. Given that team members from diverse cultural backgrounds have different communication styles and preferences, they can also have different ways of presenting messages. Therefore, failure to adapt communication strategies by participants in virtual workspaces can cause problems. Thus, messages must be composed or adapted in a way that readers would understand their meanings (Anawati & Craig, 2006; St.Amant, 2001; Warren, 1998). This is not easy to do however; it requires a proactive effort on the part of virtual workers. For instance, it was found that "while 80.3% of participants change their method of spoken communication during CCV [cross-cultural virtual] team interactions, only 60.7 % of respondents change their method of written communication" (Anawati & Craig, 2006, p. 47). It stands to reason that it is easier to adapt to culture differences in spoken communication than the written communication. Anawati and Craig (2006) argue that when one hears other people's accents or level of fluency, one is reminded to adapt communication by speaking slowly and clearly to be understood.

Furthermore, there appears to be a correlation between length of time in a team and communication adaptation regardless of spoken or written (Anawati & Craig, 2006). This implies a need to provide virtual teams enough time to engage in cultural learning—where participants develop affinity for each other's culture and communication styles, and then develop a way to overcome their challenges. For example, individuals who have been in a virtual group for three months or less were found to be the least likely to adapt their communication, while those who have been in the group for six months or more were more likely to engage in adaptive communication behavior. It was also found that length of time in the group influences whether members want their colleagues to know about their cultures (Anawati & Craig, 2006). These findings provide indirect support for the principle of anticipation of future interaction as a norm-conforming mechanism. For instance, as individuals come to realization that they are more likely to meet over and over, they begin to develop communication strategies that help foster positive interactions, trust, and relational development. As a result, it is recommended that organizations should attempt to simulate or induce anticipation of future interactions in their virtual teams, and more specifically to use ongoing rather than short-term virtual teams (Olaniran, 2004).

Some ideas for adapting written communications in virtual interactions have been offered. They include: avoidance of slang and jargon, use of simple English or short words, keep to the point, change between formal and informal writing, clarification through repetition, be more descriptive, use lists or points, use short e-mail especially to non-native-English speakers, write from the recipient's point of view, and avoid humor and formulate criticism carefully (Anawati & Craig, 2006; Olaniran, 2004). Embarking on these adaptive communication behaviors would help facilitate mutual respect, understanding, and trust, which can help in achieving the relational building goal in virtual environment. Furthermore, it is important to note that cultural challenges in virtual workspaces can be overcome when members take the time to ask questions and take interest in other participants such that they could determine whether the correct intention was comprehended. After all it is true that communication in cross-cultural virtual environments, like any other cross-cultural contexts, has a lot to do with "releasing the right responses" than actually sending the right message (Hall & Hall, 1989).

FUTURE TRENDS

For the most part researchers have focused on physical characteristics of communication technologies as a predictor or determinant of communication task-technology fit. However, the argument provided in this chapter based on other research findings points to the fact that the process of communication technology task-fit is more complex than initially thought. Moreover, culture is presented as one of the underlying variables that determines which communication technology is selected for use in a virtual team, and the general outcomes or effects of their usage. For instance, it has been shown that some aspects of task factors were related or aligned with the principles of media richness theory; but the social factors were based on other variables (Sivumen & Valo, 2006; Yoo & Alavi, 2005). Sivumen and Valo (2006) found that accessibility is one of the critical factors when virtual team leaders are selecting technology for use in virtual workspaces. It is also important to note that how virtual workers learn and select communication media affect their perceived level of voice within an organization and eventually the expectations they have about other participants along with how they respond to messages (Farrell & Holkner, 2004).

Furthermore, the argument presented in this piece suggests that the proclamation that certain technologies (e.g., text-based e-mail) are not rich, as proposed by the media richness theory, may be too simplistic. The nature of tasks and members' perceptions of the necessary communication which is influenced by one's culture may in fact determine the use of media regardless of whether the media are lean or rich. Sivumen and Valo (2006) suggest that in some situations, users may benefit from using communication technology with few cues. In essence, it is important for participants to understand the need to be aware of cultural differences and adapt their communication behavior to accommodate differences. Therefore, it is suggested that virtual team training includes aspects of culture that would facilitate adaptive behavior in virtual interaction (Anawati & Craig, 2006; Olaniran, 2006, 2004).

At the same time, while organizations using virtual teams need to be aware of culture, it is important that they pay attention to the different aspects of culture. For instance, Olaniran (2004) stresses the need to identify the differences between organizational culture and local cultures, along with the interaction of the two in virtual workspaces. It is important to know the extent to which organizational culture influences the local (national) cultures and vice versa.

As research in virtual workspaces continues, it would help our understanding to focus on how individuals use the new media at the moment of interaction. For the most part, the literature is rife with laboratory experiments and *post facto* self-report questionnaires that may or may not relate to actual experiences. Similarly, it would help to heed the call for designers to figure out how to create a communication medium that is easily accessible to all users regardless of participants' geographical locations (Sivumen & Valo, 2006). Along the same line of thinking is the need to create groupware that gives options to users regardless of cultural communication preference. The groupware would represent a one-stop suit where virtual members are able to adapt their communication relatively. It would also help for researchers to address the issue of virtual participants' propensity to use or adopt communication technologies. This chapter concurs with other researchers by calling for other approaches that do not examine mediated communication or interactions using face-to-face medium or "talk" as the baseline. Rather, mediated interactions need to be viewed as a unique mode of interaction, part of which is significantly based on written text (e.g., e-mail and bulletin boards). After all, the business memos and letters upon which some of the text-based communication technologies share similarities provide precedence (e.g., Churchill & Erickson, 2003; Yates & Orlikowski, 1992). Such an approach would allow researchers to engage varieties of methodology that allow for taking into account historical and material culture.

CONCLUSION

Virtual collaboration is an idea driven by globalization and sophistication in communication technologies such that virtual interaction and collaborations are made possible regardless of workers' geographic or national boundaries. Notwithstanding, as individuals collaborate globally in virtual workspaces, they are faced with the issue of culture that compounds the complexity of communication interactions and in return influences communication outcomes. Nonetheless, culture represents a phenomenon that is seldom taken into consideration, potentially leading to increased conflicts and interactions that fail to accomplish the set goals. This chapter explained culture and its implications in virtual workspaces. Hofstede's dimension of cultural variability, especially the idea of individualistic-collectivistic and the power distance, explore implication for media selection and usage while addressing social and relational development in virtual workspaces. From the presentation of culture implications, suggestions were offered for overcoming challenges created by culture. The last portion of the chapter attempts to provide general direction for future research in virtual interactions.

GLOSSARY

Computer-mediated communication: Involves communication process where computers and other electronic media serve as the medium of interaction.

Culture: Involves collections of symbols and norms that affect how humans gather, organize, and process information.

Globalization: A social, political, and economic trend in which denationalization of markets, politics, and legal systems occur, and from an organizational standpoint, corporations or companies attempt to participate in the global economy by maintaining their presence in foreign markets through incorporation of communication technologies (i.e., the Internet) for virtual presence on the international marketplace or e-business.

Individualism-collectivism: Individualism speaks to the cultural approach where people look after themselves and their family—in essence, the emphasis is on self. Collectivism emphasizes cultures where people belong to in-groups or collectives that are supposed to look after them in exchange for loyalty; thus the emphasis is on others.

Power distance: Identifies cultural norms or beliefs where a society comes to accept how power is distributed (either equally or unequal).

Virtual team: A collaboration approach in which members interact via communication technologies for the purpose of accomplishing a group goal.

Virtual workspace: Involves organizational environments in which coordination process and communication activities exist with the aid of electronic-mediated technologies.

REFERENCES

Anawati, D., & Craig, A. (2006). Behavioral adaptation within cross-cultural virtual teams. *IEEE Transactions on Professional Communication, 49*(1), 44-56.

Barnett, G. A., & Sung, E. (2005). Culture and the structure of the international hyperlink network. *Journal of Computer-Mediated Communication, 11*(1). Retrieved July 15, 2006, from http://jcmc.indiana.edu/vol11/issue1/barnett.html

Carlson, P. J., & Davis, G. B. (1998). An investigation of media selection among directors and managers: From self to other orientation. *MIS Quarterly, 22*(3), 335-362.

Churchill, E. F., & Erickson, T. (2003). Introduction to this special issue on talking about things in mediated conversation. *Human-Computer Interaction, 18,* 1-11.

Connolly, T., Jessup, L. M., & Valacich, J. S. (1990). Effects of anonymity and evaluative tone on idea generation in computer-mediated groups. *Management Science, 36,* 97-120.

Crampton, C. D. (2002). Attribution in distributed work groups. In P. Hinds & S. Kiesler (Eds.), *Distributed Work* (pp. 191-212). Boston: MIT Press

Daft, R. L., & Lengel, R. H. (1986). Organizational information requirements, media richness and structural design. *Management Science, 32,* 554-571.

Daft, R. L., Lengel, R. H., & Trevino, L. K. (1987). Message equivocality, media selection, and manager performance: Implications for information systems. *MIS Quarterly, 11,* 355-366.

De Mooij, M. (2003). Convergence and divergence in consumer behavior: Implications for global advertising. *International Journal of Advertising, 22*(2), 1-20.

Desanctis, G., & Poole., M. S. (1994). Capturing the complexity in advanced technology use: Adaptive structuration theory. *Organization Science, 5*(2), 121-147.

Devereaux, M. O., & Johansen, R. (1994). *Global work: Bridging distance, culture, & time.* San Francisco: Jossey-Bass.

Dube, L., & Pare, G. (2001). Global virtual teams. *Communications of the ACM, 44*(12), 71-73.

Egginton, B. (1993). Cultural complexity and the improved performance of a large international projects. In *Proceedings of the International IEEE Engineering Management Conference,* India (pp. 53-58).

Ess, C. (2002). Cultures in collision philosophical lessons from computer-mediated communication. *Metaphilosophy, 33*(1-2), 229-253.

Ess, C., & Sudweeks, F. (2005). Culture and computer-mediated communication: Toward new understandings. *Journal of Computer-Mediated Communication, 11*(1). Retrieved July 17, 2006, from http://jcmc.indiana.edu/vol11/issue1/ess.html

Farrell, L., & Holkner, B. (2004). Points of vulnerability and presence: Knowing and learning in globally networked communities. *Discourse: Studies in the Cultural Politics of Education, 25*(1), 133-144.

Fulk, J., Schmitz, J., & Steinfield, C. W., (1990). A social influence model of technology use. In J. Fulk & C. Steinfield, (Eds.), *Organizations and communication technology* (pp. 117-140). Newbury Park, CA: Sage.

Garton, L., & Wellman, B. (1995). Social impacts of electronic mail in organizations: A review of the research literature. In B. R. Burleson (Ed.), *Communication yearbook* (Vol. 18, pp. 434-453). Thousand Oaks, CA: Sage.

Gimenez, J. (2002). New media and conflicting realities in multinational corporate communication: A case study. *IRAL, 40,* 323-343.

Gudykunst, W. B., Chua, E., & Gray, A. J. (1987). Cultural dissimilarities and uncertainty reduction processes. In M. McLaughlin (Ed.), *Communication yearbook* (Vol. 10, pp. 457-469). Beverly Hills, CA: Sage.

Gudykunst, W. B., & Kim, Y. Y. (2003). *Communication with strangers: An approach to intercultural communication* (4th ed.). Boston: McGraw-Hill.

Hall, E. (1959). *The silent language.* New York: Doubleday.

Hall, E., & Hall, M. (1989). *Understanding cultural differences.* Yarmouth, ME: Intercultural Press.

Hall, E. T. (1976). *Beyond culture.* New York: Doubleday.

Heaton, L. (1998). Talking heads vs. virtual workspaces: A comparison of design across cultures. *Journal of Information Technology, 13,* 259-272.

Hermeking, M. (2005). Culture and Internet consumption: Contributions from cross-cultural marketing and advertising research. *Journal of Computer-Mediated Communication, 11*(1). Retrieved July 15, 2006, from http://jcmc.indiana.edu/vol11/issue1/hermeking.html

Hiltz, S. R., Johnson, K., & Turoff, M. (1986). Experiment in group decision making communication process and outcome in face to face vs. computerized conference. *Human Communication Research, 13,* 225-252.

Hofstede, G. (1980). *Culture's consequences.* Beverly Hills, CA: Sage.

Hofstede, G. (1996). *Cultures and organizations: Software of the mind.* New York: McGraw-Hill.

Hofstede, G., & Bond, M. (1984). Hofstede's culture dimensions: An independent validation using Rokeach's value survey. *Journal of Cross-Cultural Psychology, 15,* 417-433.

Hofstede, G. H. (2001). *Culture's consequences: Comparing values, behaviors, institutions, and organizations across nations.* Thousand Oaks, CA: Sage.

Holton, J. A. (2001). Building trust and collaboration in a virtual team. *Team Performance Management, 7*(3/4), 36-47.

Jarvenpaa, S. L., & Leidner, D. E. (1999). Communication and trust in global virtual teams. *Organizational Science, 10,* 791-815.

Kawachi, P. (1999). *When the sun doesn't rise: Empirical findings that explain the exclusion of Japanese from online global education.* Retrieved August 26, 2005, from http://www.ignou.ac.in/Theme-3/Paul%20%20KAWACHI.html

Keesing, R. (1974). Theories of culture. *Annual Review of Anthropology, 3,* 73-97.

Kroeber, A. L., & Kluckholn, C. (1952). *Culture: A critical review of concepts and definitions.* Cambridge, MA: Mass Papers Peabody Museum.

Lee, O. (2002). Cultural differences in e-mail use of virtual teams a critical social theory perspective. *Cyberpsychology & Behavior, 5*(3), 227-232.

Lewis, R. D. (2000). *When cultures collide: Managing successfully across cultures.* London: Nicholas Brealy.

McConnell, S. (1998). NGOs and Internet use in Uganda: Who benefits? In C. Ess & F. Sudweeks (Eds.) *Cultural attitudes towards communication and technology* (pp. 104-124). Australia: University of Sydney.

McSweeney, B. (2002). Hofstede's model of national cultural difference and their consequences:

A triumph of faith—a failure of analysis. *Human Relations, 55*(1), 89-117.

Olsen, M. (1978). *The process of social organization* (2nd ed.). New York: Holt, Rhinehart, & Winston.

Olaniran, B. A. (1994). Group performance in computer-mediated and face-to-face communication media. *Management Communication Quarterly, 7,* 256-281.

Olaniran, B. A. (1995). Perceived communication outcomes in computer-mediated communication: An analyses of three systems among new users. *Information Processing and Management, 31,* 525-541.

Olaniran, B. A. (1996). A model of satisfaction in computer-mediated and face-to-face communication. *Behavioural and Information Technology, 15,* 24-36.

Olaniran, B. A. (2001a). The effects of computer-mediated communication on transculturalism. In V. Milhouse, M. Asante, & P. Nwosu (Eds.), *Transcultural realities* (pp. 83-105). Thousand Oaks, CA: Sage.

Olaniran, B. A. (2001b). Computer-mediated communication and conflict management process: A closer look at anticipation of future interaction. *World Futures, 57,* 285-313.

Olaniran, B. A. (2004). Computer-mediated communication in cross-cultural virtual groups. In G.M. Chen & W. J. Starosta (Eds.), *Dialogue among diversities* (pp. 142-166). Washington, DC: NCA.

Olaniran, B. A. (2006). Challenges to implementing e-learning in lesser developed countries. In A. Edmundson (Ed.), *Globalized e-learning cultural challenges (pp. 18-34).* Hershey, PA: Information Science Publishing.

Pauleen, D. J., & Yoong, P. (2001). Relationship building and the use of ICT in boundary-crossing virtual teams: A facilitator's perspective. *Journal of Information Technology, 16,* 205-220.

Potter, R. E., & Balthazard, P. A. (2002). Virtual team interaction styles: Assessment and effects. *International Journal of Human-Computer Studies, 56,* 423-443.

Rice, R. E., & Love, G. (1987). Electronic emotion: Socioemotional content in a computer-mediated network. *Communication Research, 14,* 85-108.

Roach, K. D., & Olaniran, B. A. (2001). Intercultural willingness to communicate and communication anxiety in international teaching assistants. *Communication Research Reports, 18,* 26-35.

Sivunen, A., & Valo, M. (2006). Team leaders' technology choice in virtual teams. *IEEE Transactions on Professional Communication, 49*(1), 57-68.

Smith, P. B. (2002). Culture's consequences: Something old and something new. *Human Relations, 55*(1), 119-135.

Solomon, C. M. (2001). Managing virtual teams. *Workforce, 80*(6), 60-65.

St.Amant, K. (2001). Cultures, computers, and communication: Evaluating models of international online production. *IEEE Transactions on Professional Communication, 44*(4), 291-295.

Vroman, K., & Kovacich, J. (2002). Computer-mediated interdisciplinary teams: Theory and reality. *Journal of Interprofessional Care, 16,* 161-170.

Walther, J. B. (1992). A longitudinal experiment on relational tone in computer-mediated and face-to-face interaction. In J. F. Nunamaker & R. H. Sprague (Eds.), *Proceedings of the Hawaii International Conference on System Sciences* (Vol. 4, pp. 220-231). Los Alamitos, CA: IEEE Computer Society Press.

Walther, J. B. (1994). Anticipated ongoing interaction versus channel effects on relational communication in computer-mediated interaction. *Human Communication Research, 20,* 473-501.

Walther, J. B. (1995). Relational aspects of computer-mediated communication: Experimental observations over time. *Organization Science, 6*(2), 196-203.

Walther, J. B. (1997). Group and interpersonal effects in international computer-mediated collaboration. *Human Communication Research, 23,* 342-369.

Walther. J. B. (2002). Time effects in computer-mediated groups: Past, present, and future. In P. Hinds & S. Kiesler (Eds.), *Distributed work* (pp. 235-257). Boston: MIT Press.

Walther, J. B., & Burgoon, J. K. (1992). Relational communication in computer-mediated interaction. *Human Communication Research, 19,* 50-88.

Warren, T. (1998). National cultures in international communication. In *Proceedings of the Professional Communications Conference*, Quebec City, Canada (pp. 305-307).

Yates, J., & Orlikowski, W. J. (1992). *Control through communication: The rise of system in American management.* Baltimore: John Hopkins University Press.

Yoo, Y., & Alavi, M. (2001). Media and group cohesion: Relative influences on social presence, task participation, and group consensus. *MIS Quarterly, 25*(3), 371-390.

Chapter VII
The Language of Leaders: Identifying Emergent Leaders in Global Virtual Teams

Simeon J. Simoff
University of Technology, Sydney, Australia

Fay Sudweeks
Murdoch University, Australia

ABSTRACT

Virtual teams and their leaders are key players in global organizations. Using teams of workers dispersed temporally and geographically has changed the way people work in groups and redefined the nature of teamwork. Emergent leadership issues in computer-mediated communication are vital today because of the increasing prevalence of the virtual organization, the flattening of organizational structures, and the corresponding interest in managing virtual groups and teams. This chapter examines the communication behaviors of participants in two different case studies to determine if number, length, and content of messages are sufficient criteria to identify emergent leaders in asynchronous and synchronous environments. The methodology used can be embedded in collaborative virtual environments as a technology for identifying potential leaders in organizational and educational environments.

INTRODUCTION

Identifying competent leaders is a crucial component in building high performing teams that operate in global, cross-functional, and cross-cultural environments. During the different phases of team development, leaders may need to take different roles (Kolb, 1999). For instance, in an early stage, leadership skills may involve understanding individual differences, work styles, and cultural nuances. In later stages, leaders may assume a peer relationship with team members. In any of the roles, leaders communicate with the other team members. If a team collaborates over the Internet, then we can observe that communication. Our assumption is that emergent leaders can be identified through their communication patterns. We address our assumption focusing on *communication behaviors* of participants in two different case studies to determine if number, length, and content of messages are sufficient criteria to identify emergent leaders in both synchronous and asynchronous environments. We pose the following research questions:

- How will leadership be reflected in communication patterns and communication style among team members; in other words, what is the language of leaders?
- Are there differences in these patterns/trends in different scenarios?
- Can we facilitate technologically the identification of leaders in virtual teams, based on the patterns of their communication?

In this chapter, we initially provide background research that frames the context of the problem. Then we present our approach, based on a complementary explorative data analysis (CEDA) research methodology developed by the authors for conducting Internet research (Sudweeks & Simoff, 1999). We examine the patterns of communication of leaders in two different case studies of online teamwork. The case studies complement each other in terms of activity scenarios and text-based communication modes. One case study is a group of autonomous and diverse individuals using an asynchronous communication medium over a relatively long period of time, while the other case study is a group of individuals using a synchronous communication medium and bound by the communication network for a short period of time. Finally, we present technological solutions for enabling leader identification in virtual environments along with some concluding discussions.

BACKGROUND

The formation of global virtual teams has changed the way people work in groups and redefined the nature of teamwork (Lipnack & Stamps, 1997; Mabry, 2002; Meier, 2003). A global virtual team is defined as teams of workers dispersed temporally and geographically which are assembled using a combination of telecommunications and information technologies to accomplish an organizational task (Townsend, DeMarie, & Hendrickson, 1998). Members of such teams work and interact in various modes, using a diverse set of computer-mediating technologies (Maher, Simoff, & Cicognani, 2000). In the climate of enterprise globalization, such virtual teams are essential components in the enterprise "toolbox" to remain competitive (Maznevski & Chudoba, 2000). Research on communication in virtual teams is less well documented (Furst, Blackburn, & Rosen, 1999). However, understanding the elements of group dynamics of virtual teams is of crucial importance in facilitating and managing these teams.

Leadership is acknowledged as a key element in virtual team dynamics and is well researched (Cascio & Shurygailo, 2003; Zaccaro & Bader, 2003; Zigurs, 2003), yet less is studied about how leadership activities influence group collaborative processes (Avolio, Kahai, & Dodge, 2000; Cascio & Shurygailo, 2003; Pauleen & Yoong, 2001; Zaccaro & Bader, 2003; Zigurs, 2003). Emphasizing the paucity of research on leadership in virtual teams, Misiolek and Heckman (2005) provide a broad and up-to-date overview of the literature on virtual teams in organizational context.

Addressing leadership in virtual teams is an essential part of an in-depth study of virtual team dynamics and analysis of their development (Sudweeks, 2004). Leadership issues in virtual teams remain vital today because of the increasing prevalence of the virtual organization, the flattening of organizational structures, and the corresponding interest in managing virtual groups and teams. In this chapter, we distinguish between *assigned* leadership and *emergent* leadership. An assigned leader is an individual who is assigned to a position of leadership. An emergent leader is an individual who is not assigned to a leadership position, who has the same status as other team members initially, but who gradually emerges as a leader through the support and acceptance of the team over a period of time (Guastello, 2002). The establishment of emergent leaders is a result of their actions and their communication behaviors, which include being involved, informed, firm but seeking the opinion of others, and initiating new ideas (Fisher, 1974). Leaders emerge according to the needs of the group (Myers, Slavin, & Southern, 1990) and usually exhibit the following characteristics: (1) participate early and often; (2) focus on communication quality as well as quantity; (3) demonstrate competence; (4) create social structures (Avolio et al., 2000); and (5) help build a cohesive team (Hackman & Johnson, 2000).

Earlier research in leadership in face-to-face environments shows that leaders are identified by high participation rates in discussions (Mullen, Salas, & Driskell, 1989; Regula & Julian, 1973; Sorrentino & Boutillier, 1975). McCroskey and Richmond (1998) related effective leadership to "talkativity." However, Yoo and Alavi (2002) proposed that, because of the reduced awareness of social presence and social context, the receiver of a message via computer-mediated communication (CMC) pays more attention to the message than the messenger. This observation led Yoo and Alavi to study emergent leaders in virtual teams. They found that, in asynchronous communication, emergent leaders could be identified by the number, length, and content of messages. Not only did emergent leaders send more messages and longer messages, their messages were more task oriented than other team members.

Misiolek and Heckman (2005) studied patterns of emergent leadership behavior in distributed virtual teams based on an analysis of interactions between college seniors, captured during a two-week course of a virtual collaboration exercise. The authors conducted a content analysis of the interactions using a coding scheme derived from behaviorally based leadership theory. The two-week period may not necessarily be sufficient time for the development of emergent leaders. However, the analysis is an interesting attempt to employ behaviorally based and functional theories of leadership to the analysis of leadership in virtual teams.

CASE STUDIES

The first case study (Case Study 1) was a two-year collaborative research project conducted by an international group of volunteer researchers, most of whom had never met either online or offline. This is different from the majority of the studies of leadership in virtual teams, where usually there has been at least one face-to-face session (Avolio et al., 2000; Misiolek & Heckman, 2005; Pauleen, 2003). The collaborative activity of the group was the collection and analysis of data from electronic discussion groups. Computer-mediated *asynchronous communication* was used for coordination, participant recruitment, distribution of information, formulation and discussion of policies, decision making, encouragement, and technology transfer. The number of members varied at any one time, but 143 people were consistently involved in the project. Two participants were assigned leadership roles, and they took on the facilitating task of encouraging the group to work together interdependently in a collaborative manner.

The second case study (Case Study 2) involved a group of 18 students engaged in collaborative learning in nine one-hour workshops over a four-month period. The workshops required substantial preparation, both individually and collaboratively. The workshops were held in a WebCT chat room

Table 1. Feature summary of the two case studies

Feature	Case Study 1	Case Study 2
Medium	E-mail	Chat room
Mode	Asynchronous	Synchronous
Duration	2 years	4 months
Leadership	Assigned	Appointed
Formation	Spontaneous	Predefined
Meetings	Unstructured	Structured
Purpose	Research project	Workshop series
No. of participants	143	19
Location of participants	Global	Mostly Australia
Age group	20-65	Mostly 20-30
Process	Unstructured	Structured

and were part of a unit of study. Although the participants lived within a 100 km range of the university, the majority of the students had never met either online or off-line. For each workshop, a tutor participated and a different student was appointed as a moderator. Moderators were required to lead the group discussions and facilitate learning through discussion of set readings.

The two case studies therefore differed in the features listed in Table 1. To some extent they complement each other, allowing an exploration of various facets of emergent leadership both in asynchronous and synchronous online scenarios.

RESEARCH METHODOLOGY

Research in virtual teams is part of what constitutes Internet research. The replication of Internet field research is difficult, if not impossible, for two main reasons (apart from the usual problems of the environment and human nature itself constantly changing). On a *technological* level, the Net is perpetually changing its configuration and supporting technology. On an *interaction* level, the difficulties in replication come from the creative aspect of language use and the evolution of computer media. Apart from standard clichés, sentences are rarely duplicated exactly, yet each variation is generally comprehended. It follows that experiments involving text generation can rarely be repeated. Furthermore, studies of virtual teams have been usually conducted under controlled experimental conditions which may not present an accurate picture of the reality of virtuality. The problems can come from the following factors: (1) subjects are an atypically captive audience who would probably behave differently in a laboratory than they would in a real-world setting; (2) groups studied in experiments tend to be unrealistically small; and (3) an almost natural inclination of experimental design is to compare CMC with a face-to-face standard (Rafaeli & Sudweeks, 1997, 1998), a comparison that may be misleading.

The CEDA methodology—developed by Sudweeks and Simoff (1999), further extended by Riva and Galimberti (2001) as complementary explorative multilevel data analysis (CEMDA), and revised in Sudweeks (2004)—has developed multi-method research design principles in order to address the abovementioned issues. Hence, the methodology applied to each case study follows the

Figure 1. Research methodology and communication model: (a) the composition principles of a CEDA-compliant study (Adapted from Sudweeks & Simoff, 1999); (b) representing activities in virtual environments (Adapted from Simoff & Maher, 2000)

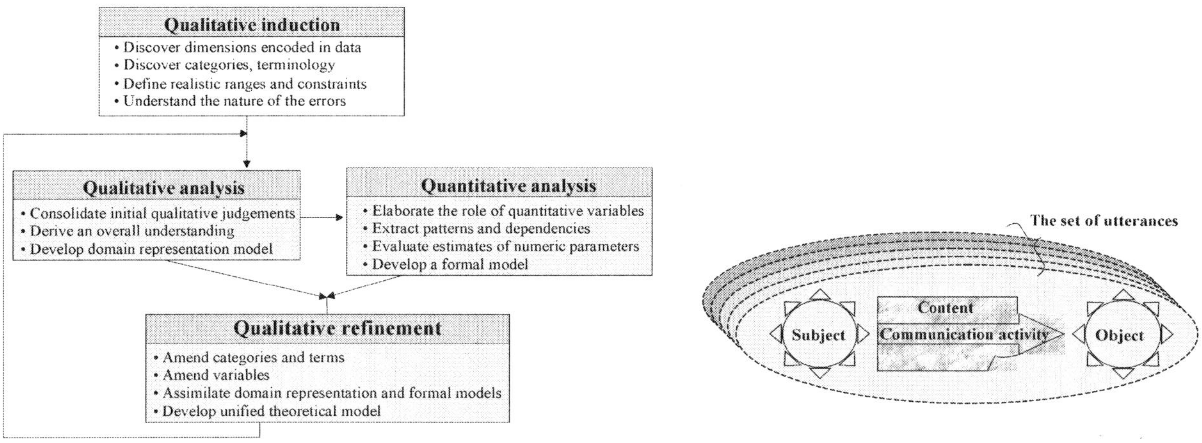

(a) (b)

CEDA approach. The methodological composition principles are illustrated in Figure 1a.

The integrated data sets include a variety of sources including participant observations and archived electronic discussions. Throughout the duration of each study, one of the authors was participating and observing communication processes within both case studies, thus providing richness of data (Witmer, 1997). Observation assisted in discovering the underlying assumptions and dimensions of which group participants may have been unaware (Hammersley & Atkinson, 1983).

In Case Study 1, team members posted more than 1,000 e-mail messages while collaborating on their project. In Case Study 2, students engaged in nine hours of synchronous discussions, which were automatically logged and downloaded by one of the authors.

To be able to compare the results of two dissimilar case studies, the communication was viewed as *sets of utterances*. The communication model for the analysis of virtual design teams, proposed by Simoff and Maher (2000), is based on the premise that each communication activity is composed of: (1) a subject who performs the communication event; (2) the content of the communication event; and (3) an object(s) to whom the communication event is addressed. In other words, in an utterance, a subject is communicating content to an object. Formally, each utterance can be represented as ⟨Subject, Object, Content⟩ which is defined as a SOC-triple. Hence an utterance can be denoted as: $u_i = \langle s_i, o_i, c_i \rangle$, where s_i, and o_i denote the corresponding subject and object in utterance u_i, respectively, and c_i is the content of utterance u_i. This model of an utterance provides a common timeless representation for both asynchronous and synchronous communication as illustrated in Figure 1b. Communication within a particular time window is represented as a sequence U of SOC-triples u_i, $U = u_1, u_2, \ldots, u_n$, where n denotes the *length* of the sequence of utterances. A *communication pattern* in our terms is a subsequence $U_{kl} = u_k, \ldots, u_l$, where $U_{kl} \subset U$ and $k < l$. In this framework, communication patterns can be grouped into three classes: (1) *content-independent patterns*—these patterns are statistics or utterance sequences that can be derived without consideration of the content of the

Figure 2. Representing the interactions via listserv and via a chat room as a set of utterances, (a) interactions via listserv media, (b) interactions via a chat room

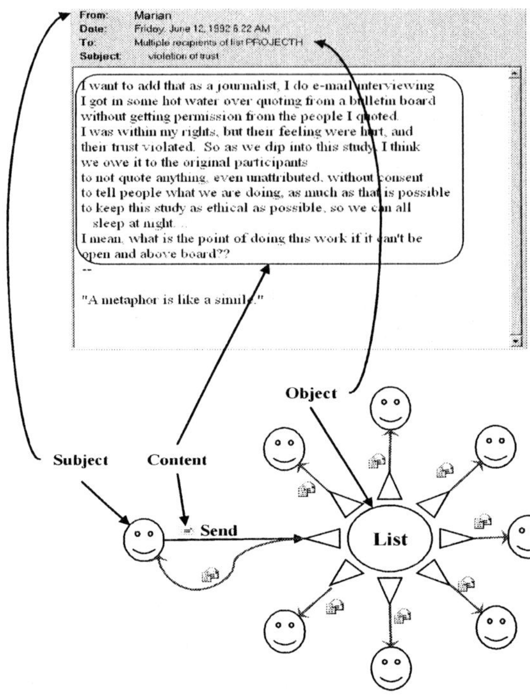

(a)

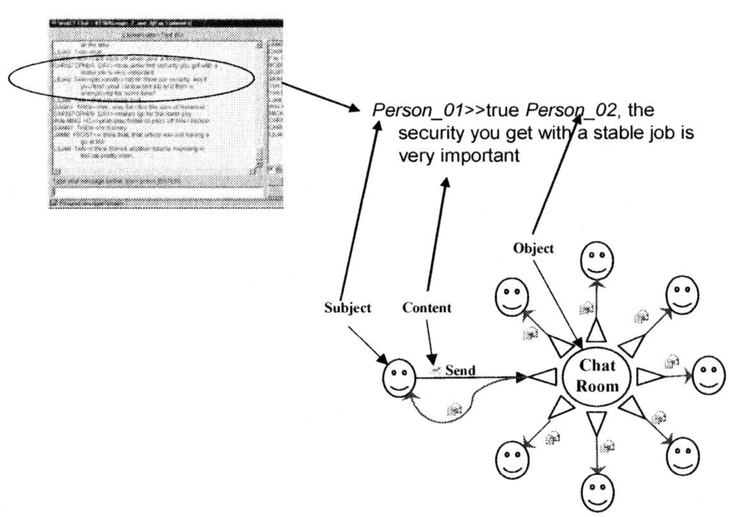

(b)

utterances; (2) *content-dependent patterns*—these patterns are statistics or utterance sequences that can be derived from an analysis of labeled communication sequences, where the labels belong to a predefined coding scheme, and each utterance is labeled according to its content; and (3) *content-based patterns*—these patterns consist of various text statistics, term clusters, contingency analyses, and taxonomies of words, which are derived from a text analysis of the utterance content.

This general model is applicable in text-based CMC to communication activities in both asynchronous environments and synchronous environments. In an initial analysis, the data was segmented into utterances—that is, each communication event (e-mail message or chat turn) was converted into utterances on the basis of one object (receiver) per utterance. The object in a communication utterance could be the whole group, part of the group, or an individual. The approach applied to both case studies is illustrated in Figure 2. Thus the e-mail messages and chat room logs were converted to 1,345 utterances for Case Study 1 and 4,547 utterances for Case Study 2.

To enable content-dependent analysis, all utterances were coded using an open hierarchical coding scheme (Table 2) designed to investigate increasing levels of detail with the possibility for consistent extension. The coding features included five categories: management, reflection, content, style, and interactivity. Coding of the data was performed by three independent coders using Excel spreadsheets. Each coder was given a copy of the coding scheme with examples of each variable and then trained on a sample data. A level of accuracy was set and coders began coding when that standard of accuracy was attained.

ANALYSIS

The descriptive statistics of the utterances in both case studies is presented in Table 3. In Case Study 1, where each of the utterances represents

Table 2. Open hierarchical coding scheme

Level 1	Level 2	Level 3
Management	Formal Management	
	Informal Management	
Reflection	Awareness	
	Environment	
Content	Social	Chat, Agree, Disagree
	Emotional	Argumentative, Supportive
	Conceptual	Idea, Clarification, Acceptance, Rejection
	Task	Clarification, Acceptance, Rejection, Instruction
Style	Negative	
	Humor	
	Asking	
	Positive	
Interactivity	All	
	Part of a Group	
	Person	

a communication act via an e-mail message, the average length of an utterance is 776 characters (~120 words) whereas the average length of an utterance in Case Study 2 is 45-50 characters (~9-11 words). The distribution of utterances in both cases contains a number of extreme cases far from the average, which is indicated by the differences between the mean and the other measures of location—the median and the mode. The data sets in both case studies are positively (right) skewed. In Case Study 1, the range of utterance length is from 3 characters (1 word) to almost 16,000 characters (2,818 words), whereas the range of utterance length in Case Study 2 varies from 1 character to 909 characters. The maximum range across the groups is fairly consistent, varying between 663 and 909 (112-163 words). The distributions of utterance lengths in both case studies are heterogeneous, as indicated by the relatively large value of the heterogeneity factor.

Case Study 1

The activity levels of 143 participants were initially analyzed in terms of: (1) number of utterances; (2) total number of words; (3) average utterance length; and (4) task-related utterances sent. Figure 3 illustrates the total number of utterances over the entire period of Case Study 1. The utterance level is organized into five intervals. The first bin [1, 10] of the lowest number of utterances accommodates the levels of activities of typical participants—that is, 78% of the group members. The remaining 22% of the group are spread across the other five bins. The two bins of the highest activity (more than 40 utterances), representing only eight participants (6% of the group), are highlighted.

Rather than using the whole data set of 143 participants, the 31 participants who were the most active on any of the four activity criteria (number of utterances, total number of words, average utterance length, activity-related utterances) were selected. These 31 participants generated 78% of

Table 3. Descriptive statistics of utterances for Case Studies 1 and 2

	Case Study 1	Case Study 2
Total number of utterances	1,343	4,547
In terms of characters		
Average utterance length	776	48
Median	401	34
Mode	37	3
Average deviation	679	35
Standard deviation	1,218	58
Range of the length	15,970	908
Minimum length	3	1
Maximum length	15,973	909
Characters (total)	1,0417,99	218,950
Characters (without spaces)	865,414	183,298
Heterogeneity	13	16
In terms of words		
Average utterance length	120	10
Median	67	7
Mode	7	1
Average deviation	117	7
Standard deviation	214	11
Range of the length	2,817	163
Minimum length	1	1
Maximum length	2,818	164
Words (total)	177,932	40,185
Heterogeneity	13	15

Figure 3. Activity levels of different participants

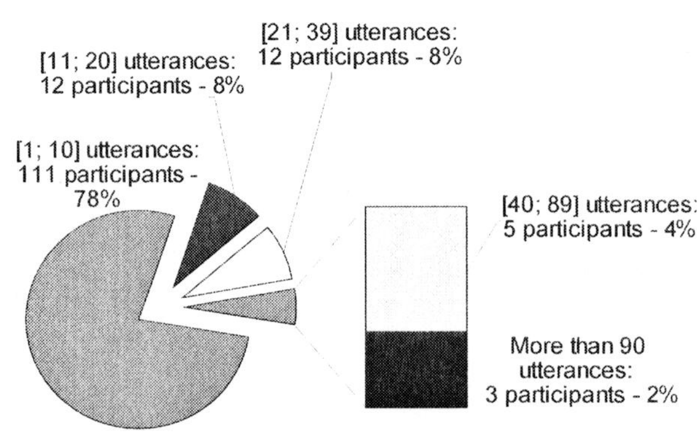

Table 4. Comparison of eight participants who rated highest on (a) the number of utterances, (b) density of utterances, and (c) activity-related content criteria

Number	
Participant	*No. of utterances*
Fay	160
Sheizaf	101
Jeff	90
Barbara	47
Catherine	47
Deborah	42
Jonathan	42
Eric	40

(a)

Density			
Participant	*Total no. of words*	*Participant*	*Average utterance length*
Sheizaf	28,408	Jonathan	291
Fay	22,994	Sheizaf	281
Jeff	16,770	Jeff	186
Jonathan	12,211	Jamie	183
Barbara	6,863	Daniel	149
Eric	5,675	Barbara	146
Nadia	3,960	Fay	144
Catherine	3,958	Eric	142

(b)

Content	
Participant	*Activity-related utterances*
Fay	111
Jeff	71
Sheizaf	61
Jonathan	35
Barbara	33
Catherine	31
Eric	25
Deborah	21

(c)

the utterances throughout the project. The measures for the participants who rated highest on the four activity criteria are given in Table 4.

From these activity measures, participants were tentatively classified as one of three types:

- **Assigned leader** (participants who have been assigned as leader explicitly or implicitly at the beginning of the project);
- **Emergent leader** (participants who are identified as potential emergent leaders using the number of utterances, total number of words, average utterance length, and task-related utterance criteria); or
- **Participant** (participants who are identified as non-leaders).

With an appropriate combination of inductive techniques, a collection of attributes is used to ascertain which of these attributes are most important in characterizing the three participant types. The collection of attributes include the four activity criteria plus the number of utterances received by an individual and the number of task-related utterances received by an individual. Table 5 lists the set of six attributes which were used as candidates for defining *Participant Type*. In our classification problem *Participant Type* is the target ("dependent") variable, and the six attributes listed in Table 5 are the "independent" variables.

The analysis included two inductive techniques: (1) decision (classification) tree induction (Witten & Frank, 2000), that was run in an exploratory mode; and (2) visual clustering. First, the CART (Classification and Regression Trees) (Lewis, 2000) technique produced a classification tree of *Participant Type*. Guided by the derived classification tree, the second step, visual clustering (Miner3D), was performed. The major goal in looking at a decision tree model is to understand the attributes that are responsible for the phenomenon. The derived tree offers a description of the concept of *Participant Type* in terms of the six attributes.

Figure 4 shows the derived classification tree which isolates each of the three participant types: *assigned leaders*; *emergent leaders*; and *participants*. This induction technique shows that *Utterances* (number of utterances sent) is the primary attribute that splits the sample of participants into Assigned Leaders and the rest. At the next level,

Table 5. Attributes used for defining participant type

Attribute	Description
Utterances	Total number of utterances
Total Number of Words	Total number of words posted by an individual
Average Length in Words	Average length of utterances in words of an individual
TSK+CON(U)	Number of activity-related utterances sent by an individual
Addressed	Number of utterances of any variable addressed to an individual
TSK+CON(A)	Number of activity-related utterances addressed to an individual

Figure 4. The decision (classification) tree for Participant Type in Case Study 1

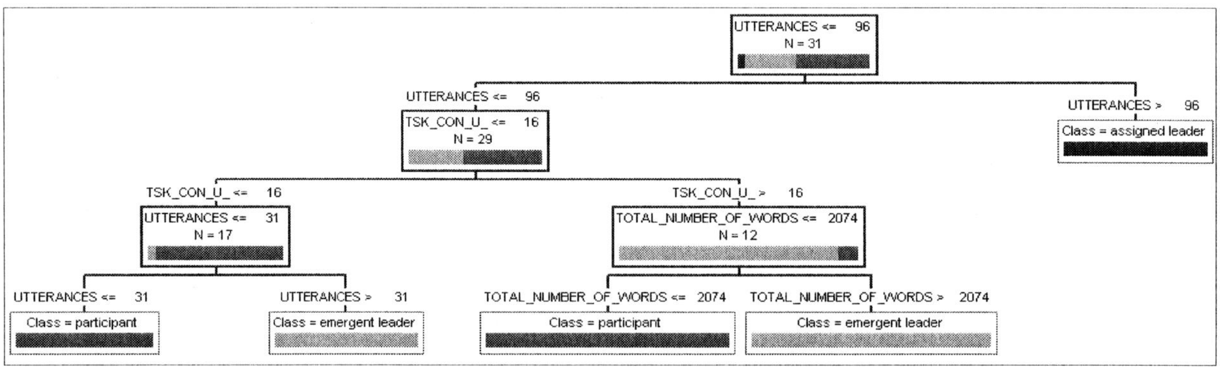

the attribute *TSK+CON(U)* (activity-related utterances sent by an individual) captures a significant portion of the leadership characteristics. At the next level, the *Utterances* and *Total Number of Words* attributes split the sample into emergent leader and participant classes. These three attributes partitioned the data to cover all participant types in Case Study 1.

The CART technique is then complemented by visual clustering. Visual clustering is the process of finding a partitioning of the data set into homogeneous sub-sets (clusters) (Keim & Ward, 2003). The key element in this technique is the mapping between the attributes and the corresponding visual features; in other words, this technique looks for groups of instances (individuals) that "belong together." Once the mapping is done, the visual clustering is an interactive procedure. In our case, the procedure is guided by the results of the decision tree induction. As this is an unsupervised technique, the clusters are not known in advance.

Figures 5 and 6 show the results of visual cluster analyses performed on the data set of 31 participants and the six attributes listed in Table 5. Figure 5 shows the initial visualization of the data set. The X, Y, and Z axes are *Utterances, TSK+CON(U),* and *Total Number of Words* respectively. The value of the *Average Length in Words* attribute has been used to define the size of the spheres. Guided by the classification tree (Figure 4), in which the *Utterances* attribute splits the data at >31, *Utterance* is set to '32'. This setting filters out a cluster of 23 participants. The remaining nine individuals are shown in Figure 6. Again, guided by the classification tree, in which

Figure 5. Initial visualization of the data set

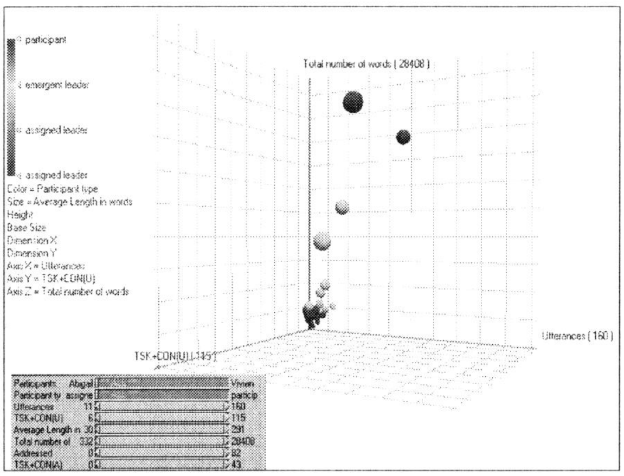

Figure 7. Enlargement of the emergent leaders identified in Figure 6

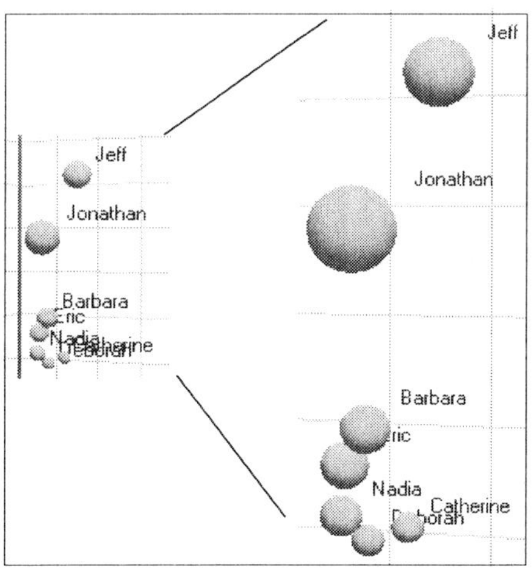

the *TSK+CON(U)* attribute splits the data at >16 and the *Total Number of Words* attribute at >2,074 (see Figure 4), these attributes are set at '17' and '2075' respectively in Figure 6. The same nine individuals remain—that is, two assigned leaders and seven emergent leaders.

Figure 7 is an enlargement of a section of Figure 6 showing the seven emergent leaders identified by name—that is, Jeff, Jonathan, Barbara, Eric, Catherine, Deborah, and Nadia. Figure 8 shows that when the *Utterances* attribute is set to '97'

Figure 6. Clustering on Utterances attribute at value '32', TSK+CON(U) *attribute at '17', and* Total Number of Words *attribute at value '2075'*

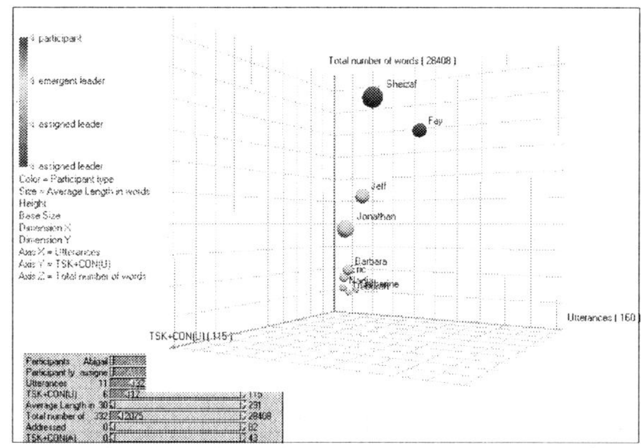

Figure 8. Clustering on Utterances *attribute at value '97' identifies assigned leaders*

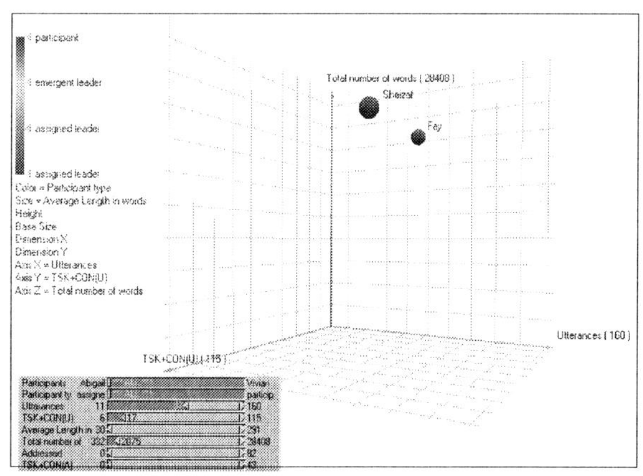

(see Figure 4, which indicates that the *Utterances* attribute splits the data again at >96), Fay and Sheizaf are identified as assigned leaders.

Hence, the classification tree in Figure 4, visualized as clusters in Figures 5 to 8, shows that the attributes *Utterances*, *TSK+CON(U)* and *Total Number of Words* were able to split the sample of 31 active participants into three *Participant Types* as listed in Table 6.

Table 6. Assigned leaders, emergent leaders, and participants

Assigned Leaders	Emergent Leaders	Participants	
Fay	Jeff	Donna	Michael
Sheizaf	Jonathan	Marian	Daniel
	Barbara	Ben	Stuart
	Eric	David	Nicola
	Catherine	Vivian	Brad
	Deborah	Brent	Jamie
	Nadia	Sally	Marie
		Chloe	Andy
		Tom	Clive
		Sarah	Peter
		Carleen	Abigail

Figure 9. Engagement level of participants

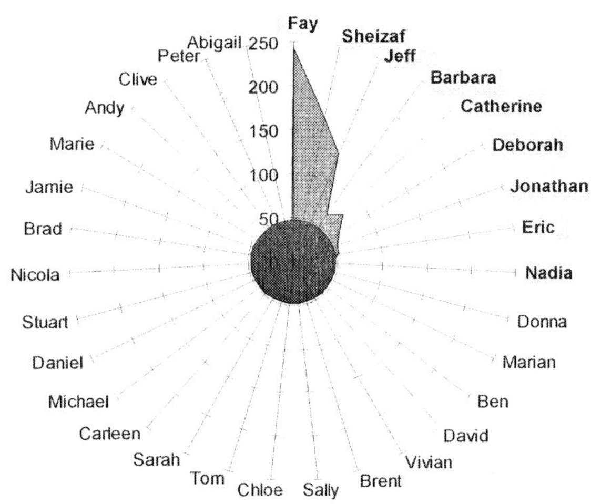

The descriptive statistics indicated four strong emergent leaders (Jeff, Barbara, Jonathan, and Eric) and two weaker candidates (Catherine and Deborah). When additional criteria were added and the data were analyzed with the CART and cluster tools, it was found that the *TSK-CON(U)* attribute (number of activity-related utterances) contributed to the identification of the same four strong emergent leaders (Jeff, Barbara, Jonathan, and Eric) and another three (Catherine, Deborah, and Nadia). The engagement graph confirms the set of six emergent leaders identified by the descriptive statistics (Jeff, Barbara, Jonathan, Eric, Catherine, and Deborah), with Nadia at the highest end of the mean engagement level.

The combination of two attributes—*Utterances* and *Addressed*—also gives a measurement of the intensity of engagement for any participant. Figure 9 illustrates the engagement level for the 31 participants examined in the classification tree model. The circle in the middle of Figure 9 corresponds to the mean level of engagement across the data set of 31 participants. The graph illustrates the findings of CART (and visualized in Minder3D); that is, the two assigned leaders (Fay and Sheizaf) have the highest level of engagement, while six of the emergent leaders (Jeff, Catherine, Barbara, Deborah, Jonathan, and Eric) are above the mean level of engagement. Nadia has the next highest level of engagement and was identified in the CART procedure.

Thus, it has been demonstrated that the criteria used for descriptive statistics (number of utterances, total number of words, average utterance length, and activity-related utterances), classification tree and clustering (number of utterances sent, total number of words, average utterance length, activity-related utterances sent, number of utterances received, and activity-related utterances received), and the radar chart on engagement level (number of utterances and utterances received) all point to a set of leaders that emerged during the life of the group.

Case Study 2

As discussed earlier, the participants in Case Study 2 formed a community of learners in a series of nine workshops using synchronous interaction in a chat room. Each week a different student was appointed to moderate the discussions. Given that each student moderated just one workshop, each student would be expected to dominate discussions in one workshop only. Hence, it would be expected that the participation pattern of any one student would be a large number of utterances in one workshop and a smaller number of utterances in the remaining eight workshops. Therefore the contributions to the discussions from each participant were potentially equalized across the period of the nine workshops.

The number and density of utterances are effective criteria for measuring verbosity in participants. Figure 10 illustrates the activity level of different participants, measured as the total number of utterances over the entire period of Case Study 2. The appointed leader and Gail communicated most intensively (bin [500; 800]), with Doug and Lorna communicating more than the other 15 participants (bin [300; 499]).

Density of utterances is measured by total number of words throughout the workshop series and the average utterance length in words. Note that average number of words per utterance is not as informative as for Case Study 1 since the range for Case Study 2 is 1-10 words. What this measure does highlight is the very different style of communication in a synchronous vs. asynchronous environment. Utterances in a synchronous environment are short, acronyms or abbreviated.

Figure 10. Activity levels of different participants

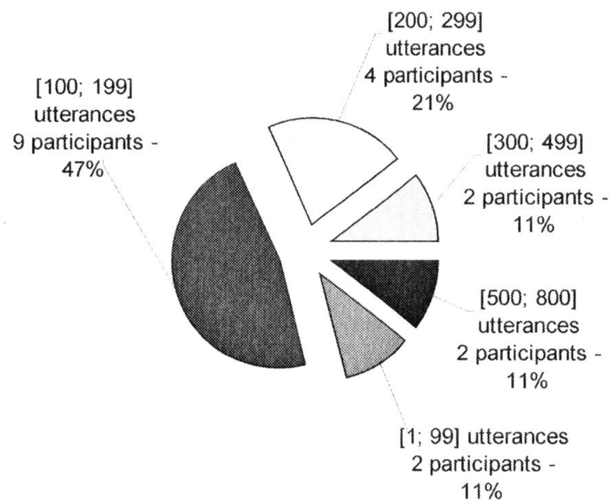

When using the number of utterances criteria, the potential emergent leaders were Gail, Doug, Lorna, Henry, Kirk, Leah, and Joe, in order of most frequent utterances (see Table 7). When using the total number of words criteria, the potential emergent leaders were Gail, Doug, Henry, Duncan, and Lorna, in order of most words. When

Table 7. Comparison of eight participants who rated highest on different criteria: (a) the number of utterances, (b) density of utterances, and (c) activity-related content criteria

Number	
Participant	*No. of utterances*
Fay	743
Gail	626
Doug	410
Lorna	317
Henry	256
Kirk	225
Leah	209
Joe	205

(a)

Density			
Participant	*Total no. of words*	*Participant*	*Average utterance length*
Gail	7,039	Gail	10
Fay	5,743	Henry	10
Doug	3,834	Duncan	10
Henry	2,688	Fay	8
Duncan	2,583	Leah	7
Lorna	2,328	Donald	7
Joe	1,849	Kirk	6
Kirk	1,807	Louis	6

(b)

Content	
Participant	*Activity-related utterances*
Fay	471
Gail	432
Lorna	228
Doug	225
Henry	180
Leah	166
Kirk	158
Susan	133

(c)

Figure 11. The decision (classification) tree for participant type in Case Study 1

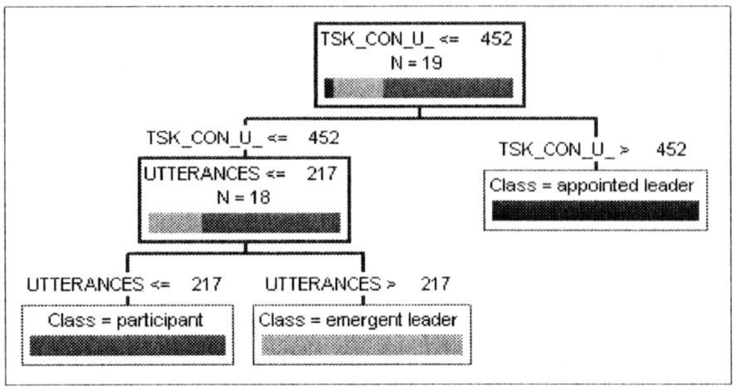

Figure 12. Visual clusters of data set of appointed leader, emergent leaders, and participants for Case Study 2

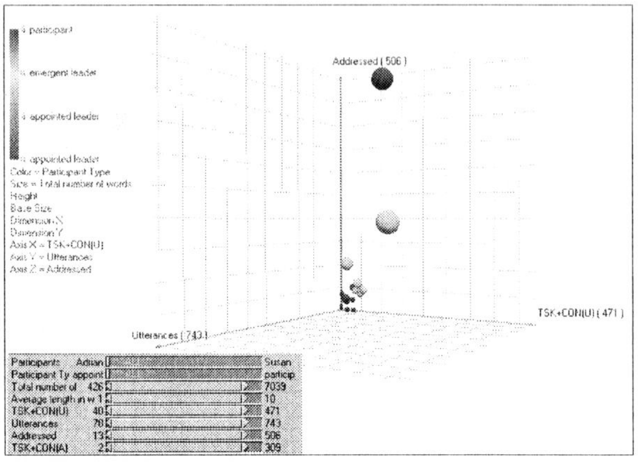

Figure 13. Clustering on Utterances *attribute at value '218'*

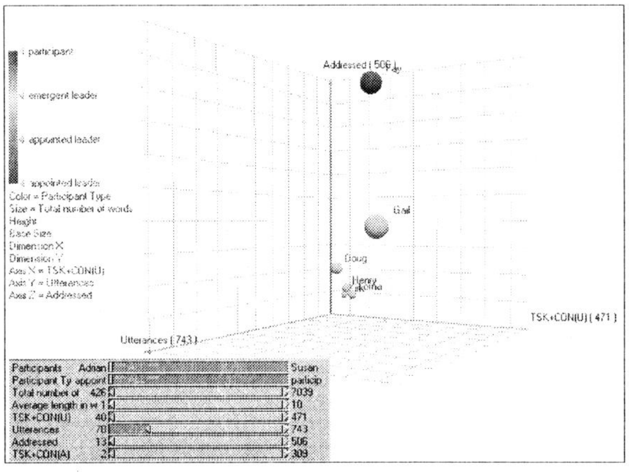

Figure 14. Engagement level of participants

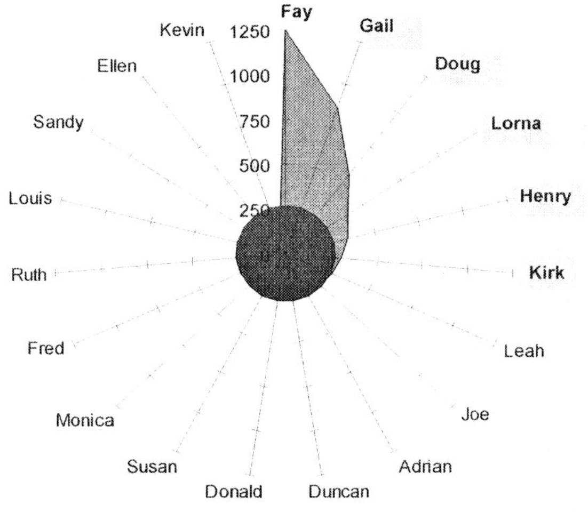

using the average utterance length, the potential emergent leaders were Gail, Henry, Duncan, Leah, and Donald, in order of longest average utterance. Adding task-related content as a criteria for identifying leadership characteristics refines the set of emergent leaders. Apart from the appointed leader, only Gail, Henry, and Kirk show evidence of emergent leadership. If any three of the four criteria are taken into account, then Doug, Lorna, and Leah are also contenders.

An expanded set of criteria was used to explore the emergent leadership within the group. Similar to Case Study 1, we reframed the problem as a classification task, classifying the group members as one of the participant types—that is, (1) appointed leader; (2) emergent leader; or (3) participant. Note that in the case of synchronous communication, the total number of words was not depicted as a parameter in the emergent leadership classification model. The classification tree model is presented in Figure 11, and the results of the visual clustering are shown in Figures 12 and 13.

The descriptive statistics indicated three strong emergent leaders (Gail, Henry, and Kirk) and three weaker candidates (Doug, Lorna, and Leah). When additional criteria were added and the data were analyzed with CART and the visual cluster tools, it was found that the *Number of Utterances* attribute contributed to the identification of the same three strong emergent leaders (Gail, Henry, and Kirk) and two of the weaker candidates (Doug and Lorna). The engagement graph in Figure 14 confirms the set of five emergent leaders identified by the descriptive statistics, and the classification tree model and clustering.

DISCUSSION

The methodology for studying emergent leadership that has been used in both case studies led to the development of an overall approach and technology for facilitating the identification of emergent leadership from project and organizational scenarios where there is a record of the communication among the individuals involved. The operationalization of the approach that can be embedded into CSCW systems is illustrated in Figure 15. Currently, the approach is focused on text-based communication data, including e-mail, chat transcripts, and communication transcripts generated from virtual environments. The collected communication data is segmented into utterances (as an utterance is considered as a data unit in the data set). In a collection of e-mail messages, the e-mail message is usually considered the organizing unit of the data. In our approach, during the data pre-processing, the sequence of e-mail messages is converted into a sequence of utterances. Each e-mail message may include one or more utterances. For example, a single message may include two or more utterances—one that addresses all team members on the list and a few utterances that address particular individuals. In practice, these and similar types of messages require segmentation into the corresponding separate utterance. The segmentation is implemented using rule-based techniques. The first set of rules is applied to the selection of messages that may potentially have several utterances.

Figure 15. Operationalization of leader identification approach into an embedded technology: (a) depicting the language of leaders, (b) depicting leaders via the three analysis streams

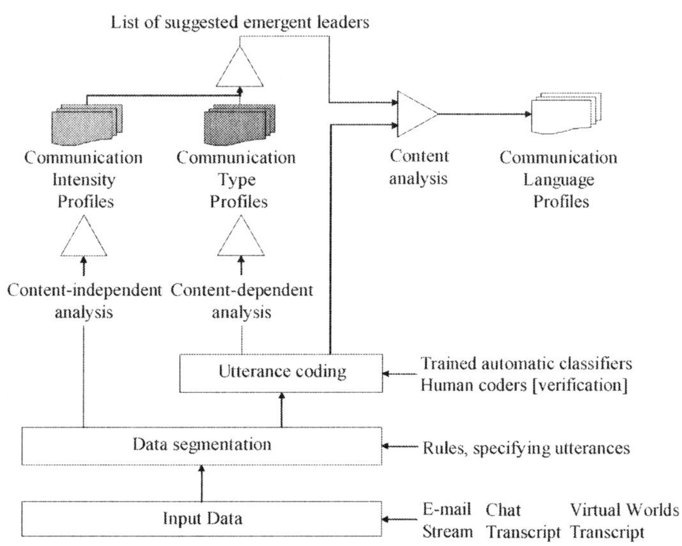

(a)

Figure 15. continued

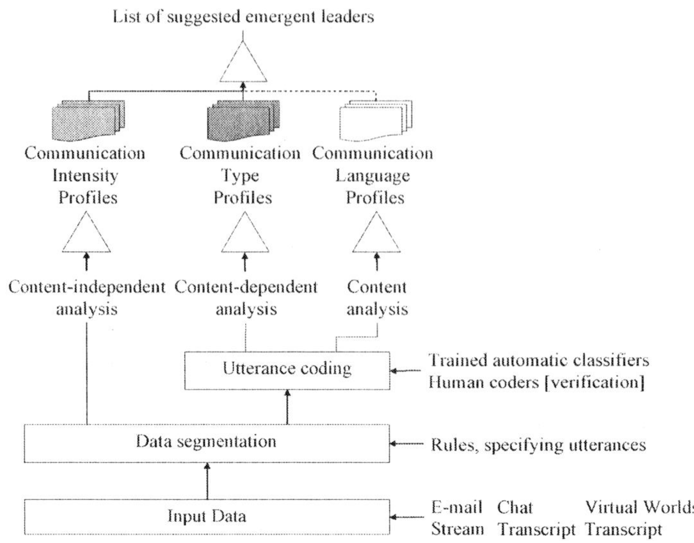

(b)

The rules applied at this stage are targeting specific constructs that identify separate utterances. Selected messages are then parsed to identify the split points. Then the splits are repacked as time-stamped SOC-triples. Technically, this is sufficient for the content-independent analysis algorithms (which include various descriptive statistics based on the presence of an utterance) and algorithms for depicting regular sequences of communicating subjects. Note that at this stage we do not consider whether high participation is a result of a high percentage of social utterances (in which case we may be looking for potential coordinator) or a high percentage of topic-related utterances (in which case we may be looking for a potential expert).

The content-dependent analysis requires labeling of the SOC-tripples following the coding schema presented in Figure 1. The automation of the process includes training of a classifier and then using it for labeling the utterances. Currently, a sample of the data set is selected as a training data, it is labeled manually and used for training a classifier for the labels at a particular level of the coding schema. Classifiers are trained for coding at different levels, and the consistency between the levels is part of the selection process of the appropriate classifier. Then the rest of the data set (the unseen part) is run, and the classification output is used for the content-dependent analysis.

FUTURE TRENDS

The content analysis described in this chapter is currently being developed and incorporated in the system in a way that it caters for two research scenarios. Figure 15a shows the system which identifies the leaders based on the outcome of content-independent and content-dependent analysis. The list of leaders and the list of coded messages is then used to filter the leadership messages of different categories and investigated through text mining the "language" of leadership (including statistical features of the sentences, use of emoticons, spectrum across word length, the dictionaries for different categories, keywords in context analysis, and a semantic net of the concepts used). Figure 15b shows the other scenario for the system where the abovementioned text features are used to identify the list of leaders, complementing the other two methods.

The study reported here provides parameters that can be used to monitor CMC of team members in collaborative virtual environments for predicting emergent leaders within the groups. Emergent leaders may impact the group dynamics, and the prediction of potential leaders may assist smart project management in virtual organizations.

The findings of the research have been embedded in a technology that facilitates the identification of leaders based on transcripts of CMC. The application is useful in both the educational and organizational setting. Researchers in organizational science have increased their efforts in group development in organizations. These studies have been motivated by the increased flattening of organizational structures, which leads to the emergence of informal groups. Knowing the structure of such emergent groups and their emergent leaders is invaluable for company management. The development of methods that assist in identifying such structures and emergent leaders is directly related to the research work presented in this chapter.

Emergent leadership also plays an important role in collaborative learning. Group development is one of the key components of social constructivism in online (flexible) learning strategies. The approach presented in this chapter is appropriate for conducting a detailed study of social learning in flexible (computer-mediated) learning environments.

Current work is focused on the incorporation of the analysis of the activities in the virtual environment in the leadership identification methods. The work is based on pilot developments in mining collaborative virtual environments reported in Simoff and Biuk-Aghai (2003a, 2003b).

CONCLUSION

The increasingly global nature of business is associated with an increased role of virtual distributed teams that communicate electronically. If such virtual organizations are to be successful, they will have to ensure that aspects of team organization transfer. We have focused on emergent leadership, as it has received little attention within the literature when compared to other models of leadership (Kickul & Neuman, 2000).

We looked at the patterns of communication that indicate the emergence of leaders. Three criteria were added to the verbosity criteria to identify emergent leaders: number of utterances addressed to an individual, number of activity-related utterances sent by an individual, and number of activity-related utterances addressed to an individual. In both case studies, a non-parametric technique and a visual clustering procedure identified a small group of participants who emerged as leaders. The findings therefore suggest that frequency, density, content, and engagement level of communication contribute to identifying emergent leadership within virtual teams.

In Case Study 1, three attributes were instrumental in categorizing assigned leaders, emergent leaders, and participants: number of utterances sent, number of activity-related utterances sent, and total number of words. In Case Study 2, two attributes categorized appointed leaders, emergent leaders, and participants: number of activity-related utterances sent and number of utterances sent.

In both asynchronous and synchronous environments, there were several people who emerged as leaders—that is, emergent leadership functions were shared. Emergent leaders send more messages, but the messages are more likely to be task related. In other words, sheer volume of words does not make an emergent leader, but frequent messages with topic-related content does contribute to leadership qualities. These findings demonstrated that emergent leadership patterns were consistent in both synchronously and asynchronously mediated virtual teams.

REFERENCES

Avolio, B., Kahai, S. S., & Dodge, G. E. (2000). E-leadership: Implications for theory, research, and practice. *The Leadership Quarterly, 11*(4), 615-668.

Cascio, W. F., & Shurygailo, S. (2003). E-leadership and virtual teams. *Organizational Dynamics, 31*(4), 362-376.

Fisher, B. A. (1974). *Small group decision making: Communication and the group process.* New York: McGraw-Hill.

Furst, S., Blackburn, R., & Rosen, B. (1999). Virtual team effectiveness: A proposed research agenda. *Information Systems Journal, 9*(4), 249-269.

Guastello, S. J. (2002). *Managing emergent phenomena.* Mahwah, NJ: Lawrence Erlbaum.

Hackman, M. Z., & Johnson, C. E. (2000). *Leadership: A communication perspective.* Prospect Heights, IL: Waveland Press.

Hammersley, N., & Atkinson, P. (1983). *Ethnography: Principles in practice.* New York: Tavistock.

Keim, D., & Ward, M. (2003). Visualization. In M. Bertold & D. J. Hand (Eds.), *Intelligent data analysis* (2nd ed., pp. 403-427). Heidelberg, Germany: Springer.

Kickul, J., & Neuman, G. (2000). Emergent leadership behaviors: The function of personality and cognitive ability in determining teamwork performance and KSAS. *Journal of Business and Psychology, 15,* 27-51.

Kolb, D. C. (1999). *Team leadership.* Durango, CO: Lore International Group.

Lewis, R. J. (2000). An introduction to classification and regression tree (CART) analysis. In *Proceedings of the Annual Meeting of the Society for Academic Emergency Medicine,* San Francisco.

Lipnack, J., & Stamps, J. (1997). *Virtual teams: Working across space, time and organizations.* New York: John Wiley & Sons.

Mabry, E. A. (2002). Group communication and technology: Rethinking the role of communication modality in group work and performance. In L. R. Frey (Ed.), *New directions in group communication* (pp. 285-298). Thousand Oaks, CA: Sage.

Maher, M. L., Simoff, S. J., & Cicognani, A. (2000). *Understanding virtual design studios.* Heidelberg, Germany: Springer.

Maznevski, M. L., & Chudoba, K. M. (2000). Bridging space over time: Global virtual team dynamics and effectiveness. *Organization Science, 11*(5), 473-492.

McCroskey, J. C., & Richmond, V. P. (1998). Willingness to communicate. In J. C. McCroskey, J. A. Daly, M. M. Martin, & M. J. Beatty (Eds.), *Communication and personality: Trait perspectives* (pp. 119-131). Cresswell, NJ: Hampton Press.

Meier, C. (2003). Doing 'groupness' in a spatially distributed work group: The case of videoconferences at Technics. In L. R. Frey (Ed.), *Group communication in context: Studies of bona fide groups* (2nd ed., pp. 367-397). Mahwah, NJ: Lawrence Erlbaum.

Misiolek, N. I., & Heckman, R. (2005). Patterns of emergent leadership in virtual teams. In *Proceedings of the 38th Annual Hawaii International Conference on System Sciences (HICSS'05),* HI.

Mullen, B., Salas, E., & Driskell, J. E. (1989). Salience, motivation and artifact as contributions to the relation between participation rate and leadership. *Journal of Experimental Social Psychology, 25,* 545-559.

Myers, M. R., Slavin, M. J., & Southern, W. T. (1990). Emergence and maintenance of leadership among gifted students in group problem solving. *Roeper Review, 12*(4), 256-260.

Pauleen, D. J. (2003). Leadership in a global virtual team: An action learning approach. *Leadership and Organization Development Journal, 24*(3), 153-162.

Pauleen, D. J., & Yoong, P. (2001). Facilitating virtual team relationships via Internet and conventional communication channels. *Internet Research: Electronic Networking Applications and Policies, 11,* 190-202.

Rafaeli, S., & Sudweeks, F. (1997). Net interactivity. *Journal of Computer-Mediated Communication, 2*(4).

Rafaeli, S., & Sudweeks, F. (1998). Interactivity on the nets. In F. Sudweeks, M. McLaughlin, & S. Rafaeli (Eds.), *Network and Netplay: Virtual groups on the Internet* (pp. 173-190). Cambridge, MA: MIT Press.

Regula, C. R., & Julian, J. W. (1973). The impact of quality and frequency of task contributions on perceived ability. *The Journal of Social Psychology, 89,* 115-122.

Riva, G., & Galimberti, C. (2001). Complementary explorative multilevel data analysis—CEMDA: A socio-cognitive model of data analysis for Internet research. In G. Riva & C. Galimberti (Eds.), *Towards cyberpsychology: Mind, cognitions and society in the Internet age* (pp. 19-35). Amsterdam: IOS Press.

Simoff, S. J., & Biuk-Aghai, R. (2003a). Generation, transfer and utilisation of knowledge: Knowledge discovery and visualisation. In H. Hasan & M. Handzic (Eds.), *Australian studies in knowledge management* (pp. 184-238). Wollongong, Australia: University of Wollongong Press.

Simoff, S. J., & Biuk-Aghai, R. (2003b). Multimedia mining of collaborative virtual workspaces: An integrative framework for extracting and integrating collaborative process knowledge. In O. Zaiane, S. J. Simoff, & C. Djeraba (Eds.), *Mining multimedia and complex data* (pp. 168-186). Heidelberg, Germany: Springer.

Simoff, S. J., & Maher, M. L. (2000). Analyzing participation in collaborative design environments. *Design Studies, 21,* 119-144.

Sorrentino, R. M., & Boutillier, R. G. (1975). The effect of quantity and quality of verbal interaction on ratings of leadership ability. *Journal of Experimental Social Psychology, 11,* 403-411.

Sudweeks, F. (2004). *Development and leadership in computer-mediated collaborative groups.* PhD thesis, Communication Studies, Murdoch University, Australia.

Sudweeks, F., & Simoff, S. (1999). Complementary explorative data analysis: The reconciliation of quantitative and qualitative principles. In S. Jones (Ed.), *Doing Internet research* (pp. 29-55). Thousand Oaks, CA: Sage.

Townsend, A. M., DeMarie, S. M., & Hendrickson, A. R. (1998). Virtual teams: Technology and the workplace of the future. *Academy of Management Journal, 12*(3), 17-29.

Witmer, D. F. (1997). Communication and recovery: Structuration as an ontological approach to organizational culture. *Communication Monographs, 64,* 324-349.

Witten, I. H., & Frank, E. (2000). *Data mining: Practical machine learning tools and techniques with Java implementations.* San Francisco: Morgan Kaufmann.

Yoo, Y., & Alavi, M. (2002, Summer). Electronic mail usage pattern of emergent leaders in distributed teams. *Sprouts: Working Papers on Information Environments. Systems and Organizations, 2.*

Zaccaro, S. J., & Bader, P. (2003). E-leadership and the challenges of leading e-teams: Minimizing the bad and maximizing the good. *Organizational Dynamics, 31*(4), 377-387.

Zigurs, I. (2003). Leadership in virtual teams: Oxymoron or opportunity? *Organizational Dynamics, 31*(4), 339-351.

Chapter VIII
Researching Online Intercultural Dialog in Business: Using Established Methods to Create New Tools

Pamela Estes Brewer
Murray State University, USA

ABSTRACT

Within a single business day, technical communicators may be challenged to work with colleagues located in a variety of nations. In such interactions, they use several kinds of communication technologies and must constantly adjust in order to effectively communicate with colleagues from other cultures. This is the realm of online intercultural dialog (OID), or online textual conversation between interlocutors from different cultures. Preparing individuals to work effectively within such contexts, however, is no easy task. This chapter will discuss the importance of doing research on such situations, present a methodology for gathering evidence based on widely accepted practices in technical communication, and show how the results of the research can be applied to business and academic practices worldwide.

INTRODUCTION

Within a single business day, technical communicators may be challenged to work with colleagues in the UK, France, Germany, India, Canada, New Zealand, the U.S., and Australia. They may use many types of technology as they work with these colleagues from all over the world, and the use of online interaction technology is increasing. Technical communicators find that they must adjust in order to effectively communicate with colleagues from other cultures, and they must do so within the parameters of advanced technology. While they can easily go to a Web site, take a seminar, or read one of hundreds of texts for intercultural face-to-face tips, how does the technology itself affect the intercultural communication? Much research has been done into the rhetoric of face-to-face communication and into intercultural communication, but little has been done at the intersection of online and intercultural, particularly regarding dialogic communication.

Online intercultural dialog (OID) is online textual conversation between interlocutors of different cultures which can take place synchronously or asynchronously; this type of interaction supports much of the business communication that takes place globally. Research is needed that helps us better understand these online intercultural dialogs. Do interlocutors of different cultures perceive online communication differently in virtual office contexts? Where do misunderstandings most often occur? What causes breakdowns? Ulijn's (1996) research indicates that people continue to screen the effectiveness of communication based on the norms of native culture. Business people have long been prepared for the constantly evolving etiquette of communicating interculturally in face-to-face situations, but how should business people be prepared for the "netiquette" of communicating interculturally online?

In order to better prepare people for global online communication, evidence is needed which can guide effective training. This chapter will discuss the importance of doing such research, present a methodology for gathering evidence based on widely accepted practices in technical communication, and show how the results of the research can be applied to business and academic practices worldwide. Research using these methods will most importantly seek to discover how specific characteristics of online dialog affect intercultural understanding in the virtual offices which OIDs support.

DEFINING ONLINE INTERCULTURAL DIALOG

In his work, St.Amant has foregrounded the need to study this type of communication. The definition of online intercultural dialog (OID) is largely based on St.Amant's (2004) definition of intercultural online interactions as:

> *... direct online contact between individuals from different cultures. This interaction can be immediate/synchronous—such as an online chat—or it can be time delayed/asynchronous—such as an e-mail. The key factor, however, is that individuals are interacting directly with one another in a two-way/back and forth/give and take communication relationship.* (pp. 321-322)

This definition and label can be further refined by identifying the communication as text based. For example, real-time video connections would present a different context altogether. On the other hand, while OIDs are text based, visual rhetoric (e.g., emoticons and the arrangement of the text) is an intrinsic part of the message. Because "dialog" is more specific to conducting conversation and exchanging words than "interaction," "online intercultural dialog" more specifically identifies this type of communication.

Studying online intercultural dialog presents a unique challenge—particularly studying this type of communication in business settings. The greatest part of intercultural communication research has been done regarding face-to-face communication, and there is nothing to indicate that once taken online, the characteristics of intercultural communication remain the same. While one can probably assume that participants in OIDs will seek to create a context in which they can communicate and will call upon familiar face-to-face cues as a starting point, these cues vary with culture, and we do not know how they have evolved online. For example, ethos plays a significant role in face-to-face and online communication. "[I]t is only through ethos that other presentation factors become effective" (St.Amant, 2004, p. 318). St.Amant further identifies two levels of credibility: attention and acceptance. A communication must be credible both in order to draw consideration in the first place and at a deeper level in order to be accepted. However, ethos conditions are neither random nor uniform—they are context dependent. In particular, ethos conditions vary from culture to culture. But how is ethos established? In face-to-face communication, ethos may initially be established by a person's clothing or speech pattern or any other number of characteristics, but online interlocutors must seek ethos differently. It is perhaps the wittiest person or the person with the fastest keying skills who first begins to establish a strong online ethos.

As indicated by Hall (1981), context is yet another rhetorical device that is dependent on culture. Interlocutors from low-context cultures tend to include all information in a communication, while those from high-context cultures leave information out of the communication if it is assumed that the recipient should already know it. Some studies have indicated that the inclusion or exclusion of contextual information may vary online (Ulijn, Lincke, & Karakaya, 2001); thus, a person from a high-context culture who traditionally excludes much detail from a face-to-face communication might, or might not, include more detail in an online communication. In moving toward a better understanding of intercultural communication as mediated by technology, we must better understand the concrete situation of the online interaction—particularly its immediacy and elimination of face-to-face cues—as well as the contexts that feed into it. It is discourse that blends characteristics of speech and writing. Consider Figure 1; it represents the dialogic relationship between components of the computer-mediated intercultural interaction. In this model the tool is central to the communication; it is a concrete part of the communication situation and contributes to the creation of meaning. Participants both act on and are acted on by the technology. For example, they are constrained by purely textual messaging that must be entered through a keyboard, but they act on the tool by creating abbreviations and emoticons that shorten text and add meaning. On each side of the dialog, the participants draw on the norms of their native cultures to send and interpret messages, and over time adaptation online may begin to affect native cultures.

Figure 1. Components of online intercultural dialog

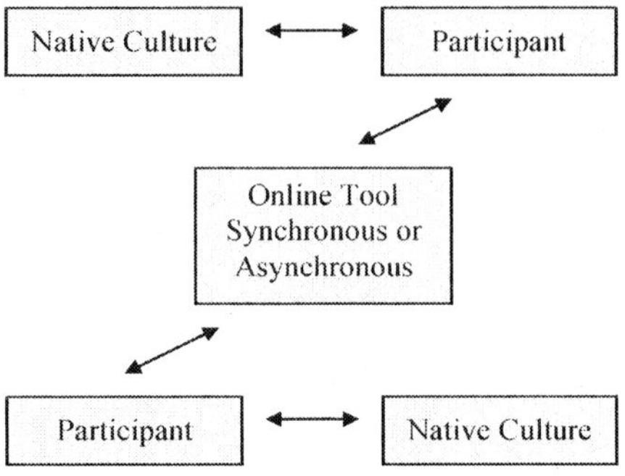

Interlocutors participating in OIDs may be particularly vulnerable to communication breakdowns because they are participating in a quick give and take that may not be fully thought out and in the absence of familiar contextual clues. We lack the terminology, taxonomies, and researched patterns that may be necessary to understand this type of communication, but as I will outline in this chapter, we do not lack methods.

METHODS AND PERSPECTIVES IN CMC AND INTERCULTURAL CMC

A review of computer-mediated communication (CMC) and intercultural CMC literature shows that a significant part of the research that has been conducted focuses on reduced and alternative cues, and on social/personal relationships (Yam & Hara, 2005; Ramirez, Walther, Burgoon, & Sunnafrank, 2002; Tidwell & Walther, 2002). In regards to methods, content analysis at both conceptual (i.e., "a concept is selected, coded, and counted for its presence in a text or corpus") and relational (i.e., "identifies a number of concepts and then examines the relationships among them") levels (Huckin, 2004, p. 14) is heavily used in the analysis of computer-mediated communication. For example, Tidwell and Walther (2002) counted occurrences of direct and indirect communication in uncertainty reduction approaches. Relationally they found "verbal interrogation (i.e., question asking) and self-disclosure as the uncertainty reduction strategies most available in CMC" (p. 323). Utz (2000) counted paralinguistic devices and their affect on developing relationships.

The October 2005 issue of the *Journal of Computer-Mediated Communication* (JCMC) contains a special theme on "Culture and Computer-Mediated Communication" with emphasis on the dimensional theories of Hall and Hofstede. The articles in this issue focus primarily on Web site communication, with some articles on intercultural marketing, media use, and education. Going back through five years of JCMC, I found that there were six intercultural articles published if I define "intercultural" as communication between interlocutors of two or more nations. In addition, a 2003 special issue on the Multilingual Internet focused on linguistics in the use of instant messaging, e-mailing, and chatting. These articles primarily reported studies of the use of various languages in specific situations, for example, the use of Japanese in bulletin board messages.

Consider too that much of the research in computer-mediated communication has been conducted face to face. For example, Heaton (1998) "examines the social construction of one group of technologies, systems for computer supported cooperative work" (p. 259). She used two case studies, collecting data via observation and interviews, to illustrate the argument that culture is an important factor in technology design. In another study of online intercultural communication, Lee, Tan, and Hameed (2005) used a computer-assisted telephone survey of 1,048 adult Singapore citizens and permanent residents to examine "the relationship among polychronicity, Internet skills, Internet use, and Internet perception. In addition, it investigates the hypothesis that Internet use displaces other activity." Rice, D'Ambra, and More (1998) conducted a study of managers working in four countries who rated media that supported business communication for richness and effectiveness. The managers ranked memos, telephone, e-mail, and face-to-face communication; preferences varied along collectivist and individualist cultural lines. The researchers used focus groups and judging to establish measures.

And last, Vishwanath (2003) focused on the effects of online information on German, Japanese, and North American communicators in his study using eBay auctions. Most of these studies explored intercultural online communication without focusing explicitly on dialogic communication, methods varied widely, research of online issues was often not conducted online, and articles on communication between cultures were sparse.

METHODS AND PERSPECTIVES IN ONLINE INTERCULTURAL DIALOG

As the above review of CMC literature indicates, research in the use of online intercultural dialog is limited. The few studies that have been done regarding the dynamics of OIDs (though they have not been named thus) have employed a number of methods and perspectives. Gammack (2002) proposes that we study epistemological bases rather than cultural in any attempt to better understand the communication within online intercultural communication. In proposing epistemological categories as more useful, he is recognizing the limitations of analyzing communication from a culturally averaged perspective. Any attempt to describe people by national culture, as is done by the dimensional theories of Hall and Hofstede, does not account for every individual. In fact, an averaged description may not describe any individual. For example, we know that at one point the average number of children living in U.S. families was 2.2, but no individual family had 2.2 children.

In his study, Gammack collected demographic data on 29 mature students in a graduate business information technology program in the UK. He then used the communication styles test of Casse to discover mindscape characteristics of participants "including decision making, ethics, beauty, social attitudes, and so on," and finally quantitatively analyzed the data. He suggests that dominant mindscape tendencies can be used to understand computer-mediated communication involving intercultural interlocutors more effectively than national culture.

Hewling (2005) also notes the limitations of an averaged view of culture in studying online intercultural dialog and the importance of understanding what individuals bring to the communication; her proposed solution is to see "culture as a process of ongoing negotiation." Using discussion board postings from two online classes, Blackboard tracked data, and interviews, Hewling found that cultural characteristics are not as significant as individual differences in mediating these online exchanges.

A study conducted by Kim and Bonk (2002) employed a mixture of qualitative and quantitative methods, specifically content analysis of transcripts of online conferences, a questionnaire, and data gathered from a videoconference. The researchers wanted to study the intercultural differences in "online collaborative behaviors among undergraduate preservice teachers." They found averaged differences in online communication consistent with native cultural boundaries. For example, they reported that "Korean students were more social and contextually driven online, Finnish students were more group-focused as well as reflective and, at times, theoretically driven, and U.S. students more action-oriented and pragmatic in seeking results or giving solutions."

Yum and Hara (2005) have also contributed to current research in OIDs with a study meant to compare cross-cultural relationship development. They focused on self-disclosure in relationship development, comparing three cultures: Korea, Japan, and the United States. Their study relied exclusively on data gathered from a survey of 361 college students in the three cultures. They used several scales that had been used in previous Internet and face-to-face relationship studies to analyze the data, particularly the depth and breadth measures used by Parks and Floyd (1996). They found that for "Americans, Japanese, and Koreans, self-disclosure was directly associated with online relationship development. However, the relationship between self-disclosure and trust was positive only for Americans."

In his computer-mediated conversations study, Ringo Ma's (1996) purpose was to "examine the differences between the two versions of intercultural communication [face-to-face and computer-mediated] and claim that some current theories of intercultural communication, when applied to communication via computer networks, are subject to modification" (p. 173). In this study he defines computer-mediated communication much as I define OID here. Ma proceeded to make five propositions based on the current face-to-face intercultural research and then empirically tested them on CMC experiences of college students. His study focused on East Asian and North American students, as that is where the bulk of face-to-face research had been done at that time.

U.S. students in junior-level college classes in the northern U.S. filled a course requirement by participating in relay chats with students attending college in East Asia and, for comparison, with students in North America. They then wrote a report on their experiences with some specifically defined content areas. East Asian students were interviewed via synchronous chat and e-mail about their experiences. The empirical data largely supported the propositions. Ma also reported two additional phenomena:

First, participants in synchronous CMC do not seem to have as high a commitment as when they engage in FTF [face-to-face] communication. They tend to be more direct and self-disclosing, but the reason they do so is not the same as that for similar behavior in FTF situations. In FTF communication, being more direct and self-disclosing usually indicates a more serious commitment to a close relationship, while in synchronous CMC participants self-disclose simply because the chance to meet the other is "rare", or there is little risk involved ... Second, there seemed to be a discrepancy between how U.S. students perceived East Asian students and how East Asian students perceived themselves with regard to directness and self-disclosure. (p. 184)

In summary, recent OID scholarship has focused on collaboration, relationship development, comparisons between face-to-face and computer-mediated communication, and the factors most affecting OIDs. Research in each of these areas is only a beginning to what is needed. In all of the studies mentioned here, which represent a significant portion of the extant research in online intercultural dialog, research was conducted among participants from academic settings—students and teachers. Significantly, we do not have research devoted to how the use of online intercultural dialogs looks in the business/professional setting of virtual and hybrid offices.

Methods used have been both quantitative and qualitative, and include collection of data from discussion board postings, Blackboard-tracked data, face-to-face and online interviews, online conferences, questionnaires and surveys, chats, written reports, and videoconferences. An analysis of the findings and discussions in these studies reveals widely differing perspectives on the communication that takes place. Some scholars believe that cultural averaging (or modified cultural averaging in Ma's case) is an effective predictor of communication patterns in OIDs. Others believe that epistemological perspectives and individual differences may be greater contributors to communication patterns. Research is needed to clarify the dynamics of online intercultural dialog and to prepare interlocutors for success.

A PROPOSED METHODOLOGY FOR STUDYING OIDs

In moving toward a better understanding of intercultural communication as mediated by technology (see Figure 1), it is important to understand the concrete situation of the online dialog as well as the contexts that feed into it. In order to study the use of online intercultural dialog in business settings, we can call upon methods widely applied in technical communication to yield both data and tools that are efficacious in the study of OIDs. The methodology outlined here will have three phases, have its foundations in real business practice, and produce tools that can be used in ongoing global studies of OIDs. The first phase will be conducted face to face, the second online (focused), and the third online (global).

Phase 1: Grounded Study

Established patterns for the use of OIDs have not yet been explained in theory, and thus, a grounded approach will offer the opportunity to gather contextual data on their use and to identify patterns. An evolved definition of grounded theory and its goals is presented by Corbin and Strauss (1990) as follows:

The procedures of grounded theory are designed to develop a well integrated set of concepts that provide a thorough theoretical explanation of social phenomena under study. A grounded theory should explain as well as describe. It may also implicitly give some degree of predictability, but only with regard to specific conditions. (p. 5)

In particular, this approach will allow a focus on potential causes of breakdowns, while at the same time allowing other content areas to emerge—as manifest data from real business settings. That is, what people say and what they do are often two different things (Blomberg, Giacomi, Mosher, & Swenton-Wall, 1993, p. 130). Instead of relying on data gathered in educational settings or on ideal data gathered from business professionals, the foundations of this research will be based on what business people are actually doing rather than what anyone theorizes they are doing. This naturalistic approach ensures that the data will not be substantially altered because people have been separated from their social resources (Doheny-Farina & Odell, 1985), increasing validity; additionally, observing the context as well as people will contribute to the collection of rich data.

The concept of a "breakdown" has been adopted from user-centered design theory—a research theory widely applied in technical communication. Consistent with Bødker and Grønbæk's (1991) definition, I use "breakdown" to refer to "when work is interrupted by something, e.g. that the tool behaves differently from what was anticipated. In these situations the tool as such becomes the object of our actions" (p. 458). Their study in prototyping led them to conclude that much can be learned about a process from breakdowns. This theory complements the study of OIDs in business well because they are constructed of people using technology tools to communicate. Breakdowns in the process of communication using these tools can illuminate communication mismatches that are a product of the communication tools (i.e., online technology).

Grounded research will allow characteristics of OIDs such as those listed below to emerge as patterns that can then be studied further and, perhaps, be linked to breakdowns:

- Communication is primarily textual.
- Non-verbal cues characteristic of face-to-face communication are absent.
- Common abbreviations are used by those familiar with online communication and can be international or language specific.
- Customs may be expressed textually (e.g., a German interlocutor may key "knock knock," in order to applaud).
- Emoticons are used as "superfixes" to provide expressions that go above the utterance and affect its meaning (term coined by Trager, in Hall, 1981, p. 127). For example, one participant uses the ;-) emoticon to let another participant know that she is joking in a statement just made.
- All caps are used to emphasize.
- Participants often write in sentence fragments.

For example, does the use of superfixes prove confusing to some interlocutors of varying cultures? What characteristics of OID discourse correlate with breakdowns interculturally?

Phase 1, as proposed here, will be conducted as one or several case studies. (A case study is a type of grounded study that "examines a 'bounded system' or a case over time in detail, employing multiple sources of data found in the setting" (McMillan & Schumacher, 2001, p. 36). In order to conduct a grounded study, researchers will observe workers within U.S. businesses who use OIDs to conduct a significant portion of their business. Thus, the first phase of this research will be conducted face to face, which is necessary to gain a contextual overview of virtual and hybrid office patterns regarding OIDs. During these observations, participants will be encouraged to think aloud. This protocol is widely used in usability testing and can prove valuable in a research situation such as this because a study of OIDs, similar to usability testing, finds the researcher observing real people doing real things in order to better understand how the people perceive the interaction. Think-aloud protocol will provide fertile ground for making out-of-awareness cultural screening (Hall, 1981) in OIDs surface to a technical (conscious) level where it can be observed. Think-aloud protocol should be conducted consistent with current usability practice as reported by Boren and Ramey (2000); that is, the researcher will take an active role in listening—intervening with prompts that encourage the participant to keep talking or that "proactively probe for additional diagnostic information" (p. 274).

During this phase, contextual interviews will also be conducted; the researcher will ask questions that grow from the observations, but remain aware that the more questions asked, the more likely the researcher is to influence the activity (Blomberg et al., 1993). Interviewing holds a key role throughout the phases of this methodology. It has the potential to complement artifact studies by probing motivation, satisfaction, and attitude as well as linking activities in virtual systems.

With the participants' permission, the researcher will also videotape these sessions, preserving the complexity of the communication for further analysis and comparison, and will collect transcripts of past communications as they are available. Artifacts of the OIDs will be collected whenever possible, but the success of this research cannot depend on such artifacts, as companies may be unwilling to allow access to the specifics of their professional online dialogs. Observation, think-aloud protocol, interviewing, taping, and artifact collection will increase the validity of the study through triangulation of data collection.

A grounded approach will serve to organically isolate communication issues within OIDs, which may or may not reflect extant patterns of intercultural communication as presented in current scholarship. For example, the dimensions of Hall and Hofstede may emerge as significant and aid in identifying patterns and content areas of importance. Do dimensional characteristics account for online communication behaviors and perceptions? On the other hand, patterns closely tied to issues of face and ethos may emerge as most significant. Most importantly, the data will have the opportunity to emerge contextually.

Phase 2: Textual Analysis, Focused Questionnaire, and Interviews

Once patterns of importance have been isolated, they will be refined for use in the next phase of research—a focused analysis—which will be conducted online. Via textual analysis of videotapes and other artifacts gathered during the grounded study, and consistent with Ma's method, a series of propositions will be formulated. These propositions will inform the development of a set of questions that may, depending upon their nature, be put to participants via questionnaire and synchronous interview. Participants who use OIDs as a significant part of their workday to accomplish significant work goals will be selected.

Each participant will be given the questionnaire and then be interviewed using the questions informed by the textual analysis—specifically addressing the content areas that have been isolated in the first phase of this research. The study will include participants from both U.S. and international offices. These interviews will be arranged over a period of time immediately following the part of the workday that most focuses on OID communication so that participants' recall of the OIDs will be most clear. For example, organizations based in the U.S. may conduct most OID communication during the morning hours in order to work with colleagues in other time zones. They would therefore be asked to fill out the questionnaire and participate in a brief online synchronous interview during the late morning. Each participant will fill out the questionnaire and be interviewed several times over a period of time in order to provide data that reflect on average communication rather than the particular communication of a single day, and to support the validity and stability of the data collection.

The interviews, while structured, may also include probing questions which have been formed by the information gathered from the questionnaires. Such questions will be particularly important in this phase of the research which will include participants from various cultures unlike research in the first phase. In all phases of the research, we seek to understand the perceptions of interlocutors in OIDs. The entire methodology is recursive in nature, each part informing others. Consequently, the tools used in this phase must enable the perceptions of interlocutors from all cultures to emerge; if breakdowns occur, the questionnaire and interview questions can and should be further refined during this phase so that it evolves as a reliable tool for analyzing the communication.

For example, I once participated in educational OIDs via synchronous chat that included members of several cultures. After each synchronous session, I interviewed both a North American and a German participant as to their perceptions of the chat. While the North American noted no breakdowns during one particular session, the German participant noted that moments of silence, unacknowledged comments, and the use of evidentials such as "I believe" and "I think" caused uncertainty in the communication. "In linguistics, evidentials are defined as words that express a writer's attitude toward knowledge" (Barton, 2004, p. 72). In fact, these educational chats revealed a very high incidence of evidentials among students. Rich data such as this are critical to our understanding of OID communication, and using recursive methods will increase the likelihood of discovering it.

This methodology varies from Ma in that all participants will complete a questionnaire and a synchronous interview, whereas Ma asked North American students to complete a report but interviewed East Asian students via synchronous chat and e-mail. In a study of businesses, the process must be as efficient as possible (in order to gain support from business) while at the same time collecting rich data. In addition, the time it would take participants to write a report would be an unrealistic expectation, and reliability is increased by collecting data using the same methods for all participants.

The data gathered during this focused phase of the study should reveal both individual attitudes most likely to affect OIDs, as well as a somewhat stable profile of the communication issues most likely to affect OIDs. Data may identify dimensions of significance in OIDs—that is over-arching concepts that characterize a culture (St.Amant, 2004). In addition to its value in understanding communication within the specific OIDs studied, this data can also be used to further refine the questionnaire and interview questions which can then provide the foundation for a wider study of OIDs.

Phase 3: Global Application of Study Tools

It is inadequate to consider the effects of technology on "intercultural" communication. We must study the individual cultures themselves as they are mediated by new forms, and we need reliable and efficient methods for doing so. The questionnaire and interview questions recursively developed in the first and second phases of this research can be applied to a global study of intercultural online communication similar to Hofstede's study of intercultural face-to-face communication. Data can be collected with such tools culture by culture to provide a worldview comparing the responses of many cultures via a defined set of OID characteristics using much the same process as outlined in phase two. That is, the questionnaire and interview questions developed as a part of this research can be administered online to participants engaged in OIDs over a period of time and immediately following concentrated use of this type of communication. Though such data would be ideal, the questions themselves would be grounded in manifest communication, thus providing a more valid (real-world) understanding of this key communication in business.

CONCLUSION

Scholars are beginning to research online intercultural dialog, while practitioners in business have been learning as they go. Technical communication research methods provide a rich means for understanding business communication in an age of global technology. While the methodology proposed here has not been used in the study of computer-mediated communication, the individual methods have been applied to past studies in CMC and, more widely, in technical communication. Using this methodology—a combination of grounded research, think-aloud protocol, breakdown identification, questionnaires, and interviewing—online intercultural dialogs and the virtual office systems in which they occur can be better understood. The tools that emerge from this study can be applied to gain a culture-by-culture, or perhaps profession-by-profession, understanding that will enable professionals in business and education to prepare people to communicate more effectively online. In addition, these methods can be used to develop other tools that enable the study of intercultural online communication. Even as technology itself evolves, such recursive methodology can provide a means by which we can study intercultural communication as it is mediated by that technology.

REFERENCES

Barton, E. (2004). Linguistic discourse analysis: How the language in texts works. In C. Bazerman & P. Prior (Eds.), *What writing does and how it does it* (pp. 57-82). Mahwah, NJ: Lawrence Erlbaum.

Bødker, S., & Grønbæk, K. (1991). Cooperative prototyping: Users and designers in mutual activity. *International Journal of Man-Machine Studies, 34*, 453-478.

Boren, T. M., & Ramey, J. (2000). Thinking aloud: Reconciling theory and practice. *IEEE Transactions on Professional Communication, 43*(3), 261-278.

Blomberg, J., Giacomi, J., Mosher, A., & Swenton-Wall, P. (1993). Ethnographic field methods and their relation to design. In D. Schuler & A. Namioka (Eds.), *Participatory design: Perspectives on systems design* (pp. 123-155). Hillsdale, NJ: Lawrence Erlbaum.

Corbin, J., & Strauss, A. (1990). Grounded theory research: Procedures, canons, and evaluative criteria. *Qualitative Sociology, 13*(1), 3-21.

Doheny-Farina, S., & Odell, L. (1985). Ethnographic research on writing: Assumptions and methodology. In L. Odell & D. Goswami (Eds.), *Writing in nonacademic settings* (pp. 503-535). New York: Guilford.

Gammack, J. (2002). Mindscapes and Internet-mediated communication. *Journal of Computer-Mediated Communication, 7*(3). Retrieved March 10, 2006, from http://jcmc.indiana.edu/issues.html

Hall, E. T. (1981). *The silent language.* New York: Anchor Books.

Heaton, L. (1998). Talking heads vs. virtual workspaces: A comparison of design across cultures. *Journal of Information Technology, 13*(4), 259-272.

Hewling, A. (2005). Culture in the online class: Using message analysis to look beyond nationality-based frames of reference. *Journal of Computer-Mediated Communication, 11*(1). Retrieved

March 10, 2006, from http://jcmc.indiana.edu/issues.html

Huckin, T. (2004). Content analysis: What texts talk about. In C. Bazerman & P. Prior (Eds.), *What writing does and how it does it* (pp. 13-32). Mahwah, NJ: Lawrence Erlbaum.

Kim, K., & Bonk, C. (2002). Cross-cultural comparisons of online collaboration. *Journal of Computer-Mediated Communication, 8*(1). Retrieved March 10, 2006, from http://jcmc.indiana.edu/issues.html

Lee, W., Tan, T. M. K., & Hameed, S. S. (2005). Polychronicity, the Internet, and the mass media: A Singapore study. *Journal of Computer-Mediated Communication, 11*(1). Retrieved March 10, 2006, from http://jcmc.indiana.edu/issues.html

Ma, R. (1996). Computer-mediated conversations as a new dimension of intercultural communication between East Asian and North American college students. In S. C. Herring (Ed.), *Computer-mediated communication: Linguistic, social, and cross-cultural perspectives* (Vol. 39 of the series "Pragmatics and Beyond New Series"; pp. 173-185). Amsterdam: John Benjamins.

McMillan, J. H. & Schumacher, S. (2001). *Research in education* (5th ed.). New York: Longman.

Ramirez, A. Jr., Walther, J. B., Burgoon, J. K., & Sunnafrank, M. (2002). Information-seeking strategies, uncertainty, and computer-mediated communication: Toward a conceptual model. *Human Communication Research, 28*(2), 213-228.

Rice, R. E., D'Ambra, J., & More, E. (1998). Cross-cultural comparison of organizational media evaluation and choice. *Journal of Communication, 48*(3), 3-26.

St.Amant, K. (2004). International digital studies: A research approach for examining international online interactions. In E. Buchanan (Ed.), *Readings in virtual research ethics: Issues and controversies* (pp. 317-337). Hershey, PA: Idea Group, Inc.

Tidwell, L. C., & Walther, J. B. (2002). Computer-mediated communication effects on disclosure, impressions, and interpersonal evaluations. *Human Communication Research, 28*(3), 317-348.

Ulijn, J. M. (1996). Translating the culture of technical documents: Some experimental evidence. In D. C. Andrews (Ed.), *International dimensions of technical communication* (pp. 69-86). Arlington, VA: Society for Technical Communication.

Ulijn, J. M., Lincke, A., & Karakaya, Y. (2001). Non-face-to-face international business negotiation: How is national culture reflected in this medium? *IEEE Transactions on Professional Communication, 44*(2), 126-137.

Utz, S. (2000). Social information processing in MUDs: The development of friendships in virtual worlds. *Journal of Online Behavior, 1*(1). Retrieved March 10, 2006, from http://www.behavior.net/JOB/v1n1/utz.html

Vishwanath, A. (2003). Comparing online information effects: A cross-cultural comparison of online information and uncertainty avoidance. *Communication Research, 30*(6), 579-598.

Yum, Y. O., & Hara, K. (2005). Computer-mediated relationship development: A cross-cultural comparison. *Journal of Computer-Mediated Communication, 11*(1). Retrieved March 10, 2006, from http://jcmc.indiana.edu/issues.html

Section III
Aspects of Online Design in International Environments

Chapter IX
Web Site Analysis Across Cultures:
An Inquiry into Intercultural Values and Web Site Design

Barry Thatcher
New Mexico State University, USA

Edgar Barrantes
New Mexico State University, USA

Amy Dalzell
New Mexico State University, USA

Qiumin Dong
New Mexico State University, USA

Christopher Ford
New Mexico State University, USA

Rachel Gallagher
New Mexico State University, USA

Sipai Klein
New Mexico State University, USA

Marohang Limbu
New Mexico State University, USA

Lisa Ramirez
New Mexico State University, USA

ABSTRACT

The countries and cultures of the world are becoming much more interdependent. At the same time, the development and use of the World Wide Web has dramatically increased cross-cultural contact in many areas. This growth in cross-cultural contact must be influencing the rhetorical and cultural patterns at the local, national, and international level. Researchers, however, simply do not know how and what is happening in such exchanges. This chapter presents a theoretical approach to Web site design, cultural values, and rhetorical traditions. The chapter then describes the methods of researchers who examined the Web site designs in nine cultures, and finally discusses the results of comparing Web site designs to local cultural and rhetorical traditions.

INTRODUCTION

It seems apparent that the countries and cultures of the world are becoming much more interdependent, due in large measure to global economics, politics, environment, and immigration. In addition, the development and use of the World Wide Web has dramatically increased cross-cultural contact in many areas, including academia, industry, governmental and non-governmental organizations, and many others (e.g., Collis, 1999). This dramatic increase in cross-cultural contact must be influencing the rhetorical and cultural patterns at the local, national, and international level, but researchers simply do not know how and what is happening.

As a long established line of research has documented, cultural assimilation and change due to communication technologies is never straightforward (Grossberg, 1992; Kauffer & Carley, 1993; Bolter & Grusin, 1999). In other words, just because there is cross-cultural contact does not mean that the cultures in contact are simply blending or assimilate the influences. As Stuart Hall (1980) argued long ago, cultures adopt, resist, appropriate, refract, and refashion cultural and rhetorical patterns for a variety of social, economic, political, personal, and aesthetic reasons. Further, local cultures often not only reject assimilation, but re-trench themselves in their local values against cross-cultural influences (Adler, 1991). However, are the rules of this game (Bourdieu, 1992) changing because of the greater global interdependence and the World Wide Web? In other words, how, if at all, is the World Wide Web influencing local, regional, and even national cultural and rhetorical values, and does this influence follow similar patterns of other communication media such as television and film (Grossberg, 1992). And if the Web is influencing cultural change, what kinds of changes are occurring and why?

These questions are critical for intercultural researchers working on rhetoric and technology because the rapid move towards global interdependence requires effective cross-cultural communications and cooperation, much of which is mediated by the World Wide Web. These questions about the Web and culture, however, are broad and comprehensive, and researchers are only beginning to address them. Thus, this chapter seeks to lay a foundation for exploring the Web and culture by examining Web site designs across cultures. The hypothesis to be tested is the following: *If the World Wide Web (and global interdependence) is indeed influencing cultural and rhetorical patterns at the local, regional, and national level, this influence should be reflected in Web site designs. That is, local Web site designs should reflect a mixture of their local culture as well as global or other cultures. However, if local Web site designs show less mixture and more local patterns, then the Web and global interdependence is perhaps not influencing as much as expected.*

This chapter first lays out the theoretical approach to Web site design, cultural values, and rhetorical traditions. Next, it describes the methods of researchers who examined the Web site designs in nine cultures. Finally, it explores the results of comparing Web site designs to local cultural and rhetorical traditions.

CONNECTING INTERCULTURAL VALUES TO WEB SITE DESIGNS

The scholarly debate about the relations of cultural patterns and communication technologies has been a long and fruitful one. Plato originally argued that writing—the newest and most significant communication technology—would have damaging cultural and rhetorical effects because of its inherent weaknesses as compared to orality. According to Plato (1956), writing would deteriorate memory, wreak havoc on logical constructions, and create an artificial reality. As a result, argues Plato, cultures that embrace writing would become inferior mentally, socially, and logically. Later, Ong (1987) traces the evolution of oral to literate societies, arguing that the development of writing radically changed cultural and rhetorical patterns. Pre-literate societies developed cultural and communicative patterns based on the oral medium: repetition to increase memory and transference; logic based on narrative organizational patterns; and persuasion based on "presence," a development of concrete sensory language that evokes context, personal relationships, and events. Ong (1987) then argued that with the advent of literacy and writing, these cultural patterns changed: concision instead of repetition; analytical organizational patterns instead of narrative; persuasion based on formal logic rather than presence. Ong labels this transformation of writing supplanting orality as perhaps the most significant cultural change in history.[1] In the 1960s, McLuhan (1962) offered similar deterministic accounts of communication media such as television and radio; in other words, the medium was the message in that the physical structure of the medium superimposed rhetorical and cultural patterns on its audiences.

In the mid 1980s and into the 1990s, many scholars, however, began to discount the simplicity of these deterministic theories, arguing instead for mutually constitutive patterns. These relationships were based on the degree of commonality and difference between the broad sociocultural patterns, the communication patterns the media reinforced, and the specific sociohistorical context in which the media are associated. For example, Martín-Barbero (1993), a Colombian communication scholar, and Grossberg (1992), a U.S. scholar, both argue that communication media such as movies, television, and print journalism do encourage or reinforce certain rhetorical and cultural patterns, but the effect of this influence varies greatly; and these scholars argue for exploring the ways local cultures appropriate, resist, and even reject cultural patterns associated with both the content and form of communication media. Kaufer and Carley (1993) similarly posit theories of mutually constitutive relationships among cultural patterns and writing and orality, and they explore the social and cultural dynamics reinforced—but not determined—by the print media. Rarely does one medium supplant another one, as Ong originally argued; rather, the rhetoric of one medium corresponds more closely to—and reinforces the rhetorical patterns of—a culture more than another medium. Kaufer and Carley (1993) did not extend this comparison to cross-cultural inquiries, but the connection is obvious. Some cultures feel more comfortable with some media over others because the patterns reinforced by these media correspond generally well with their overall cultural patterns (e.g., Thatcher, 2005).

In conclusion, each cultural and rhetorical tradition will relate to Web site designs differently, based on the concepts of rhetorical fit and reciprocity. Understanding how the World Wide Web fits rhetorical traditions around the world requires understanding the intercultural variables that underlie rhetorical traditions. For this Web site comparison, the research team selected five major variables: individual-collective, universal-particular, ascription-achievement, specific-diffuse, and power distance. We used these variables to explore the relationships between Web site design and cultures.

CONNECTING THE NEW MEDIA TO CROSS-CULTURAL TRADITIONS

As detailed elsewhere by Thatcher (2001, 2005), cross-cultural comparisons of rhetoric require a framework of inquiry that places the cultures on as equal standing as possible. To do this, researchers need to start by recognizing similarities based on shared contexts and then considering differences within the framework of these similarities (Bhawuk & Triandis, 1996; Lucy, 1996). In other words, when comparing apples and oranges, researchers should not start by looking at all the emic characteristics of the orange and then the apple, because the apple will be seen in light of the orange, an ethnocentric approach. Instead, apples and oranges are both fruit, generally about the same size, are used for similar purposes, and their textures are not remarkably different. This embedding of difference in a framework of similarities works against analyzing a second culture using the cultural constructs of the first culture (Bhawuk & Triandis, 1996. pp. 23-24), thus establishing a common ground in which to operationalize and compare the variables of both cultures in the intercultural context (Lucy, 1996).

The proposed intercultural frameworks are not the modernist conceptions of progress or other hegemonic narratives that have been used, often unethically, when comparing cultures. They are shared contexts or thresholds of interaction that are common to humans, as perhaps compared to other species on this planet. For many intercultural scholars, the shared contexts tend to be time, relationships to each other, relationships to nature, and a conception of humanity (Stewart & Bennett, 1991). This chapter will focus on five constructs: I/other relations, defining and applying rules, status, degree of involvement, and authority. All cultures usually exhibit a variety of approaches to these common human thresholds, but most cultures have a yin/yang tension between two contradictory, yet complementary approaches. And one side of the yin-yang dyad usually predominates. Thus, in this chapter, we will focus on the two most likely approaches because we do not have space to cover others.

First, all cultures share the I-other threshold of human interaction. For some cultures, the most likely approach is individualistic, while in others it is collective. Individualism exists when people define themselves, see the world through, and negotiate life based on individual identities and efforts. Hofstede (2004) and Hampden-Turner and Trompenaars (2000) argue that individualism is the strong default approach in the United States, Australia, South Africa, and other western European-Protestant countries. In the Web sites, the researchers operationalized individualism as large personal space; isolation of lone individuals; and an emphasis of personal achievement, self-creation, and individualized navigation design.

Collectivism predominates when people define themselves, see the world, and negotiate life based on social or family groups. Hofstede (2004) and Hampden-Turner and Trompenaars (2000) argue that collectivism is the most likely approach in many countries in Mexico, Central and northern South America, Asia, and the Middle East. In the Web sites, the researchers identified as collective patterns an emphasis of interpersonal relationships, social hierarchy, social leveraging, group identities, and close personal space.

Second, all cultures establish norms or rules, but their approaches and applications greatly vary. The two most common approaches are universal and particular. In universalist cultures, the default approach is to establish rules that define what is good and right regardless of the social standing of the individuals. The ideal of a "level playing field" is an apt analogy for universalist cultures. For the Web sites, the researchers operationalized universalism as invoking strategies of fairness, justice, equality, parallelism in Web design, and strong use of templates or Web branding.

In particularist cultures, the default approach is to define what is moral and right depending on relations and context. Exceptions are the rule, based on social relationships. Instead of "the level playing field," the particularist playing field is overtly structured. The researchers operationalized particularism as uniqueness in page designs across each Web site, emphasis of context and particular relationships, exceptional circumstances, and social prestige and power relations. According to Hofstede (2004) and Hampden-Turner and Trompenaars (2000), universalism is the default approach in many western European countries, the United States, Canada, and South Africa, while Latin American, Arab, and Asian cultures show the strongest particular approaches.

The third common threshold of human interaction is status—how it is developed and communicated. The two most common approaches are ascription and achievement. Ascriptive-oriented cultures derive status from their social group, race, gender, age, ethnicity, and language. These variables provide the communicator with the power to get things done in complicated social structures. Ascription orientation says: I am great because of my ascribed background; therefore, I can do. The researchers operationalized ascription as overt signs of ascribed status, power through people, and importance of context and history to the university.

In achievement-oriented cultures, it does not matter where you come from, who your family is, your age, language, and so forth, as long as you can achieve your goals. Achievement orientation says: I have achieved many things; therefore, I am great. The Web researchers operationalized achievement by looking for emphasis of accomplishment such as enrollment, number of scholarships, graduate placement, and faculty honors. Hofstede (2004) and Hampden-Turner and Trompenaars (2000) argue achievement is the most likely approach in the United States, Canada, Australia, and the United Kingdom, while ascription is the most likely approach in Asia, former communist countries, and Mexico and Central and Andean South American countries.

The fourth common threshold of human interaction is degree of involvement across different spheres of life. Researchers have defined the two most likely approaches as specific or diffuse. In diffuse cultures, friendships, social relations, and work environments often cross boundaries—you work with your friends and close social acquaintances, thus sharing the relationship across these spheres. These acquaintances also cross over into your more private life including religion, family, recreation, and business. The hierarchy of the relationship also crosses over into other spheres.

On the other hand, specific cultures easily separate spheres of life with friendship belonging to one or two spheres only, with little or no crossover of authority. In other words, you have a boss at work, but she is your boss only at work; at the beach, she is just another acquaintance. Thus, you have school friends, work friends, beach friends, and bar friends, with little crossover.

The Web researchers operationalized diffuse patterns by looking for mentoring approaches that address the whole person in all her aspects and dimensions. Specific communication patterns easily section off people into different groups: students, teachers, administrators, alumni, staff, and so forth. One major data for specific-diffuse is smiling in public photographs—almost no smiling in diffuse cultures. As Hofstede (2004) and Hampden-Turner and Trompenaars (2000) have argued, the specific approach is most likely in the United States, Canada, and western Europe, while the diffuse approach predominates in Asian, Latin American, and Arab countries.

The final common human threshold addresses the handling of inequality in interpersonal communication or what Hofstede (2004) calls power distance. Power distance measures the ability for two people of different power and authority to influence the other. The researchers operationalized low power distance as an emphasis on collaborative learning between teachers and students, disdain for close supervision, teachers not afraid to show they do not know the answers, and students not afraid to disagree with teachers. High power distance was evident in one-way, top-down communication and teaching styles, hierarchical collaboration, students afraid to disagree, and teachers afraid to show they do not know the answers. In Hofstede's (2004) research, the United States, northern Europe, and Israel were most likely to address inequality through low power distance means, while high power distance approaches predominated in Mexico, parts of Asia, and Andean South America.

These five variables provide the frame for understanding the relationships between Web site designs and cultures. The research team focused on how closely the Web sites from nine countries or cultures reflected the cultural patterns attributed to them by the research on intercultural values and why. If differences between the cultural values were attributed to a country and the Web sites, the researchers hypothesized as to the reasons for those differences.

RESEARCH METHODS

The research methods can best be described as a rhetorical analysis of Web sites using qualitative sampling (see Flicke, 2002) and intercultural empirical methods (Thatcher, 2001). Nine graduate students in a graduate course in intercultural rhetoric at New Mexico State University (NMSU) first selected a target culture or country for their analysis, and each researcher needed to be either a native of that target culture or possess significant knowledge of its culture and language. Then the researchers selected public university Web sites from that culture, using New Mexico State University as a benchmark. They used the following criteria to select possible universities for analysis:

- Public university of significant profile but not nationally recognized in any certain area
- Student population between 10,000 and 30,000
- Offering undergraduate and graduate degrees in humanities, engineering, social science, and agriculture
- Land-grant or affiliated strongly with community
- University Web site comparably developed

For China, South America, the United States, Canada, and Mexico, the researchers developed a list of public universities following these criteria, and then they randomly selected three universities for analysis. For Israel, Nepal, and Ireland, there were not sufficient universities to randomize effectively, so the researchers selected the universities that most closely followed these criteria. Table 1 displays the researchers and Web sites.

As outlined by Thatcher (2001), in order to be valid, intercultural empirical research needs to establish a comparative and neutral approach to both data analysis and data collection. The research team decided that the five intercultural variables provided the comparative analysis and the criteria provided the neutral framework for data collection. In selecting each university to analyze, the researchers also used randomization or sampling logic described by Flicke (2002) to develop as generalizable results as possible.

For the data analysis, the research team compared the Web sites of their three universities to the communication characteristics of each intercultural variable. The units of analysis of the Web sites that were compared to the communication characteristics included: screen layout, links, language style, content of Web site, graphics, artwork, navigation, use of flash and audio components, color, photographs, maps, icons, and other Web site components. To develop the results, the research team moved back and forth between inductivee (open) coding to deductive (theoretical) coding (Flick, 2002, pp. 176-195). Moving back and forth between open and theoretical coding ensures validity in the data analysis and construction of results.

At the beginning of the study, many on the research team openly acknowledged some or much doubt as to the accuracy of the intercultural variables attributed to the local cultures. Thus, they analyzed very critically any data that supported the attributions and welcomed exceptions

Table 1. Researchers and university Web sites

Researcher Name and Country of Origin	Target Culture/ Nation	Web Sites Selected for Analysis
Edgar Barrantes, Costa Rica	Central and South America	• Universidad de Costa Rica • Universidad Nacional de Colombia • Universidad de Chile
Amy Dalzell, USA	Canada	• University of Calgary (Alberta) • University of Ottawa (Ontario) • Concordia University (Quebec)
Qiumin Dong, China	China	• Hebei University • Zhengzhou University • Hunan University
Christopher Ford, USA	Ireland	• University College, Dublin • National University of Ireland, Galway • National University of Ireland, Cork
Sipai Klein, Israel	Israel	• Hebrew University of Jerusalem • Tel-Aviv University • Ben Gurion University of the Negev
Rachel Gallagher, USA	USA	• New Mexico State University • Texas Tech University • Washington State University
Marohang Limbu, Nepal	Nepal	• Kathmandu University • Tribhuvan University • Mahendra Sanskrit University
Lisa Ramirez, Belize	Mexico	• Benemérita Universidad de Puebla • Universidad Autónoma de Campeche • Universidad Autónoma de San Luis Potosi

to the attributions made by researchers to the target culture.

RESEARCH RESULTS

The most surprising result of this study is that in most cases the Web site designs of the universities strongly reflected the cultural patterns attributed to that local culture by the intercultural researchers such as Hofstede (2004) and Hampden-Turner and Trompenaars (2000). The Canadian and U.S. Web sites reflected almost exactly the values attributed to these countries in the intercultural research literature. The Web sites from the other countries reflected at least four of the five values attributed to them. In other words, despite the growing interdependence and cross-cultural contact through the World Wide Web, the Web sites simply reflected local culture in most areas. This results section will follow the previous section that outlined the intercultural variables for comparison: individual-collective, universal-particular, ascription-achievement, specific-diffuse, and power distance. It first describes NMSU, Texas Tech (TTU), and Washington State University (WSU) as the starting point because NMSU was the benchmark used for comparison. And then the other Web site characteristics are discussed.

Individual-Collective

All three U.S. Web sites follow individualist patterns. The use of diverse student and faculty photos (i.e., diverse in gender and ethnicity), often showing people alone and smiling as opposed to in a group setting, and their well-stratified content design tailored to specific audiences (prospective students, current students, faculty staff, alumni, etc.) are all hallmarks of individualism. These characteristics noticeably set the individual, as opposed to the group, as the lens, displaying self-reliance or even self-interest. Uniqueness or novelty is highly valued in this framework, again by the focus on diversity through the use of design and rhetoric. Writing is generalized and inclusive for a wider audience, yet is also directed to specific groups in certain instances. For example, prospective students can access information on how to apply to the university, while faculty and staff would most likely not need this information. Instead, they can check on faculty and staff policies or benefits by going to specific links. It is interesting to note that the Canadian sites showed the same Western proclivity for individualism over collectivism, using very similar methods.

The three universities in Costa Rica, Colombia, and Chile have a clear orientation towards collectivism, following the values attributed to these countries by the intercultural researchers. Given the convulsive history of Latin America, and specifically in Colombia and Chile, all three universities explicitly state their role in reaching/maintaining social stability and peace. Individuals are expected to achieve a high level of critical thinking and independence, but at the same time they are to become pillars of civil stability, hence the dual nature of these institutions: on one hand as autonomous entities centered on the individuals, on the other as providers of social stability and a center of collaboration. The heritage of public service and concern for national problems emerges in the history pages. The foundation of all three institutions is depicted as a social and historic landmark in the life of these young countries. For the University of Chile, the university is an indispensable condition of social progress and cultural development. The University of Colombia depicts itself as a legacy from forerunners and founding fathers. The University of Costa Rica advances the idea of higher education as a form for achieving social unity, and the mission of offering educational options for those with economic disadvantages. Also, individuals are portrayed in all sites as sharing space in communal settings or activities, as opposed to individual settings with delimited personal space. Pictures in all Web sites show closer distances in terms of proxemics. Pictures of all activities depict academic events as a reason for social gatherings.

Among the Irish universities, of all the traits being examined, a clear preference for either individualism or collectivism seems to be ambivalent, again following the mixed scores on the individual-collective continuum (see Hofstede, 2004). On the one hand, all three universities mention their role within the National University of Ireland system and seem to regard their membership as an important element of their identities. Yet each site also takes pains to differentiate itself as a unique institution within the larger whole. Interestingly, however, none of the Web sites makes obvious mention of how the individual universities cooperate with the other institutions comprising the larger university system. This is interesting since the National University of Ireland Web site describes cooperation and working together as underlying reasons for federalizing Ireland's public university system. While the formation of the National University of Ireland seems communitarian in intent, actual practice may differ considerably.

In contrast, traits indicating individualism can also been seen on the different Irish university Web sites as well. In their descriptions about educational opportunities and degree programs, the overall emphasis seems to be on what the individual can gain by attending. The word "choice" (and its synonyms) is highlighted prominently in each of the school's descriptions of its programs. The clearest example of this can be found on the University College, Dublin Web site, which dedicates a separate sub-Web site to details about the various educational opportunities available to its students. This individualistic emphasis on the choices available to each student is also evidenced by the links for prospective and current students to a wide variety of clubs, sports, recreational activities, and societies. Each of the universities highlights how there is "something for everyone" to be found at its school.

In Mexico, the three universities' emphasis on public service and social concern reflects the collective environment that prevails in Mexican culture. In all the Web sites, there is a focus on the student's opportunity to interact with and serve the surrounding communities; on the Web sites there is even a link that features cultural and social involvement in the community. In most of their mission statements, this commitment to cooperation and community spirit is also explicitly evident. The comments of alumni also indicate the level of community involvement that surrounds each university community, a feature prevalent in the first page of all three Web sites. Additionally, the pictures of students portray them in groups, working together to accomplish various projects. The Web sites rarely depict students in an individual setting, except on a few occasions where students are involved in learning new technology or using lab equipment. But even in some of these settings, they are seen working in groups. The universities are all seen as an integral part of the community in which they reside, and even the presidents' messages establishes a firm commitment to social development.

In China, collectivistic orientation is one of the most conspicuous characteristics of all three Web sites. Each of the three Web sites uses much space describing and introducing the university to the audience, and the introduction focuses on its history, heritage, and the outstanding status it enjoys, rather than what students learn from there and will be able to do after they graduate. Two long paragraphs in "Brief Introduction" trace Hunan University's "... origin to ... 976 A.D., and [it is] regarded as one of the four most prestigious academies of the Song Dynasty" and list outstanding alumni including generals, thinkers, political reformers, the first Chinese who landed on Antarctic, the Giant Galaxy Computer designer, and the number of academicians and innovative entrepreneurs it has graduated. Zhengzhou University claims in its long "Brief Introduction" to "be one of the top 100 universities in China chosen for the national '211 Project'". Zhengzhou University does not have a long history, so its Web description emphasizes its grants received from the central government, its national and provincial "key disciplines," "research centers," "state centers for ... research," "state research base for ...," "key laboratories," and so forth. Hebei University combines the feature of both Hunan University and Zhengzhou University in its brief introduction in that it includes its past and present.

The dimension of collectivism is also reflected in the use of graphics on the three Chinese Web sites. Neither Hunan University nor Zhengzhou University has many photos of individuals on its Web site. The photos are about groups of people at a conference or a leader wearing ties and suits speaking to his audience. In the pictures you seldom see people smiling, and the reader will see pictures of campus buildings more often than people, among which the graphics on the Hunan University Web page are framed in the same fashion as a Chinese painting, with traditional design pattern in the margins.

Although Nepalese culture is generally communitarian, the Web site of Kathmandu University tends to be individualism. For example, in one picture the students are shown working and reading individually, which distinctly portrays that the students have to draw themselves and create themselves to compete with others, which are the characteristics of individualistic culture. On the other hand, the Mahendra Sanskrit University Web site does not have a single picture of students; in addition, the language used in the different links depicts that it is more communitarian. For example, the language used in the message from the vice chancellor emphasizes the origin of value, the importance of Sanskrit, Sanskrit as a universal treasure, the story of Rhishis, and so forth, in an attempt to nurture and educate the audiences and make them feel responsible. For instance the Web site says: "Our forefathers having spent thousands of arduous years formulated this incomparable treasury of wisdom and knowledge, which has now been passed on to us. It must not perish." This Web site gives a sense of cooperation and social concern, which is strongly communitarian.

Universal-Particular

The Web sites for NMSU, TTU, and WSU are overall more universal in nature rather than particular. Interestingly enough, each site has undergone a brand-identity initiative to help present a more unified look throughout each site—that is, every page has a similar look and feel, treatment of photos, and overall design. This branding, along with the parallel, uniform, linear, and stable design and structuring of each site, makes for simple and quick navigation for a generalized audience. In essence, it allows for a wider group of people to view it and discern information easily. There is no "in" or "out" group with these sites, which would be apparent in particular cultures. Instead, the NMSU, TTU, and WSU sites attempt to be as inclusive as possible, reaching a variety of people in a variety of locations, while maintaining a uniform identity.

The use of photos on all the U.S. sites is also similar, capturing a diverse student body and faculty with a variety of ethnicities represented. This focus on diversity is clearly a universal feature, indicating an equal opportunity or "level playing field" institution. The diversity focus also establishes a generalized audience—no matter who may view the site, their ethnic group or even gender will be represented. In addition to the diversity focus, the overall use of language for these Web sites is also geared toward a general audience. Writing is relatively active, short, concise with direct sentence and paragraph structure—a key universal trait. This direct use of language allows audiences to view a site and learn what they need quickly. The explicit rules and policies of each university are also evident on the respective sites. The fact that these policies are actually written and easily accessed (usually on the second or third levels of the sites) is clearly a universal mark. Universal cultures are known for their explicit rules, which tend to be in writing.

The Web site organization of the three universities in Costa Rica, Colombia, and Chile displayed both universal and particular patterns. All sites essentially present the same information in the front page with varying levels of detail. Links for current and prospective students are readily available in the front pages, the "about" and "contact" links are present, as well as search features. Although these design choices seem to point to the universalism mental model, there is a clear contradiction between the way the sites volunteer information (universalistic) and the universities' missions and objectives (particularism). At the discursive level all three sites show particularism tendencies as they link the individuals to the body of society (individuals are but a cell of a larger being), and individual achievement is superseded by social or communitarian achievement. The University of Costa Rica makes the future professional responsible for attaining social justice and common welfare. The University of Colombia strives to offer a wide selection of academic programs in order to form socially responsible students who will help to change the direction of the national project. The University of Chile brings to the forefront a similar program for its graduates by emphasizing that individuals should contribute to the solution of major national issues and in the realization of the nation's most important projects.

Unlike NMSU's Web site where each page is branded, the universities in Mexico exhibit a tendency toward the unique, making the pages different, particularly if the pages are links to the various academic departments. Whether they are structured differently, or use different degrees of animation, it is rare that the pages mirror each other exactly. Often times, the format will remain, but some aesthetic feature is either added or altered in some way to preserve uniqueness. Each department has its own unique logo and its own unique way of presenting its information, which is indicative of a particular environment.

In Mexico, uniformity does not seem to be the norm in most of these Web sites, and even the rules of the university do not seem to apply to every individual equally. Because of the distinctly structured hierarchy, rules and regulations are specific to administration, faculty, or students. Even the manner in which they are written reflects the specificity of their application to the different members of the university community. The rules are sectioned off into articles, and each set of articles pertains to the different individuals that comprise the university. Rules are clearly role based, and there is emphasis placed on the contextually based roles exhibited throughout the university community. Additionally, it is evident that there is a clear "in group-out group" orientation in all the university Web sites, as particular relationships define the level of integration and participation of both students and faculty. Even the provision of information is particular as there is clearly a level of academic orientation expected in accessing the various links. The information on the pages is clearly not intended for the average individual, and the more specialized the faculties become, the more highly specialized the rhetoric becomes. Information is clearly not written for a general audience, and the majority of it is written in very formal academic language.

In China, while the design of the three university Web sites is simple and clear to follow, they all include information that attempts to attract the attention of a particular audience other than ordinary teachers and students. For example, despite the linear pattern, a closer look at the homepage of the reader will find him or her such information as "Keep Communist Party's progressiveness," "Party Secretary Zhao (made-up name) inspected campus," and "Unification Office (political organization) met for the new year's plan." This type of news, typical in China because of China's political system, on the one hand will draw attention of the central and local government and thus make the authority happy. On the other hand, it demonstrates to the government the university authority's commitment to their work and devotion to the Party.

Ascription-Achievement

For the Canadian Web sites, there is no question on ascription-achievement; the emphasis is entirely on achievement of status through the acquisition of education. The purpose of the school is to help students achieve personal goals through solving specific problems by ensuring universal access to information that will allow an even playing field. The University of Calgary's Senior Administration page features a picture of the campus with the offices listed in order of importance, followed by department listings. President and Vice Chancellor Dr. Harvey P. Weingarten's picture can be accessed with his personal information, but importance is ascribed to the office, not the individual—an office anyone can achieve through effort and problem solving. The viewer must traverse three pages to discover Dr. Huguette Labelle on the Ottawa site, even if her picture is slightly reserved; there is no reference to administration on the homepage. Again, Dr. Lajeunesse graces the Concordia homepage, but his message and demeanor are entirely genial and neighborly His picture on the homepage is small, dwarfed by the message: "Be part of the fastest growing university in Quebec. Find out more>."

The three Web sites of Costa Rica, Colombia, and Chile show a clear tendency towards ascribed status. By enrolling in one of these institutions, the individual will enjoy the social stature that comes from being a part of a national tradition. All universities promote the pride that comes from being part of this larger group. By becoming a student of a university that fosters research and outreach programs, individuals participate from institutional prestige. In other words, the individual's track record is important only once he or she has become part of the group. In order to achieve status, each student must begin by sharing from the ascribed status pool provided by each institution. Here, the student's potential is developed by becoming part of a larger social body. The ascribed status is also evidenced in the Web sites by the pictures of individuals engaged in social activities and social contexts. Unlike achieved status Web sites, persons are depicted as part of a larger group and as mutually dependent.

An orientation toward ascribed or achieved status is not easily or quickly discerned on the Irish university Web sites. While prominent and notable persons are noted only on the homepage of the University College, Dublin Web site, the brief notations describe only what their contributions are as scholars or as a guest speaker—and only in the context of this being "current news." This would seem to indicate achieved status is what is valued, particularly since neither the National University of Ireland, Cork or the National University of Ireland, Galway make any such mention on their Web pages. Brief notices about awards, grants, or other noteworthy activities by administrators, faculty members, and students are relegated to news and current events pages, which are linked to the universities' homepages. Even on these pages, only short blurbs telling about a recent award or honor is included. Ascribed status plays an even lesser role. With the exception of the universities presidents' Web pages already discussed, little attention seems to be given to the role of individuals by any of the universities. When matters of status are evidenced, it is primarily in terms of the role a person plays within the institution or the contributions the person has made to the university, the nation, or his or her academic discipline. This tendency may be more evidence of collectivism traits at work or perhaps even an expression of universalism's preference for wholes instead of parts than a clear preference for achieved status.

In Israel, achievement is most clearly seen at Hebrew University of Jerusalem, where on its homepage there are both news of recent scientific discovery done at the university and a link to a faculty member who recently became a Nobel Prize winner. On a similar tone, on the Ben Gurion University homepage, there is a notice of a faculty member's recently earned award. On the Tel-Aviv University site, little information is given to show weather or not achievement is as important of a role as at Hebrew University and Ben Gurion University. While both portray a strong sense of achievement value, there are also strong strains of the ascribed, status-based value. In both universities' history pages, there is a discussion of university founders and their prominent role in Israel culture. These founders are directly linked to each university's mission statement and therefore assert the importance of ascribed value. Tel-Aviv University's ambiguous standing on the issue can be resolved by examining how the absence of ascribed value indicators relate to the description of the university's current and previous rectors. The rector is one of the few people who does have a picture linked to the person, along with a list of accomplishments and publications during his or her career. The long detailed list implies that even though the site as a whole does not prescribe to either achieved or ascribed status values, the administration values people who have based their careers on their achievements.

The achievement-ascription dimension is complicated with the three Chinese university Web sites. They tend to have both. They all devote much of their space describing the success of their graduates in their specific field, listing/describing academy members, other prestigious professors, and announcing their achievements in research and teaching. All this indicates that individual achievement is highly valued and cherished. However, this feature also makes sense for the three Web sites to be defined as ascription-oriented on the rationale that these prestigious people lend credit to the schools; therefore, the description here constitutes rhetorical appeals to the audience. But at the same time, it is also evidence of individual achievements.

The ascription orientation of the Chinese culture reflected on the Web site is also exemplified in the "Brief Introduction" of the schools. Usually the "Brief Introduction" is long, consisting of information about the history, the presidents, prestigious alumni, leaders of the country who visited campus, academy members, and well-known professors; they also include how many graduates they have graduated, number of enrollment and degree programs, and grants and projects received. These statistics and facts are intended to impress the reader with its status.

Another example of the ascription orientation is reflected by the fact that all three Chinese Web sites quote from some historic and important figure, and/or use his calligraphy as the school motto on the homepage. This historic person might be the first president of the university as evidenced by the motto of Hunan University, or a previous leader of the country, or a scholar who enjoyed a high reputation. The calligraphy or quote of a well-known person, to the Chinese audience, increases the school's strength and raises its status.

All three Nepalese universities tend to be achievement based. For example, Kathmandu University has a motto "Quality Education for Leadership" which distinctly shows that it is achievement based. Besides, it has many programs that have free fields to compete to achieve goals. It has many other students' activities, which clearly invites the students for competition to achieve their objectives. Tribhuvan University and Mahendra Sanskrit University also have achievement-based Web sites. They distinctly encourage students to get better opportunities through competition. However, the hierarchies of the vice chancellor, professors, assistant professors, and students' various positions reflect an ascriptive culture, too. For example, Kathamndu University, one of the highest-ranking achievement-based universities in Nepal, also has hierarchies of professors, executive members, and various titles which clearly reflect the ascriptive culture. On the other hand, Tribhuvan University also has hierarchical structures and pictures of great persons like the late King Tribhuvan, the late King Birendra, and the King Gyanendra, ministers, and leaders which clearly reflect the ascriptive culture. Therefore, the vice chancellor, deans, professors, lecturers, and other administrative officers clearly depict the ascriptive culture. Despite the fact that Kathmandu University has some characteristics of ascriptive culture, it tends to bestow equal opportunity to all people (castes and creeds) and students; for example, the Dean of Education is from a Baisya family, and this university has name lists and pictures of many successful female professors and students. On the other hand, the Sanskrit education system in Mahendra Sanskrit University clearly seems to be ascriptive, because the Sanskrit language, which is known as a dead language, has nothing to do in the practical life of Sudra, Baisya, and Chhetri; nonetheless, it is still useful for Bahuns who conduct social and religious rituals.

Specific-Diffuse

All three U.S. Web sites' content organization, writing, and design target certain audiences (prospective students, current students, faculty and staff, etc.)—an obvious trait of specific cultures. Here the focus is on the individual in a specific sphere of action. The fact that each university is a land-grant institution dedicated to serving the citizens of each respective state is a diffuse characteristic. However, the overall Web site certainly falls more in the specific realm, with some diffuse qualities, such as mission statements for the universities stating the above dedication, or serving a wider community through the institution's resources. Public service to a broader community is certainly a diffuse quality, which NMSU for example adheres to, most notably with its cooperative extension departments and offices located throughout the state to assist New Mexicans in a variety of areas.

Specificity seems to be the predominant trait among the Irish universities, and this tendency appears prominently in how the Web sites themselves are organized. Each of the sites examined provides links to the various components of the sites, and each of these pages in turn becomes increasingly more specific and detailed. Finding information on all of the Web sites was relatively simple if a person has at least some idea of what he or she is trying to find. While each of the universities' Web sites provides a search tool for its Web site from its homepage, it was not especially difficult to find information being sought without using the Web site's search function—going from general information to more specific information seemed to be a fairly linear process.

In Mexico, all three Web sites show a clear preference for diffuseness, particularly because of the attention to aesthetic design. Bright colors dominate on almost every page and call attention to the picturesque quality of the Web sites. Even the more dominant colors that would be considered loud and inappropriate by a specific culture dominate the pages of these Mexican university Web sites. Interestingly enough, in these three particular Web sites, the color yellow or orange can be found on every page. These Web sites are a stark contrast to the less obvious color scheme that defines the NMSU Web site. Even the overall design of the Web sites is more elaborate, not simply because of the bold color schemes, but because of the creativity involved in the presentation of information. Animation is a very popular feature of all three Web sites, especially if a particular page is dedicated to student life and involvement.

The availability of links to national and international news sites is also indicative of the Mexican culture's preference for the diffuse. The university community establishes a visible link to the rest of the world, crossing boundaries between the private and public spheres. On the other hand, the way in which the different departments and kinds of information have been sectioned off is reflective of a specific tendency in an otherwise diffuse culture. All three Web sites have clearly divided the information available into their respective sections, specifically outlining which kind of information pertains to students or faculty members, or which information serves the general public's interest.

Among the three Chinese university Web sites, Hebei and Zhengzhou fit more in the diffuse pattern. If the links on the homepage are pretty direct and self-evident and tell the reader exactly what they are, the information included in the links might not be that direct. For example, when the reader enters the Web pages of a certain program offered by Zhengzhou University and Hebei University, he or she usually finds a long overview of the department and its history first before he or she gets to the degree programs. The Web site also provides some information about major courses, but no detailed information about each course. The three Web sites list the faculty members there, but none of them gives the contact information, which makes it difficult for students who want to interact with professors. However, Hunan University is an exception in terms of straightforwardness, and tends to be comparatively more specific. For example, the College of Foreign Languages Web site provides detailed information from "For Student," to "Faculty," and "Policies and Procedures." The "Courses" even include syllabi for each class with clear course objectives, materials, grading criteria, and so on.

In Nepal, Kathmandu University has a specific Web site. We can very easily enter in the specific links or discipline. For example, we can get into Arts, Education, Engineering, and Medical Sciences, but once we are in one particular discipline (e.g., Education), it is difficult to get into another (e.g., Medical Sciences). This situation clearly reveals the characteristics of specific culture. On the other hand, Mahendra Sanskrit University tends to have a diffuse Web site because the links do not seem to give specific information; as the result, the audience cannot get the right information. But the audiences familiar with the Web site and links may feel comfortable gaining access to the programs, which is a characteristic of a diffuse culture. Tribhuvan University has a more specific Web site and links than Mahendra Sanskrit University, but it is less specific than that of Kathmandu University. However, Tribhuvan University still tends to have a diffuse Web site and links, because we cannot get into different departments or different sections very easily. It is quite difficult to find any specific information of what we want.

Hampden-Turner and Trompenaars (2000) argue that diffuse cultures create strong boundaries between public and private. This is probably one of the reasons why these three Nepalese universities do not have pictures of many students on their Web sites, because the pictures can be offensive to some of the cultures; consequently, the pictures of students in different activities can affect their credibility negatively. Therefore, the universities do not have pictures heavily influenced by the western cultures as well as eastern cultures because Nepalese people, especially from the middle class, are in the transitional period, and some of them like the western popular culture, universalism, individualism, and specificity, and others still want to maintain the eastern collective culture, particularism, communitarianism, and diffuseness.

Power Distance

The U.S. and Canadian Web sites imply an overall independence for the respective faculty and student body. The same theme echoed earlier about diverse students and faculty achieving, researching, and studying what they choose, and achieving based on merit rather than what authorities designate, is an obvious characteristic of a low power distance society. There is little dependence and rather more individual choice and personal fulfillment involved with these aspects—a key low power distance trait. Also, photos of students working closely with faculty members, with everyone smiling, attest to the low power distance theme—students are independent and comfortable working with, not answering to, their instructors. Independence to choose rather than adhere to what authorities designate is a huge overtone on all the sites. Dress is also of unique importance. Most photos on each U.S. site display a more informal dress pattern for both faculty and students, further emphasizing informality and a level playing field, rather than the formality of hierarchical position or role. Also the overall focus of each site appears to be on the individual, which again expresses independence and self-reliance.

For Costa Rica, Colombia, and Chile, all their sites tend to display high power distance in both the text organization and in pictures of university personalities. The University of Costa Rica has a separate site (and URL) to account for its internal organization. There is a gallery of university presidents, and all pictures look like portraits rather than individuals in their natural work environment. Dress code is formal, which contributes to the power distance, and most individuals appear behind a desk or holding books, which creates an artificial barrier between the observer and the person portrayed in each picture. The University of Chile clearly calls its internal organization "structure," and information is arranged in order of importance from the president's office to faculties and interdisciplinary institutes. The Web site features a president's "profile," which includes a vita, a list of academic achievements, and a picture of the president next to the portrait of an academic personality. The University of Colombia calls its internal organization "institutional identity." Following the same pattern, information and links are arranged by order of importance, and pictures are presented in formal settings. High power distance is also evidenced in all sites by the organization of links by order of importance, starting with the presidents in a vertical orientation, all the way down to secretaries and workers with less decision-making power.

All three Mexican Web sites show a tendency toward maintaining high power distances between the levels of individuals in any given scholastic institution. Each university's Web site delineates this with the inclusion of a flowchart that specifies the various hierarchical levels in the institution and the systems of government, pointing out specifically the degree of respect for the authority figures of the university. The inclusion of the flowchart also recognizes the boundaries between the larger university community and those who direct its academic affairs. In addition, even in the classroom settings depicted on the Web sites, the power distance established between students and faculty members is unquestionable. The academic environment is defined by formality, as is evident in the distance maintained between teachers and students in the classroom. The teacher can rarely be seen interacting with the students, in the aisles of the classroom, or offering more personal attention to any one student. Rather, he or she is seen at the front of the classroom, either writing on the chalkboard or delivering a lecture. Even his or her attire reflects the notable difference in the high power distance between teachers and students. Furthermore, the Web sites rarely portray images that show teachers and students engaged in less formal conversations or interacting in less formal settings. Consequently, in these particular academic communities, students and teachers are not equals in the least; young or old, professors command a notable degree of respect from their students.

The three Chinese Web sites tend to be on the high end of the power distance continuum which focuses more on the schools and their status rather the people. The Web sites are university/department centered. Every school has a long introduction to its history, development, and growth in the field of higher education in the country (see "Brief Introduction" to the university and separate departments) instead of what the students will learn. The brief introduction to the College of Foreign Languages at Zhengzhou University all starts with "our school is/has/undertakes/invites ..." The reader cannot find any mission statement as to what students will learn there and will be able to do after graduation. It might be easy for the reader to find what programs they offer, but difficult for him or her to locate guidelines for actions. For example, a reader who wants to know more about undergraduate education, might not get the contact information on the Web site. As a result, he or she has to figure out where to go and which person to contact on his or her own.

The Chinese Web sites also tend to be hierarchical. On the Hebei University Web site, six academy members are grouped with presidents in "About the University." In their pictures, some of them are addressing their audience at a conference, or working in the lab or in the study, which clearly indicates their high status. On the School of Foreign Languages Web site of Hebei University, the reader cannot find all the faculty members except professors and those who have doctoral degrees. The Foreign Languages Departments at Zhengzhou and Hunan universities only have "prestigious faculty" on their Web sites, entirely excluding those who they consider are not "prestigious."

In Nepal, although Kathmandu University is generally more specific, universalist, and individualist, the pictures of vice chancellor, and the clothing fo the professors create large power distance symbols. For example, the vice chancellor of Kathmandu University looks serious, bold, and oppressive. He creates his space and shows that he is a great authority. The dean of education from Kathmandu University has a big smile; however, he is dressed in quite formal clothes and seems to create his own space. Both of them do not share space; they want to create their own space, hierarchy, and authority. On the other hand, on the Web site of Tribhuvan University, the vice chancellor equally creates his authoritarian space in a quite formal Nepali dress. He does not smile, hence he creates a gap between his subordinates, students, and audience. There are also several pictures of political leaders, professors, and students, but they do not have one picture of a woman except for Queen Komal which reflects the high power distance, hierarchy, and male sovereignty.

In addition, Tribhuvan and Mahendra Sanskrit universities greatly reflect the various hierarchies. For example, Mahendra Sanskrit University has a long list of various titles or executive posts: Chancellor, Senate, Pro-Chancellor, Vice Chancellor, Executive Council, Central Office, and so forth, but it does not have any specific duties and responsibilities they carry out; similarly, Tribhuvan University has a long list of its structure—the King, Chancellor, University Council, Executive Council, Academic Council, and Planning Council—but none of the lists has the name of a female or the name of someone from the lower castes. Kathmandu University does not have a link about the structure like that of Tribhuvan and Mahendra Sanskrit universities. Therefore, the Web sites imply that Mahendra Sanskrit and Tribhuvan universities tend to maintain higher power context than Kathmandu University.

DISCUSSION AND CONCLUSION

As already mentioned, the university Web sites generally reflected the cultural values attributed to that local culture by intercultural researchers. This was the first and most obvious result, surprising many on the research team who expected to see less correlation between the attributed cultural values and the Web sites. The U.S. and Canadian Web sites very closely reflected the cultural values attributed to these countries, and all six were very similar. This strongly suggests that at least at these universities, the attributed cultural values are deeply rooted. And, despite the claims of its multicultural heritage, the United States seems strongly homogenized in its cultural orientation, and it does not seem to have changed in the last 20 or 30 years, despite the growing global interdependence and cross-cultural contact. This lack of change is especially surprising in light of the strong individualism that still predominates on these U.S. and Canadian Web sites. Since interdependence is one of the key characteristics of the global economy, and the Web as a rhetorical system seems collective—hence the metaphor, Web—it seems highly likely that U.S. and Canadian individualism should be fading, but these six Web sites did not display that.

It is also surprising to note that particularism and collectivism are still strongly rooted in the Web sites of countries that have traditionally exhibited high levels of both values. As mentioned, particularism and collectivism show a strong in- or out-group orientation, based on tight social groups, with applications of rules based on relationships and social status. The growing interdependence of nations and the cross-cultural contact of the Web should encourage more of a universal orientation;

that is, because of the great complexity of global relations, an over-arching, universal framework is often needed for international affairs, such as the United Nations. However, the Web sites of Mexico, Costa Rica, Colombia, Argentina, Chile, Nepal, and China showed very strong particularist and collective tendencies.

Only occasionally did the university Web site contradict the attributions from intercultural research, and these exceptions deserve some discussion. First, some Web sites in South America were specific, rather than diffuse, which is a common characteristic in South America. Perhaps the diffuse attribution for South America is inaccurate or changing, or perhaps the strong specific orientation of the United States and Europe is influencing this specific approach. Or perhaps university culture is more specific than general South American culture. It is hard to tell.

In addition, the South American, Mexican, and Chinese Web sites also displayed mixed characteristics of achievement and ascription orientation, while common attributions to these cultures are strongly ascription oriented. It seems quite probable that universities are a counter-culture to ascription orientation; that is, they provide perhaps the only path to status and reputation for those who are not born into ascribed categories that permit that kind of status. Perhaps the global economy is also encouraging achievement orientation because of its competitive nature, the context for achievement orientation.

Finally, this study significantly underscores the necessity to localize Web sites that are targeted for other cultures. Since the data in this study shows strong connections between local cultures and Web designs, technical communicators cannot assume a globalized or homogenized Web culture. It simply does not exist, or if it does, it is so weak that it is difficult to detect. Further, this study also suggests that technical communicators who want to localize effectively their Web sites can draw on the already attributed cultural values from Hofstede (2004) and Hampden-Tuner and Trompenaars (2000). Even though these researchers do not address Web designs in their research, the connections between intercultural variables and Web features is fairly simple to carry out.

REFERENCES

Bhawuk, D. P., & Triandis, H. C. (1996). The role of culture theory in the study of culture and intercultural training. In H. C. Landis & R. S. Bhagat (Eds.), *Handbook of intercultural training* (2nd ed., pp. 17-34). Thousand Oaks, CA: Sage.

Bolter, J., & Grusin, R. (1999). *Remediation: Understanding new media.* Cambridge, MA: MIT Press.

Flicke, U. (2002). *An introduction to qualitative research.* Thousand Oaks, CA: Sage.

Grossberg, L. (1992). *We gotta get out of this place: Popular culture and postmodern culture.* New York: Routledge.

Hampden-Turner, C., & Trompenaars, F. (2000). *Building cross-cultural competence: How to create wealth from conflicting values.* New Haven, CT: Yale University Press.

Hofstede, G. (2004). *Cultures and organizations: Software of the mind.* New York: McGraw-Hill.

Lucy, J. (1996). The scope of linguistic relativity. In J. J. Gumperez & S. C. Levinson (Eds.), *Rethinking linguistic relativity* (pp. 37-69). Cambridge: Cambridge University Press.

Martin-Barbero, J. (1993). *Communication, culture and hegemony: From the media to mediations* (Elizabeth Fox & Robert A. White, Trans.). Newbury Park, CA: Sage.

McLuhan, M. (1962). *The Gutenberg galaxy.* Toronto: University of Toronto Press.

Ong, W. (1987). *Oralidad y escritura: Tecnologías de la Palabra* [Orality and writing: Technologies of the word] (A. Sherp, Trans). Mexico City: Fonda de La cultura Económica.

Plato. (1956). *Phaedrus.* New York: Macmillan.

Thatcher, B. (2001). Issues of validity in intercultural professional communication research. *Journal of Business and Technical Communication, 15,* 458-489.

Thatcher, B. (2005). Situating L2 writing in global communication technologies. *Computers and Composition, 22*(3), 279-295.

ENDNOTE

1 Ong later acknowledges the problems of this position in response to strong critiques of writing supplanting orality. See Kaufer and Carley (1993) for a detailed description of Ong's changing approaches and the ongoing debate of orality and writing.

APPENDIX: WEB SITES REFERENCED

Ben Gurion University: www.bgu.ac.il

Concordia University: www.concordia.ca

Hebei University: www.hbu.net.cn/en/

Hebrew University of Jerusalem: www.huji.ac.il/huji/eng/

Hunan University: www.hunu.edu.cn/English/index.html

Kathmandu University: www.ku.edu.np

Mahendra Sanskrit University: www.msu.edu.np

National University of Ireland, Cork: www.ucc.ie/en/

National University of Ireland, Galway: www.nuig.ie

Tel-Aviv University: www.tau.ac.il

Tribhuvan University: www.tribhuvan-univeristy.edu.np

University College, Dublin: www.ucd.ie

University of Calgary: www.ucalgary.ca

Zhengzhou University: english.zzu.edu.cn

Chapter X
Cultural Cognitive Style and the Web:
Toward a Theory and Practice of Web Design for International Users

Anthony Faiola
Indiana University-Perdue University Indianapolis, USA

Sorin Adam Matei
Purdue University, USA

ABSTRACT

Several technological developments have altered our world in the last half-century. Among these were the new information processing and distribution platforms supported by computer-mediated communication (CMC). In 2005, Forrester Research found that 50% of Internet users were non-English speakers, and this number would increase to 66% by 2006. For many designers and Web usability researchers, addressing this situation seem limited to translating Web interfaces or content. Although early studies in usability testing have identified considerable cultural differences among users (D'Andrade, 1984; Evers & Day, 1997), a need exists for a more rigorous investigation from a cross-cultural perspective into how Web sites are designed. The authors hold that the cultural cognitive styles of Web designers ultimately affect the performance and preferences of online users. As a result, specific attention should be paid to the impact of the Web designers' culturally shaped cognitive style on the design and development of online information.

INTRODUCTION

Several technological developments have profoundly altered our world in the last half-century. Among these, of central importance were the new information processing and distribution platforms supported by computer-mediated communication (CMC). These particular developments have reduced the time lags, human and material costs, and space limitations intrinsic to the traditional flow of communication, which, until recently, supported most large-scale organizations, businesses, and social groups. A cross-cultural community of interests, commerce, ideas, and technology has emerged as a natural outcome of these developments. Its impact is omnipresent, affecting a large number of cultures, knowledge domains, and activities (Edwards, 1994).

In 2005, Forrester Research found that 50% of Internet users were non-English speakers and predicted that by 2006 this number would increase to approximately 66%. At the same time, Inktomi and NEC Research state that only 14% of Web sites are in a language other than English, Consequently, half of all Web users are able to understand less than a sixth of the Web sites. This was especially prevalent in Asia, where Internet use climbed from 38 million to 95.8 million in three years (WorldLingo, 2000).

This unprecedented proliferation of non-English-speaking e-commerce consumers suggests that a paradigm shift is imminent in the global marketplace, in response to which Web developers must formulate new strategies for effectively delivering Web content (Del Galdo & Nielsen, 1996). Specifically, communication technologists must consider the cross-cultural challenges that will inevitably confront the online world (Houston & Eckhardt, 2001; Zhang, Schmitt, & Haley; 2003).

For many designers and Web usability researchers, these challenges seem to be limited to translating Web interfaces or content. Although accurate translation of site content is imperative, there are far more important human factors related to information perception and organization, which are rarely considered by Web site developers. Chief among these factors are the subtle psychological dimensions of human-computer interaction (HCI), which come into play when a culturally diverse audience interacts with Web content (Ess & Sudweeks, 1998; Faiola & Matei, 2005b; Herring, 1996; Vishwanath, 2003). In this context, it is particularly important to investigate and elucidate the culture-specific thinking and reasoning methods that impact a user's experience.

Moreover, a recent body of research, particularly in CMC, HCI, and instructional technology, has shown the important relationship between culture's effect on information acquisition and processing strategies, and the user's experience on the Web (Chen, & Macredie, 2002; Choong & Salvendy, 1999; Ess & Sudweeks, 1998). This literature has begun to show how culture influences a variety of behaviors and preferences that are important to understanding usability issues specific to Web sites (Chau, Massey, Montoya-Weiss, & O'Keefe, 2002; Faiola & Matei, 2005a, 2005b). For this reason, when creating content for the WWW, developers should take into account the cultural differences in the ways that users think, assign value, and behave.

In brief, culture, especially that which is rooted in national and ethnic contexts, is one of several complex and problematic areas that need more attention from those who are trying to understand the factors that impact online commerce, education, and the general accessibility of online information (Simon, 2000). At the same time, Web developers with expertise in HCI, graphic design, information technology, and psychology are increasingly concerned with the cognitive processes involved in designing Web content (Faiola, 2002; Preece, Abras, & Maloney-Krichmar, 2004). Some of their conclusions and efforts have been further developed by communication scholars with the aim of identifying the most appropriate ways of improving information delivery to international users (Blanchard, 2004; Marcus, 2000; Preece, 1998). Although early studies in usability testing have identified considerable cultural differences among users (D'Andrade, 1984; Evers & Day, 1997), there is still a need for a more rigorous investigation from a cross-cultural perspective into how Web sites are designed.

As never before, it is necessary to construct a framework rooted in theory and best practices that can support strategizing for Web site construction and that addresses cross-cultural user preferences and usability issues. New research initiatives must assist Web developers in understanding Web design from a cultural cognitive design (CCD) perspective. In accordance with our own past research (Faiola & Matei, 2005a, 2005b), which continues a line of thought developed by cultural psychologists (Chau et al., 2002; Hofstede, 1991; Nisbett & Norenzayan, 2002; Vygotsky, 1979), we hold that the cultural cognitive styles of Web designers ultimately affect the performance and preferences of online users. As a result, specific attention should be paid to the impact of the Web designers' culturally shaped cognitive style on the design and development of online information.

BACKGROUND

Behavioral Perspective of Web Design

Since the early 1990s, cross-cultural Web usability research has become increasingly interesting for a number of scholars in the fields of communication and human-computer interaction (Bourges-Waldegg & Scrivener, 1998; Del Gado & Nielsen, 1996; Eveland & Dunwoody, 2000; Kim & Allen, 2002; Preece, 2001; Sears, Jacko, & Dubach, 2000). Researchers who come from a behavioral perspective have made especially valuable contributions to the literature on online user experiences that are culturally molded (Honold, 2000; Liu, Lin, & Wang, 2003; Marcus, 2000; Vishwanath, 2003; Zahedi, van Pelt, & Song, 2001). In fact, much of the current cross-cultural research on online communication and usability has been dominated by a behavioral perspective reflecting the traditions of Hofstede (1991).

Sensitive to the richness of cultural variation, Hofstede focuses primarily on cultural acts at the behavioral level. He defines culture as the collective programming of the mind, a key designator that visibly distinguishes one group of people from another, specifically through behavior-driven cultural values (Hofstede & McCrae, 2004).

Hofstede's cultural values and their relational scores helped Marcus (2000) and others recommend an array of practical cultural differences and similarities while examining various Web sites from diverse cultures. Researchers have transformed Hofstede's theory to practice by applying the five dimensions to interface elements such as metaphor, mental models, navigation, appearance, and interaction (Honold, 2000; Sheridan, 2001; Marcus, 2000).

Cultural Shaping of Cognition Through Learning

Hofstede's (1991) categorization of culture provides a platform for research that takes a more introspective approach to the cognitive development of culture. From this perspective, culture is primarily developed cognitively, and only secondarily expressed through behavior. From a cultural-psychological perspective, behavioral diversity is the product of cultural traits, which are deeply embedded in cognitive processes. This leads to a mode of inquiry that can be most appropriately labeled as cultural psychology, which according to Schweder (1990) is defined as an examination of the way cultural traditions and social practices create differences due to cultural context and the constant dynamics of those cultural environments.

The key point here is that human experiences are molded in the matrix of their social values, ethical concerns, and narrative imagination, and these experiences constitute the main framework of the mind. The cultural perspective on cognition assumes that humans are biologically prepared not only with a variety of physiological and psychological components, but also with a number of sociocognitively shaped propensities. Hence, the human psychological profile is formed through biological and sociocultural processes, and is able to adapt and adjust to its own sociocultural environment (Greene, 1995).

The intellectual roots of this framework can be traced back to Lev Vygotsky's (1934/1979, 1932/1989) cultural psychology and to Whorf's (1956) research on the relationship between language and social processes. Vygotsky's sociocultural theories emerged as an early response to the behaviorist approach to psychology. They addressed learning as a shared process situated in the context of social interaction and discourse. Vygotsky's (1934/1979) early 20th century research in child psychology brought forth the idea that cognitive processes emerge from culturally constrained practical activities that in turn develop higher mental processes. In essence, his assertion is that culture fundamentally shapes thought (Mishra, 1996; Nisbett, Peng, Choi, & Norenzayan, 2001).

Vygotsky's sociocultural theories addressed learning as a shared process, while assuming that social interaction is grounded in culture. He placed an emphasis on social settings by arguing that: (1) knowledge is not individually constructed, but co-constructed in a collaborative setting between two or more people, and (2) problem solving and abstract thinking have a social origin. In sum, the social cognition learning model developed by Vygotsky argued that culture is the primary determinant of individual mental development, because culture influences the way we categorize our thoughts.

Despite the fact that Vygotsky's ideas were for many decades isolated from the theoretical debate on the nature of mind and thinking, they were still convergent with other early 20th century researchers. For example, Vygotsky argued that the structure and function of a group's language affects an individual's engagement and understanding of his or her world. Similarly, Whorf (1956) argued that language patterns are influenced by cultural norms. He advanced the idea that, since cultural patterns are explicitly coded in language, the grammar of a specific language is resistant to change in a manner consistent with the cultural conservatism of the group that uses that language (Brown, 1994). At the same time, however, he noted that "correlations or diagnostic correspondences between cultural norms and linguistic patterns" cannot be strictly applied (Whorf, 1956, p. 159).[1]

Both the traditions of Vygotsky and Whorf have found a new synthesis in Nisbett's recent exploration of the role of culture in shaping thinking. His learning model, which is part of a larger cultural framework, suggests that language and quantitative abilities are culturally inflected and that they profoundly shape the act of cognition (Nisbett & Norenzayan, 2002). This hypothesis was further developed by a number of researchers, such as Roberson (2005), who concurs with Nisbett and Norenzayan by stating that "recent findings, using a range of techniques, have broadly supported the tight links between language and thought in other fields" (p. 3). Davies (1997) also argues for a linkage between the lexical and cognitive categorization, while a wealth of data comes out of multiple studies of lexical variations in number systems and numerical reasoning (Choi, McDonough, Bowerman, & Mandler, 1999; Gumperz & Levinson, 1997; Levinson, 1996).

Cultural Cognitive Style Differences Between Holistic and Analytic Thinking

Further studies have shown ways that specific cognitive styles are connected to culture (Chen & Ford, 1998; Chen & Macredie, 2002; Riding & Rayner, 1998; Wood, Ford, Miller, Sobczyk, & Duffin, 1996; Zebian & Denny, 2001). Riding and Rayner (1998) define cognitive style as an individual's unique personal approach to organizing and representing information. Ford, Wood, and Walsh (1994) add that a cognitive style is a collection of strategies by which individuals differ when processing information, while Goldstein and Blackman (1978) state that cognitive styles are the characteristic ways in which people "conceptually organize their environment" (p. 79).

As culturally influenced processes of learning develop over time, the mind forms particular styles of planning, strategizing, problem solving, and organizing information (Goldstein & Blackman, 1978). Farnen (1993) suggests that this process provides a "mediating function for information transformation" (p. 378), the primary focus being on structure rather than content—that is, the cognitive organization of knowledge.

Recently, some very significant distinctions in terms of cognitive styles have been demonstrated between the holistic and analytic modes of thought. Research has identified specific cognitive strategies that could deeply impact the manner in which language and logical inferences are structured. This means that, in a Whorfian sense, languages and knowledge could be described as analytic or holistic in nature.

In general terms, a holistic perspective sees the world as a collection of overlapping, interpenetrating, and continuous entities that make up the "wholeness" of a field of experience. People who see the world from a holistic perspective prefer to organize the objects of their world via relationships of contiguity, similarities, and metaphors (Nisbett & Norenzayan, 2002). Their understanding of a course of events and objects may be intuitive and instantaneous, resulting in a focus on particular instances.

Conversely, analytic thinkers view the world as a collection of discrete objects, which can be categorized by reference to some subset of universal properties that characterize the object (citations). Those who embrace this view are inclined to decontextualize objects from their environment. They explain time-related interactions and abstract or isolated events, rather than in the context of sequential arrangements. At the same time, they may understand the behavior of the object as independent from its context. This view focuses on formal logic.

Asians, for example, because of heightened contextual sensitivity, would have an advantage over North Americans in performing tasks that require incorporating contextual information in the process of estimating quantities (line segment lengths). Kitayama, Duffy, Kawamura, and Larsen (2003) tested this hypothesis in a series of "framed-line tests." Two groups (Japanese and American) were presented with two square frames, one containing a vertical line, and the other empty and of a different size. Subjects were either asked to draw a line in the second frame of the same exact length as the line in the first frame (the absolute task) or in a ratio to the height of its frame similar to that of the already framed line (the relative task). Japanese participants were more accurate in the relative task, whereas North Americans were more accurate in the absolute task. Hence, according to Kitayama et al. (2003), the Japanese used a holistic method of appreciating size, the framing element being an essential element in the assessment of the length of the line segment. Conclusions suggested that when each culture interacts with other cultures, "individuals tended to show the cognitive characteristics common in their host culture" (Kitayama et al., 2003, p. 201).

Choong and Salvendy (1999) took these conclusions further in an HCI related study by examining the role of mental models in software design. After asking North American and Chinese subjects to describe the contents of a house, they found that the Chinese subjects tended to emphasize relationships, contexts, and classification based on interdependence within the whole more than the North American subjects. In their classification activities, the Chinese subjects also relied on their subjective experience, and did not make sharp differences between self and others or between facts and concepts. The North American users, however, tended toward classification by functions and analysis of components, and used inferences based on common features. When the researchers gave Chinese subjects the mental models favored by the North Americans, they found that the Chinese required longer performance times and were more likely to err. Similar results were found for North Americans attempting to use Chinese models of thinking.

Differences between the holistic and analytic perspective were also detected in the realm of language processing. Ji, Peng, and Nisbett (2000) found that Chinese adults and children were more likely to find strong associations between words in a functional (e.g., pencil-notebook) or contextual (e.g., sky-sunshine) relationship. At the same time, North Americans were more likely to find association between words belonging to the same category (e.g., notebook-magazine)" (p. 25).

In summary, the studies suggest that differences between holistic and analytic cultures consist mainly in the manner in which information is processed and organized. Holistic cultures seem to be more inclined to connect information to its concrete contexts, while analytic cultures emphasize functional-abstract relationships.

COGNITIVE STYLE AND ONLINE INFORMATION DESIGN

Web Designer Cognitive Style

As outlined above, past research has provided a better understanding of how culture can inform distinct cognitive strategies. Currently, most HCI researchers that have addressed cognitive style and Web design have done so from a usability perspective (Graff, Davies, & McNorton, 2004). Consequently, a significant connection has not been made between cultural cognitive style and the design of online information. One way to connect usability with information architecture research is to understand that cognitive styles are contextually shaped and impact the building of online information. The authors recommend a framework rooted in theory and best practices that can support Web site construction strategies by addressing cross-cultural user preferences with respect not only to the interface, but also to the content presented on a Web site.

Since cultural context has been identified to have an influence on thinking patterns and information processes, we argue that Web design should be observed from a CCD perspective, as illustrated in Figure 1. In this figure, CCD begins with the designer's own cognitive development, which has been largely influenced by his/her cultural or contextual influences of sociocultural, language, and educational environments. Hence, when they attempt to assemble and build information for the Web, there is a transfer of implicit thinking patterns and processes that influence how and in what way that information will be developed. Once the information is disseminated online, global users access it in a common way through the Internet, but they understand the information through a broad range of cultural cognitive variables, which are personal, distinct, and multifarious.

Contextual influences on Web designer cognitive style have an indirect impact on a Web site design that is distributed to a multicultural user base (A-D).

We have tried to elucidate the CCD process empirically in our own research, which focuses explicitly on Web design strategies and online information use. Our studies posit that cognitive style can play an important and direct role in online information design by affecting designers' problem-solving and design strategies.

In what follows, we will present some of the assumptions and findings of our research that highlight the relationship between cognitive style and Web design. We have directed our attention to national/ethnic cultural cognitive styles, and we have investigated some of the possible implications for their relationship with Web design and online interface usability. Of particular interest were questions, such as: Do online users have a preference for Web sites created by Web designers

Figure 1. The relationship between cultural cognitive style and Web development style

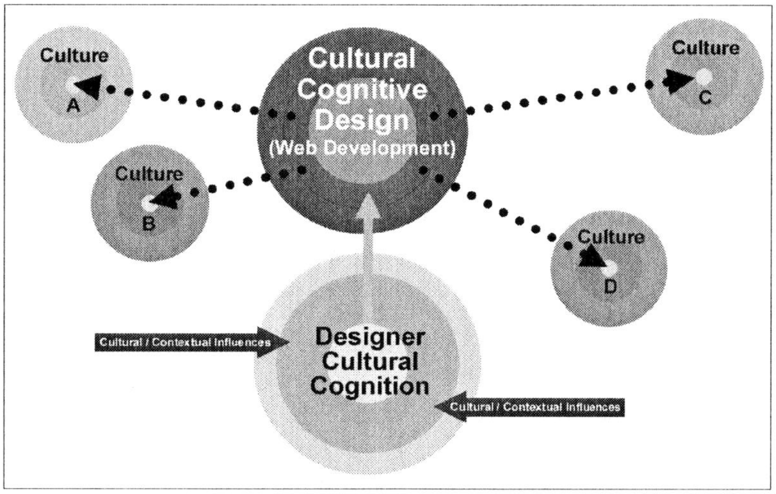

with whom they share a specific cultural cognitive style? Furthermore, do preferences rooted in cognitive development impact the performance of users?

One expected outcome was that Web sites created by designers from cultures other than those of the research participants would cause difficulty in finding information. Of course, performance difficulties experienced by participants could be assigned to the general design quality of the Web site, rather than interaction clashes between the cultural cognitive style of the Web designer and that of the Web users. To investigate these questions, we carried out a number of online and in-lab studies.

TWO CCD STUDIES

Our studies asked designers recruited from three cultures (U.S., Chinese, and Russian) to produce Web sites, which were then offered for interaction to users recruited from the same three cultures. Overall, participants displayed preferences for and better interaction with Web sites designed by designers of their own culture. Our conclusions support the previously mentioned research, showing a direct connection between cultural cognitive style and Web design.

The First Study

The purpose of the first exploratory study (Faiola & Matei, 2005a) was to ascertain whether or not the participants' preferences were aligned with a specific Web design from a particular culturethat is, designed or not designed by a Web designer of their own national culture. Hence, the hypothesis was that participants would prefer and perform better on Web sites created by designers from their own national culture. In this study, we were particularly interested in a correlation between the designers' cognitive style and cultural-specific content produced for Web sites. This included aspects related to visual communication or the graphics and typographic elements, and their relation to any peripheral factors that may affect the user's interaction or understanding of the virtual space of the site. Also, we were concerned about the hierarchical aspects of the sites' information architecture relative to site navigation and other interaction design factors.

In the first study (Faiola & Matei, 2005a), 18 participants from the U.S., Chinese, and Russian cultures were exposed to six Web sites in a lab setting. All the participants were volunteers from a general invitation via e-mail to students from Indiana University-Perdue University Indianapolis (IUPUI) and Purdue University. The Russian and Chinese participants had been studying in the U.S. for approximately two years. All students were between the ages of 18 and 27, with three males and three females from each culture. Each culture (six participants) was exposed to six Web sites.

We asked two Web designers from each culture to design one Web site each according to their culture-specific style, in their respective national language, giving us a total of six Web sites. We used two Web sites from each culture in order to offset variability in design styles at the individual level. In each case, we commissioned one average skilled Web designer and one semi-professional to professional Web designer for each culture. Each Web designer was provided the exact same content (the text) in their native language. In other words, the substantive information is common across Web sites, while the graphic elements and the information structure is cultural specific. We determined that a small business Web site might be the most non-controversial and least content specific.

Once the six Web sites were developed, each was translated into the other two languages, so that each participant was exposed to and interacted with six different sites, all in his/her national language. However, two of the sites were created by designers from their own national culture, while the remaining four where designed by the other cultures. Also, the Web designers were given the freedom to choose their own interface layout, typography, graphic elements, and information architecture.

During the experiment, the participants had 15 minutes to observe and interact with the six Web sites in their own national language simultaneously, clicking and navigating at will. After 15 minutes, they were given 30 minutes to fill out a questionnaire composed of eight items for each of the six sites. The questions asked the users how much they preferred each site in terms of eight variables: usability, visual appeal, layout and design, color, symbols and logos, menus and button choice, navigation system, and information organization. Six sets of eight variables were thus produced. These were reduced to three sets, by averaging the results for each pair of two sites produced by designers of the same cultural background (Russian, Chinese, and American). The three sets where then reduced into three "preference indices" by summating each set of eight variables for each type of site.

To test our hypothesis that cultural background is associated with user preference for Web site designs, we employed a median test. This determined if the subjects had a preference (i.e., had higher scores on the preference index) for sites produced by individuals with a cultural background similar to theirs. The results showed that the participants consistently had higher preference scores for the sites created by designers from their own culture.[2]

The Second Study

The findings of a second experiment (Faiola & Matei, 2005b) also supported the hypothesis that online task time performance will be faster when using Web sites created by designers from one's own national culture.

Twenty-seven Chinese and 26 U.S. students were compared in terms of time spent on performing a number of information retrieval and Web use tasks. Due to political and accessibility issues, contacting Chinese students in China proved to be very difficult. We decided to select a sample of newly arrived Chinese students studying at American universities. The students were contacted via e-mail, using a convenience sampling technique for the American sample and a more rigorous methodology for the Chinese students that consisted of e-mailing a broad range of Chinese student associations at 31 different American universities. The American students were recruited from the IUPUI student population in the School of Informatics, that is, 150 invitation letters were handed out in three separate classes. Subjects from both cultures received the same invitation letters containing the exact same recruiting information. They were asked to visit the test site and once arrived at the site were given instructions in their own national language as to how to proceed with the test.

We asked one Web designer from each culture (Chinese and U.S.) to design one Web site, each according to their culture-specific style, in their respective national language. The standards and content applied were the same as in the study mentioned above. The two Web sites were translated into the other language, and the participants were randomly assigned to the four resulting sites, so that half of the Chinese subjects interacted with the Chinese-designed and half with the American-designed site, both presented in Chinese, while the Americans, also divided into two sub-groups, interacted with the same corresponding sites, but in English.

At the bottom of each site, the screen displayed a frame that provided navigation instructions and asked task-related questions. The six questions ranged from very simple ones, requiring minimal effort, to more complex ones, which required more extensive searching to find the answer. During the testing, the system recorded if the participants had finished the task and the performance time for each task. As such, the primary use of the "yes" or "no" function was to establish the timeframe for each task for each participant. All the questions challenged the participants in their cognitive strategy to various degrees as they searched throughout the sites. However, the main goal of diversifying the complexity of the task-based questions was to help detect variations in task performance.

The data were analyzed using an ANOVA 2x2 design with the user's cultural background as the independent variable, interaction with Chinese- vs. American-designed Web sites as conditions, and task performance times for the six tasks as dependent variables (Faiola & Matei, 2005). The results indicated that both the Chinese and American performance times were generally shorter for tasks performed on sites designed by individuals of the same cultural background as the user. Hence, statistical results supported, with some qualifications, the hypothesis that cultural-specific Web designs facilitate user performance.

Those qualifications were related to the fact that only performance times for tasks 1-4 were shorter, while for tasks 5 and 6 no significant performance differences were observed. One reason was the fact that of the 171 subjects who began the online study, only 53 finished all the tasks, the sharpest drop off in user activity being recorded after task 4. Critical incidents caused by system errors may have played a role in distracting the online testers, causing a general lack of concentration by the time they reached task 5.

A BROAD APPLICATION OF CULTURAL COGNITIVE STYLE TO WEB DESIGN

What are the practical implications of these theories and empirical findings? In what way should awareness of cultural cognitive styles be incorporated into the Web design process? What implications does CCD suggest for Web site localization and internationalization?

A first suggestion that can be made in view of the literature reviewed above and of our own research findings is that if a Web site were to fit the cognitive style of an intended culture, designers need to organize information according to the set of cognitive principles specific to that culture. The two most important sets of principles are the holistic and the analytic. Researchers have provided considerable findings that distinguish between Western (analytic) and Eastern (holistic) cognitive styles (or field-independent/field-dependent) (Armstrong, 1999; Choong & Salvendy, 1999; Gudykunst, Ting-Toomey, & Chua, 1988; Miyamoto & Kitayama, 2002; Nisbett et al., 2001; Nisbett & Norenzayan, 2002; Parkinson & Redmond, 2000). However, in terms of practical implications, a Web site designed with a holistic user in mind can anticipate content as well as structure on the very first page of a site. The site should have related groupings of text, icons, and images, and the information architecture can provide a broader range of choices. For example, see Figure 2, where the Web site on the right, in Chinese, was made for a Chinese audience. As illustrated by this site, holistic users prefer to see the whole picture—for example, key site locations displayed in a "ring" fashion force the user to understand the site in a non-hierarchical way. Also, to accommodate this preference, Web content can be designed to display the interconnection of the various parts, allowing the user to see most of the site's main and subsections at a glance. In sum, a holistic Web design template should be intuitive, while site elements should be grouped according to their phenomenological (metaphoric) relations.

Opposed to this, as illustrated by Figure 2, the Web site on the left, users must dig down from a main menu. This site was designed for Western readers who would like to visit the same university site, but in English. In the case of the site on the left, the analytic user is being targeted, who may prefer discrete choices and elements, and may show less concern for their (graphic and informational) interrelationships within the broader structure. The analytically designed site may also display more reliance on text.

Also, interface design and information architecture for Western Web sites should generally avoid a broad range of choices. Content may be structured and divided into distinct components so that users can focus upon them individually. This is clearly illustrated by Figures 2 and 4 on the Web sites on the left. In sum, design should use hierarchically organized schemata, that is, the site metaphor may have a strong hierarchical tree organization using an abstract system of categories and subcategories. The information architecture may be represented by a site map in outline format, with main headings and subheads, so that the user can navigate by "drilling down" into the internal structure of the site.

These suggestions seem to resonate with tacit practices already in use in the Chinese or English-speaking Web design communities. In Figure 3, the Chinese Nike site shows a mix of contrasting of graphic elements in the site illustration with bright color choices, interwoven with a main menu system (which is nearly transparent to the user). Here, challenges in menu accessibility and use of figure-ground elements might stumble Western users, but Chinese users take pleasure in the complexity of relationships of playful graphic elements, which appear to be intuitively depicted. At the same time, the U.S. Nike site is much more controlled and discrete, with no sense of the images' (sneaker) context—that is, nothing is relationally depicted. Instead, the precisely articulated site presents the image and text in a focused, ordered, and logical manner. Again, Western users may prefer to use a traditional menu, with a drill-down approach to navigating to site locations.

In Figure 4, the Ford U.S. and Chinese sites show differences, not in menu availability or figure-ground elements, but in particular color palette. The U.S. site (on the left) uses a mono-colored color system, but the Chinese site uses bright, high-key primary colors, primarily in the menus. Again, a contrast between the complexity and control of the U.S. site vs. the simplicity of imagery and context (environment) of the car in the Chinese site is evident.

Also, as illustrated in Figure 4, the design of the Web site on the right caters to holistic users with multiple links and the main menu being represented by a circle, perhaps going against the "conventional wisdom" of the graphic design of the West. At the same time, the English Web

Figure 2. Beijing University of Chemical Technology U.S. and Chinese sites

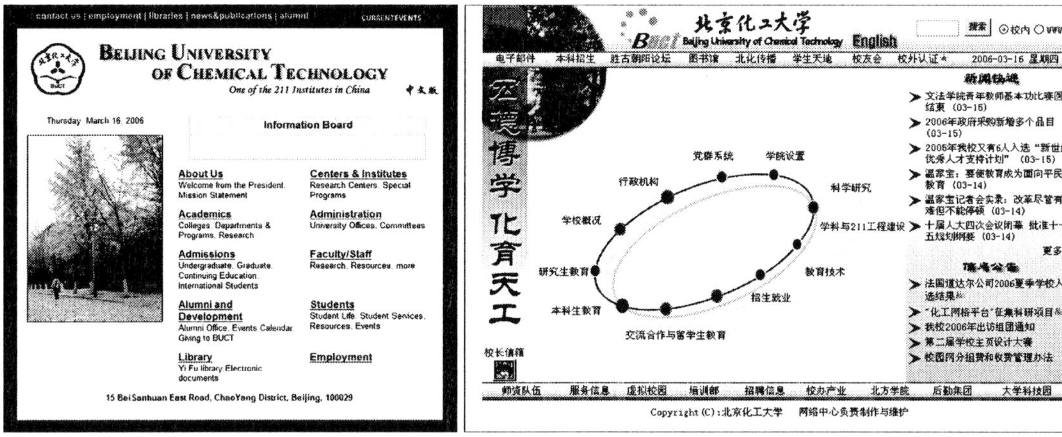

Figure 3. Nike U.S. and Chinese sites

Figure 4. Ford U.S. and Chinese sites

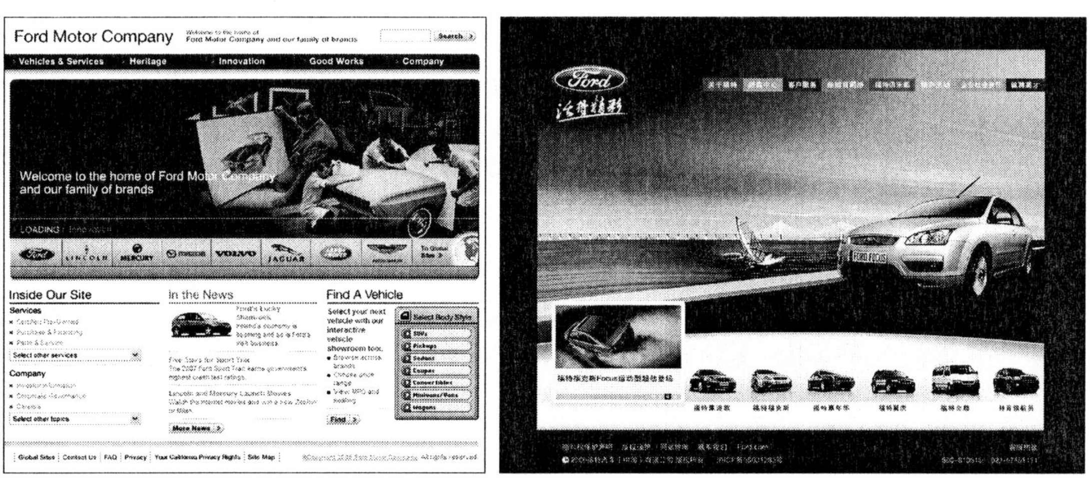

site is represented with a hierarchical design to better illustrate navigational paths, with their corresponding links.

In summary, Web site elements and their placement should be adjusted to the cognitive style of the users. The same will be true for consistency of navigational elements, that is, the proper design of menus and an information architecture that helps to identify the hierarchical order of a site. Some Web designers seem to intuitively understand the realities of these cognitive principles, as illustrated by Figures 3 to 5. For example, in Figure 2, the designers of the Beijing Technical University Web site have clearly chosen a hierarchy-driven, minimalist design for the interface presented to their English-speaking audience. At the same time, they choose a rich integrative approach for their Chinese-speaking audience, which provides a more global view of the site.

Overall, menus should allow users to freely move about within the site, but depending on how they are designed, the user will have a sense of either context and relationships or isolation and disorientation, depending on one's cognitive style. For example, general research suggests that users begin to lose their orientation with a hierarchical structure that is too deep. However, cultural style preferences should override past rules that may have been formed around Western views of information architecture. The need for instant access to various sections within the site will vary between cultures, depending on their varying cultural cognitive preferences for the interconnectedness of the site as a whole.

FUTURE TRENDS

Research that examines the influence of culture on Web design is in its inception, but it is increasing in scope, ambition, and quality (Burnett & Buerkle, 2004; Hillier, 2003; Zahedi et al., 2001). One application for this research would be online education for an international student population. In this case, Web designers will be responsible to work with educational technologists to provide distance education that is far more in tune with a broad range of cultures. To design and implement a learning environment that accommodates these cultures, they will need to understand the cognitive learning styles associated with them. The authors foresee that universities in the U.S. will take stronger measures to supplement their revenues by reaching out to a new breed of students who are living abroad and currently hold full-time positions in industry. These learners have a desire to attain an advanced degree from a U.S. university without having to disrupt their career development or uproot their families. As a result, educational technologists will investigate a broad range of convergent technologies to make portals for online education far more interactive and culturally sensitive, while maintaining the dynamics that take place in a real classroom.

For example, the faculty of the Department of Technology and Society at Stony Brook University has launched the first online executive program for Chinese MIS managers to establish a new hybrid graduate program at numerous universities in Taiwan. Currently, the program uses standard instruction techniques, with lectures being transmitted in real time to multiple campuses. However, students also have the option of reviewing an archived lecture at a later time. Instructors and students participate in the lectures using two-way voice and one-way video synchronous communication (chat), with asynchronous open forum group discussions and collaborative learning activities (Department of Technology and Society, 2006). Teng and Taveras (2005) note that the "objective of the hybrid-learning model was to help cultivate mutual dependency and cooperation among students and their instructor ... and the university" (p. 121). Programs such as this are using the most advanced technologies available, but will also need to consider the cultural cognitive design of graphic and voice user interfaces. For this, Web site developers of Web-based learning portals will need to respond to the implicit thinking patterns of cross-cultural learners.

CONCLUSION

E-commerce and educational activities are proliferating in an unprecedented way due to the Internet. As a result, more and more people are being ushered into a cross-cultural arena of global communication. Although theoretical advances have been made in psychology in the past half century, researchers have much more to explore in terms of the online behavior of cross-cultural users. The work of Hofstede (1991) and other behavioralists has helped to frame diverse cultural behavior into specific dimensions. These dimensions have provided standards of measurement for the empirical process, but lack the implicit knowledge necessary to understand the more profound and complex workings of the mind in the context of its native culture. Hofstede was well informed to note that machine software can be globalized, while the minds that use the machines cannot.

Arriving at a more comprehensive understanding of cognitive style is imperative to designing Web sites that are intended to effectively deliver content to a cross-cultural audience. Theories framing this notion were referred to as CCD theory, which suggests that particular cognitive styles of Web designers are contingent on their cultural background. The authors' intent was to show that the notions surrounding this theory can provide additional insight for Web design practitioners who play a key role in building online information for a global community of cross-cultural users. At the same time, we believe that new research opportunities will emerge that support better universal access for all international Web participants (Adams, 2004; Stephanidis & Savidis, 2001). However, more investigation by researchers in cross-cultural psychology and anthropology are also needed. As the scope of the Web continues to expand, researchers should devise better methods to measure the cognitive style of the Web site designers.

REFERENCES

Adams, R. (2004). Universal access through client-centred cognitive assessment and personality profiling. In C. Stary & C. Stephanidis (Eds.), *User-centred interaction paradigms for universal access in the information society* (pp. 3-15). Berlin: Springer-Verlag.

Armstrong, S. (1999). The influence of individual cognitive style on performance in management education. In J. Hill, S. Armstrong, M. Graff, S. Rayner, & E. Sadler-Smith (Eds.), *Proceedings of the 4th Annual Conference of the European Learning Styles Information Network* (pp. 31-50). Preston, UK: University of Central Lancashire.

Blanchard, A. (2004). Virtual behavior settings: An application of behavior setting theories to virtual communities. *Journal of Computer-Mediated Communication, 9*(2).

Bourges-Waldegg, P., & Scrivener, A. R. (1998). Meaning: The central issue in cross-cultural HCI design. *Interacting with Computers Special Issue, 9*(3), 287-309.

Brown, P. (1994). The INs and ONs of Tzeltal locative expressions: The semantics of static descriptions of location. In J. B. Haviland & S. C. Levinson (Eds.), Special issue of *Linguistics, 32,* 743-790. Berlin: Mouton de Gruyter.

Burnett, G., & Buerkle, H. (2004). Information exchange in virtual communities: A comparative study. *Journal of Computer-Mediated Communication, 9*(2), 451-459.

Chau, P. Y. K., Cole, M., Massey, A. P., Montoya-Weiss, M., & O'Keefe, R. M. (2002). Cultural differences in the online behavior of consumers. *Communications of the ACM, 45*(10).

Chen, S. Y., & Ford, N. J. (1998). Modeling user navigation behaviors in a hypermedia-based learning system: An individual differences approach. *International Journal of Knowledge Organization, 25*(3), 67-78.

Chen, S. Y., & Macredie, R. D. (2002). Cognitive styles and hypermedia navigation: Development of a learning model. *Journal of the American Society for Information Science and Technology, 53*(1).

Choi, S., McDonough, L., Bowerman, M., & Mandler, J. M. (1999). Early sensitivity to language-specific spatial categories in English and Korean. *Cognitive Development, 14,* 241-268.

Choong, Y. Y., & Salvendy, G. (1999). Implications for design of computer interfaces for Chinese users in mainland China. *International Journal of Human-Computer Interaction, 11*(1), 29-46.

D'Andrade, R. G. (1984). Cultural meaning systems. In R. A. Shweder & R. A. LeVine (Eds.), *Culture theory: Essays on mind, self, and emotion.* New York: Cambridge University Press.

Davies, I. R. L. (1997). Color cognition is more universal than color language. *Behavioral and Brain Sciences, 20,* 186-187.

Del Gado, E., & Nielsen, J. (Eds.). (1996). *International user interfaces.* New York: John Wiley & Sons.

Department of Technology and Society. (2006). *Stony Brook University, overseas program in Taiwan.* Retrieved March 15, 2006, from http://www.stonybrook.edu/est/academics/overseas.html

Edwards, P. N. (1994). From impact to social process: Computers in society and culture. In S. Jasanoff (Ed.), *Handbook of science and technology studies.* Beverly Hills, CA: Sage.

Ess, C., & Sudweeks, F. (1998). Computer-mediated communication or culturally-mediated computing? Challenging assumptions of the electronic global village. *The Electronic Journal of Communication, 8*(3-4).

Eveland, W. P., & Dunwoody, S, (2000). Examining information processing on the World Wide Web using think aloud protocols. *Media Psychology, 3,* 219-244.

Evers, V., & Day, D. (1997). The role of culture in interface acceptance. In S. Howard, J. H. & G. L. Lindegaard (Eds.), *Human computer interaction: INTERACT'97.* Sydney: Chapman and Hall.

Faiola, A., & Matei, S. (2005a). Cross-cultural computer-mediated communication and online learning: Assessing the impact of Web designer cognitive style on information design. In G. Salvendy & J. Jacko (Ed.), *Human-computer interaction—ergonomics and user interfaces, theory and practice: Volume 10. Internationalization, online communities and social computing: Design and evaluation. Proceedings of the 11th International Conference on Human-Computer Interaction,* Las Vegas, NV. Mahwah, NJ: Lawrence Erlbaum.

Faiola, A., & Matei, S. (2005b). Cultural cognitive style and Web design: Beyond a behavioral inquiry of computer-mediated communication. In C. Ess & F. Sudweeks (Eds.), *Culture and computer-mediated communication: Toward new understandings* [electronic version]. *Journal of Computer-Mediated Communication, 11*(1). Retrieved from http://jcmc.indiana.edu/vol11/issue1/faiola.html

Faiola, A. (2002). A visualization pilot study for hypermedia: Developing cross-cultural user profiles for new media interfaces. *The Journal of Educational Multimedia and Hypermedia, 11*(1), 51-71.

Farnen, R.F. (1993). Cross-national perspectives on cognitive studies, problem solving, and decision making. In R.F. Farnen (Ed.), *Reconceptualizing politics, socialization, and education: International perspectives for the 21st century* (pp. 376-460). Oldenburg: Library and Information Systems, University of Oldenburg.

Ford, N., Wood, F., & Walsh, C. (1994). Cognitive styles and searching. *Online and CDROM Review, 18*(2), 79-86.

Goldstein, K. M., & Blackman, S. (1978). *Cognitive style.* New York: John Wiley & Sons.

Graff, M., Davies, J., & McNorton, M. (2004). Cognitive style and cross-cultural differences in Internet use and computer attitudes. *European Journal of Open and Distance Learning, 2.* Retrieved from http://www.eurodl.org/materials/contrib/2004/Graff_Davies_McNorton.html

Greene, J. O. (1995). An action-assembly perspective on verbal and nonverbal message production: A dancer's message unveiled. In D.E. Hewes (Ed.), *The cognitive bases of interpersonal communication* (pp. 51 -85). Hillsdale, NJ: Lawrence Erlbaum.

Gudykunst, W. B., Ting-Toomey, S., & Chua, E. (1988). *Culture and interpersonal communication.* Newbury Park, CA: Sage.

Gumperz, J. J., & Levinson, S. C. (1997). Rethinking linguistic relativity. *Current Anthropology, 32,* 613-623.

Hara, N., & Kling, R. (1999). Students' frustrations with a Web-based distance education course. *First Monday, 4*(12). Retrieved August 1, 2005, from http://www.firstmonday.dk/issues/issue4_12/hara

Herring, S. C. (Ed.). (1996). *Computer-mediated communication: Linguistic, social, and cross-cultural perspectives.* Philadelphia: John Benjamins.

Hillier, M. (2003). The role of cultural context in multilingual Web site usability. *Electronic Commerce Research and Applications, 2*(1), 2-14.

Hofstede, G., & McCrae, R. R. (2004). Personality and culture revisited: Linking traits and dimensions of culture. *Journal of Cross-Cultural Research, 38*(1), 52-88.

Hofstede, G. (1991). *Cultures and organizations: Software of the mind.* London: McGraw-Hill.

Honold, P. (2000). Cultural and context: An empirical study for the development of a framework for the elicitation of cultural influence in product usage. *The International Journal of Human-Computer Interaction, 12,* 327-345.

Houston, M. J., & Eckhardt, G. M. (2001). Culture's consequences on consumer behaviour toward food in Asia. *Asian Journal of Marketing, 8*(2), 33-54.

Ji, L., Peng, K., & Nisbett, R. (2000). Culture, control and perception of relations in environment. *Journal of Personality and Social Psychology, 78,* 943-955.

Kim, K. S., & Allen, B. (2002). Cognitive and task influences on Web searching behavior. *Journal of the American Society for Information Science and Technology, 2,* 109-119.

Kitayama, S., Duffy, S., Kawamura, T., & Larsen, J. T. (2003). A cultural look at New Look: Perceiving an object and its context in two cultures. *Psychological Science, 14,* 201-206.

Levinson, S. C. (1996). Language and space. *Annual* review *of Anthropology, 25,* 353-382.

Liu, Y., Lin, F., & Wang, X. (2003). Education practice and analyzing behaviour of students in a Web-based learning environment: An exploratory study from China. *Online Information Review, 27*(2), 110-119.

Lucy, J. A. (1992). *Grammatical categories and cognition: A case study of the linguistic relativity hypothesis.* New York: Cambridge University Press.

Marcus, A. (2000). International and intercultural user-interface design. In C. Stephanidis (Ed.), *User interfaces for all* (pp. 47-64). Hillsdale NJ: Lawrence Erlbaum.

Miyamoto, Y., & Kitayama, S. (2002). Cultural variation in correspondence bias: The critical role of attitude diagnosticity of socially constrained behavior. *Journal of Personality and Social Psychology, 83,* 1239-1248.

Mishra, R. C. (1996). Cognition and cognitive development. In J. W. Berry, Y. H. Poortinga, & J. Pandey (Series Eds.) & J. W. Berry, P. R. Dasen, & T. S. Saraswathi (Vol. Eds.), *Handbook of cross-cultural psychology: Volume 2. Basic processes and human development* (2nd ed., pp. 1-406). Boston: Allyn and Bacon.

Nisbett, R. E., & Norenzayan, A. (2002). Culture and cognition. In H. Pashler & D. L. Medin (Eds.), *Stevens' handbook of experimental psychology: Volume 2: Cognition* (3rd ed., pp. 561-597). New York: John Wiley & Sons.

Nisbett, R. E., Peng, K., Choi, I., & Norenzayan, A. (2001). Culture and systems of thought: Holistic vs. analytic cognition. *Psychological Review, 8,* 291-310.

Parkinson, A., & Redmond, J. (2002). Do cognitive styles affect learning performance in different computer media? In *Proceedings of the 7th Annual Conference on Innovation and Technology in Computer Science Education* (pp. 39-43), Aarhus, Denmark. New York: ACM.

Preece, J. (1998). Empathic communities: Reaching out across the Web. *Interactions Magazine, 2*(2), 32-43.

Preece, J. (2001). *Online communities: Designing usability and supporting sociability.* New York: John Wiley & Sons.

Preece, J., Abras, C., & Maloney-Krichmar, D. (2004). Designing and evaluating online communities: Research speaks to emerging practice. *International Journal of Web-Based Communities, 1*(1), 2-18.

Riding, R., & Rayner, S. G. (1998). *Cognitive styles and learning strategies.* London: David Fulton.

Roberson, D. (2005). Color categories are culturally diverse in cognition as well as in language. *Journal of Cross-Cultural Research, 39*(1), 56-71.

Sears, A., Jacko, J., & Dubach, E.M. (2000). International aspects of World Wide Web usability and the role of high-end graphical enhancements. *International Journal of Human-Computer Interaction, 12*(2), 241-261.

Sheridan, E.F. (2001). Cross-cultural Web site design. *Multilingual Computing and Technology, 12*(7).

Shweder, R. (1990). Cultural psychology: What is it? In J. Stigler, R. Shweder, & G. Herdt (Eds.), *Cultural psychology: Essays on comparative human development* (pp. 1-43). New York: Cambridge University Press.

Simon, S. J. (2000). Research contributions: The impact of culture and gender on Web sites: An empirical study. *The DATABASE for Advances in Information Systems, 32*(1), 18-37.

Stephanidis, C., & Savidis, A. (2001). Universal access in the information society: Methods, tools and interaction technologies. *Universal Access in the Information Society, 1*(1), 40-55.

Teng, T. L., & Taveras, M. (2005). Combining live video and audio broadcasting, synchronous chat, and asynchronous open forum discussions in distance education. *Journal of Educational Technology Systems, 33*(2), 121-130.

U.S. Department of Commerce. (n.d.). *U.S. Census Bureau News: Quarterly retail e-commerce sales report for 2005.* Retrieved March 1, 2006, from http://www.census.gov/mrts/www/data/html/05Q4.html

Vishwanath, A. (2003). Comparing online information effects: A cross-cultural comparison of online information and uncertainty avoidance. *Communication Research, 30*(6), 579-598.

Vygotsky, L. S. (1979). *Mind in society: The development of higher psychological processes.* Cambridge, MA: Harvard University Press. [Translated from the original 1934 text.]

Vygotsky, L. S. (1989). *Thought and language.* Cambridge, MA: MIT. [Translated from the original 1932 text.]

Whorf, B. (1957). On the connection of ideas. In J.B. Carroll (Ed.), *Language, thought, and reality: Selected writing of Benjamin Lee Whorf.* Cambridge, MA: MIT Press.

WorldLingo. (2000). *Communicate in your customer's language.* Retrieved July 25, 2006, from https://worldlingo.com/en/company/pr/pr20000625_01.html

Wood, F., Ford, N., Miller, D., Sobczyk, G., & Duffin, R. (1996). Information skills, searching behavior and cognitive styles for student-centered learning: A computer-assisted learning approach. *Journal of Information Science, 22*(2), 79-92.

Zahedi, F., van Pelt, W. V., & Song, J. (2001). A conceptual framework for international Web design. *IEEE Transactions on Professional Communication, 44*(2), 83-103.

Zebian, S., & Denny, J. P. (2001). Integrative cognitive style in Middle Eastern and Western groups: Multidimensional classification and major and minor property sorting. *Journal of Cross-Cultural Psychology, 32*(1), 58-75.

Zhang, S., Schmitt, B. H., & Haley, H. (2003). Language and culture: Linguistic effects on consumer behavior in international marketing research. In S. C. Jain (Ed.), *Handbook of research in international marketing* (pp. 228-242). Northampton, MA: Edward Elgar.

ENDNOTES

1 After a period of relative obscurity, the Whorfian tradition has been recently revived. Lucy (1992), for example, has observed how linguistic differences in number-marking patterns affect thought among the Yucatee Maya, Chinese, Japanese, and English. He concluded that the Whorfian hypothesis can be supported since "the specific linguistic patterns of Yucatec and English corresponded with observable patterns of individual cognitive performance" (p. 156).

2 All differences were statistically significant with the Chinese at $p < .000$ (Chi-square= 18.000), the Russians at $p < .000$ (Chi-square= 18.000), and the North Americans at $p < .027$ (Chi-square= 7.200). The eight-item scales had high reliability scores, with Cronbach's alpha values > .9).

Chapter XI
Global Internet Usage, Web Site Design, and Cultural Communication Preferences:
Contributions from Cross-Cultural Marketing and Advertising Research

Marc Hermeking
Ludwig-Maximilians University, Munich, Germany

ABSTRACT

Though the Internet continuously gains much popularity on a global scale, marketing research reveals enormous national differences and disparities in the numbers of Internet users worldwide. To date, most users still live in North America, in the Asia Pacific Basin, and in Western Europe. In response, an increasing body of research on global Internet usage, interface design, and Web site usability has been undertaken by human-computer-interaction (HCI) and localization specialists during the past few years. Similarly, questions of global standardization vs. local adaptation are also central to international marketing, research on consumer behavior, and marketing communications. Since the Internet, including its most popular usage platform, the World Wide Web (WWW), is part of the media, its culture-specific potential and conditions for global usage—and respectively its culturally appropriate design for worldwide consumption—can be described in general by contributions from cross-cultural marketing and advertising research. This chapter presents a foundational examination of this situation and its implications for professional practices related to marketing and online consumers.

INTRODUCTION

An increasing body of research on global Internet usage, interface design, and Web site usability has been undertaken by human-computer interaction (HCI) and localization specialists during the past few years, trying to find out if people worldwide will be prepared to use the Internet equally or if cultural and other kinds of adaptation or localization will be necessary. This question of global standardization vs. local adaptation is also central to the discipline of international marketing, which includes, for example, research on consumer behavior, appropriate product design, and marketing communications. Some international marketers have predicted a final convergence of culturally different markets into a "one-world culture," which would facilitate globally standardized marketing activities (Levitt, 1983). This, however, has turned out to be an illusion—too many non-cultural hard factors and cultural soft factors still exist or arise as constraints on international marketing, so there is no real 100% standardization at all. This even applies to so-called "culture-free" products like industrial goods and high-tech digital products or information and communication technologies (ICTs) (de Mooij, 1998; Hermeking, 2000). Since the Internet, including its most popular usage platform, the World Wide Web (WWW), is part of the media, its culture-specific potential and conditions for global usage—and respectively its culturally appropriate design for worldwide consumption—can be described in general by contributions from cross-cultural marketing and advertising research.

GLOBAL INTERNET USAGE

The Worldwide Disparity

Though the Internet continuously gains much popularity on a global scale, marketing research reveals enormous national differences and disparities in the numbers of Internet users worldwide. To date, most users still live in North America, in the Asia Pacific Basin, and in Western Europe. One crucial point, however, is the definition of Internet usage. Online access quotas, which are the most commonly cited databases, are no valid indicators of any real online activity or usage. Moreover, even daily online activities may only include a retrieval or sending of e-mail messages lasting just a few seconds, especially at the workplace. A more valid indicator for Internet usage is defined by "active" Web use, which happens nearly every day, takes about one hour at least, and is typically done at home in order to meet individual needs online.

The percentage of the population using the Internet as active Web users at home in early 2005 was, for example, in the United States—48%, in Australia—46%, in Sweden—53%, in Germany—36%, in the UK—38%, in France—26%, in Spain—22%, in Japan—29%, and in Brazil—only 6%. In early 2006, this percentage was: in the United States—48%, in Australia—49%, in Sweden—53%, in Germany—39%, in the UK—40%, in France—28%, in Spain—27%, in Japan—29%, and in Brazil—still only 6% (Nielsen NetRatings, n.d.).

Some of these countries show a little increase in Web usage; others, however, seem to remain on their relatively high or very low usage level. Although those figures change continuously over time, there is a clear continuum of descent from high Internet usage in the (developed, western) North to low Internet usage in the (often less-developed, non-western) South—also known as the *digital divide*. Economic development, like technical infrastructure or GDP per capita, is only one (hard) factor. A close look at highly developed countries like France or Japan, for example, with their relatively moderate Web usage of the referenced kind, reveals that some cultural values as soft factors obviously play an even more important role.

Cultural Values as Key Factors

Culture is a complex phenomenon. As such, it is not easy to define without any simplifications, therefore hundreds of definitions of culture exist; Müller and Gelbrich (2004), for example, report more than 240 different definitions. Cultural values can be described here as collectively shared standards of orientation regarding the preferred mode of thinking, perceiving, and behaving within a specific group, society, environment, or region. Often these values result in unconscious specific mentalities and frames of reference which distinguish members of one group or culture from another in many diverse ways. In order to compare cultures, it is comfortable to choose abstract and universal value categories which are relevant to all cultures, but at different degrees.

The most popular of such categories, especially in the discipline of international marketing, are Hofstede's (originally) four "cultural dimensions": individualism (vs. collectivism), power distance, masculinity (vs. femininity), and uncertainty avoidance (Hofstede, 1991, 2001). These predominantly relate to universal aspects of social relationships (self vs. group orientation, hierarchy, gender roles, trust/risk-taking) and resulted from factorial analysis of former data of work-related values of international personnel in a multinational company (IBM) in the late 1960s. Hofstede's empirical basis with more than 116,000 questionnaires is enormously large. The possibly obsolete databases and generalizations related to people outside IBM are frequent criticisms that Hofstede continuously counters. Several replications of Hofstede's study on different samples have proved that his data are still valid (Hofstede, 2001; de Mooij, 2004). Müller and Gelbrich (2004), however, have found some statistical deficiencies like redundancies and internal correlations between the dimensions; consequently, Hofstede's model includes only three independent dimensions (individualism, masculinity, uncertainty avoidance).

These dimensions relate to complete nations only. Relational differences between (originally 40, now more than 60) national cultures are shown in a quantitative way, using index scores for those nations along each cultural dimension. Like aggregated "proxies," these index scores represent degrees of cultural values at which they are most frequently shared by individuals in a country, representing the statistical modus of such values (de Mooij, 2004). Many criticisms on such a strongly generalizing national-cultural concept exist (Holden, 2004). This approach, however, proves to be very operational for international and mass marketing especially, which often, in an equally generalizing way, deals with aggregated units like national markets and segmented clusters of thousands or even millions of individuals as one specific, homogeneous target group (Hermeking, 2005).

Universal aspects of social relationships represented by Hofstede's dimensions may also apply well to international marketing communication,

Figure 1. Internet usage in different nations, 2001 (Source: Nielsen NetRatings, n.d.)

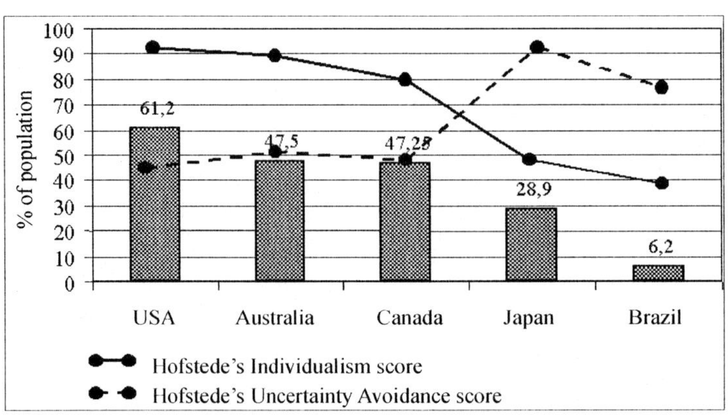

which in general has to create relationships between the supply and the demand sides, which are culturally different. The most central element of marketing is communication; individualism and uncertainty avoidance in particular have been proven to be relevant for communication processes in general and for intercultural communication in particular (Gudykunst, 2001; Schugk, 2004).

Utilizing Hofstede's model, the influence of cultural values on Internet usage can be visualized by the significant correlation of many countries' usage data with the country scores of the two cultural dimensions, individualism and uncertainty avoidance. This is shown in Figure 1 with Nielsen's total Internet usage data ("all users") of just five countries in 2001.

The general tendency of a decrease of Internet usage ("all users") from the North to the South, as well as a positive correlation with individualism and a negative correlation with uncertainty avoidance, is also evident in the majority of the highly developed countries of Western Europe. This is demonstrated in part in Figure 2.

Like all artifacts, the Internet is a specific cultural product which originates from a specific cultural environment. Accordingly, the diffusion of the Internet from its country of origin, the USA, to other countries of the world seems to depend—among other factors—on their cultural similarity to the United States, specifically with respect to cultural values of strong individualism and low uncertainty avoidance. Of course, the early dominance of the English language on the Internet, due to its origin in the U.S., and the capabilities in the English language in other countries are also relevant. Proficiency in the English language and its acceptance as a foreign language, however, may also correlate with low uncertainty avoidance.

Many marketing researchers have demonstrated that countries with low uncertainty avoidance are more open to innovations in general (Steenkamp, ter Hofstede, & Wedel, 1999) as well as the new medium Internet in particular (la Ferle, Edwards, & Yutaka, 2002; Yeniyurt & Townsend, 2003; Müller & Gelbrich, 2004; de Mooij, 2004); that is, these cultures tend to be early Internet adopters with a high diffusion rate.

With respect to increasing habituation to and language localization of the Internet, one could argue that it is only a matter of time until all high uncertainty avoidance cultures will catch up and also show high usage rates. This, however, does not seem to apply to many of them because some additional cultural influences on their Internet adoption, including individualism and other cultural values, continue to play important roles as well.

Figure 2. Internet usage in European nations, 2001 (Source: Nielsen NetRatings, n.d.)

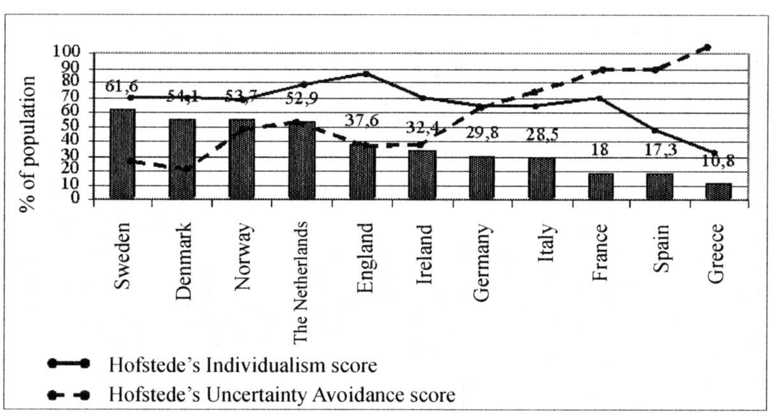

Individualism, especially, has been revealed by some marketing researchers as the most significant cultural background of global Internet usage (la Ferle et al., 2002). Individualistic cultures' high attraction to the Internet is often attributed to the very egalitarian, democratic nature of this medium (de Mooij, 2004). Additionally, the Internet in general is a relatively impersonal medium of communication between individuals, which is also closely connected with the very essential and product-specific aspect of cultural communication preferences.

Cultural Communication Preferences

Among universal cultural categories specifically related to communication, Hall's concept of "high-context" vs. "low-context" (Hall, 1977; Hall & Hall, 1990) is the most popular one. Hofstede (1991) also refers to Hall's concept of context and concludes that high individualism is very often connected to low-context communication, whereas low individualism (collectivism) very often is connected to high-context communication. According to Hall, members of low-context cultures have less personal contact with each other, therefore communication must be very detailed and very explicit. A large amount of formal information is communicated in a direct way, often by way of written texts. In contrast, members of high-context cultures have closer and more familiar contacts with each other; a lot of information is already shared among them, and therefore their preferred mode of communication is more informal, indirect, and often based merely on symbols or pictures.

Hall's model of cultural categories, of which only four (time, space, information flow, and context) gained larger popularity, is explicitly focused on the interdependence of culture and communication, and has proved to be useful in communications research.

Like Hofstede's model, Hall's model is influenced by Kluckhohn's work on basic value orientations and the theories of Functionalism (Rogers & Hart, 1998), according to which cultures are characterized by mostly unconscious and collectively shared frames of reference, "hidden dimensions," or "mental software," in order to solve universal problems of their human existence in different, culture-specific ways. Such a concept of culture is attractive to the discipline of marketing, which arranges for the supply of products and services that have to meet specific

Figure 3. Cultural communication styles according to Hall (Modified from Usunier, 1993)

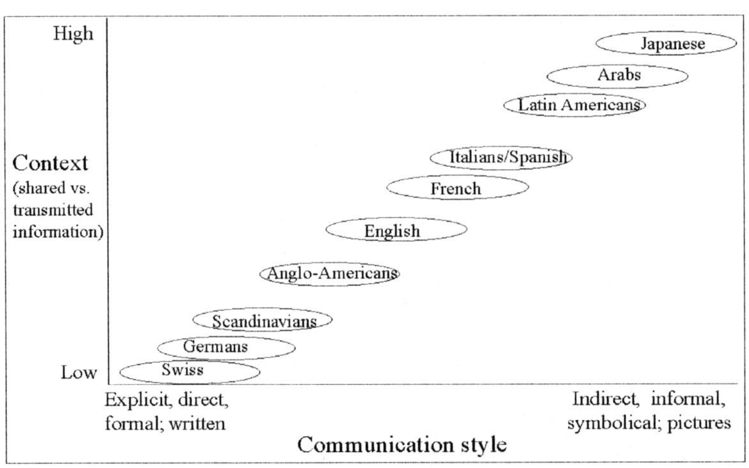

needs and demands (i.e., problems and values) of customers, target groups, or markets.

Although Hall's hermeneutical approach, which is based on anthropology, ethnography, and real-life experiences, is less criticized than Hofstede's quantified approach, it is also less operational for marketers: only a few countries or specific populations inside of them are mentioned as ideal examples; no country ratings of context exist. However, a vague ordinal positioning of some cultures with respect to their preferred communication style was made by Hall and later adapted by Usunier (1993) and other marketing researchers (see Figure 3).

According to Hall's context model and the globally different Internet usage rates, the new medium in general is more preferred in low-context cultures than in high-context cultures—with few exceptions, of course. This tendency could be explained by an obviously very information-explicit and text-heavy communication style of the Internet, which originally was developed for informative and scientific reasons in the (low-context) United States. Differences in Internet adoption among different low-context cultures over time, of course, may also depend on their different uncertainty avoidance.

This obvious impact of cultural communication preferences on global Internet usage shows interesting analogies with the global consumption pattern of the traditional media.

Analogies with Traditional Media Consumption

There are enormous and persistent disparities in worldwide consumption of traditional media such as newspapers and TV across different countries, which may well be interpreted as a result of culture-specific communication preferences. By analogy with global Internet usage, the consumption of print media shows a general decline from the (developed, western) North to the (often less-developed, non-western) South, whereas TV consumption (watching minutes per day) reveals just the opposite tendency. Various hard factors like political, economic, or technical conditions also have influence on this pattern of media consumption, but in highly developed countries, at least, they seem to be less significant.

For example, the average consumption rate of daily newspapers among the population in Western Europe between the early 1950s and the late 1990s reveals stable differences, with (low-context) Norway, Finland, Sweden, the UK, and Switzerland showing the highest consumption rates (40-65%); followed by Denmark, Germany, and Austria also showing high consumption rates (30-40%); France, Belgium, and Ireland showing only moderate consumption rates (15-30%); and finally (high-context) Portugal, Spain, and Italy showing lowest consumption rates (5-15%) (de Mooij, 2004). Newer media data continuously confirm such differences in high newspaper vs. low TV consumption rates and vice versa (Anonymous, 2005).

Equal to the different preferences in media consumption, national total advertising expenditures by media also differ. For example, total advertising expenditures for TV in (high-context) Brazil, Peru, Colombia, the Philippines, or Japan are much higher than for print media, whereas total advertising expenditures for print media in (low-context) Denmark, Germany, the Netherlands, Canada, or Sweden are much higher than for TV (Terpstra & Russow, 2000).

In general, the global disparities in traditional media consumption reveal a consistent tendency of a higher popularity of print media in low-context cultures, in contrast to a higher popularity of TV in high-context cultures. This quite evidently corresponds with the culturally preferred communication styles, as TV is mostly based on pictures, whereas print media are more or less based on written text. In addition, most newspapers are read by one individual at a time, whereas TV may be watched and listened to by a collective group.

Exceptions to this very general tendency are some highly developed East-Asian (high-context) countries like Japan, Hong Kong, and South Korea, where print media are very popular as well. This phenomenon may correspond with the nature of their written language, which is more symbolic or picture like (Usunier, 1993). Other values and phenomena of these "Confucian" cultures may also correspond to their specific two-fold media preferences (Hermeking, 2005).

In sum, global Internet usage reveals strong analogies with global print media consumption, which is in accordance with cultural communication preferences.

Implications for the Internet as a Mass Medium

The analogies with print media consumption indicate the fact that, to date, the nature of the Internet has been more similar to print media, based mainly on a text-heavy, explicit, and informational communication style, which is therefore accepted most of all in low-context cultures like the culture of origin, the United States.

Indeed, marketing research on frequent Internet users all over the world showed a continued dominance of informational needs over entertainment needs in past years (Anonymous, 2000a). Internet users in the UK, for example, are motivated primarily by information needs and also prefer informational to entertaining TV programs (Hammond, Turner, & Bain, 2000). Internet users in Germany, for example, read books and daily newspapers more often than non-users (Anonymous, 2000b). Even studies on the recent increase in Internet usage in South Korea reveal that Internet users watch TV less often, but spend more hours reading books than non-users (Rhee & Kim, 2004).

Because of their dominant information needs, Internet users worldwide may be regarded as a sub-cultural, low-context "information elite." These elites are, in terms of culturally preferred communication styles, in general proportionally larger in low-context cultures and smaller in high-context cultures (exceptions might be expected in some East Asian countries).

To broaden the basis of Internet usage in the UK, for example, Hammond et al. (2000) conclude that the Internet may need to become less information but more entertainment oriented and more comfortable to use. With respect to global Internet usage and the digital divide, this means that the communication style of the Internet should be appropriate and acceptable not only to users from low-context cultures, but also to users from high-context cultures and their specific communication preferences. This, of course, has implications for Internet communications and Web site design.

APPROPRIATE WEB SITE DESIGN

Since the WWW is the most popular usage platform or user interface of the Internet, its interface design or Web site quality in particular is of high relevance for the cultural appropriateness and acceptability of the Internet as a medium of communication. A recent study on online consumers from Germany, Brazil, and Taiwan, for example, revealed that cultural adaptation on Web sites is an important determinant of perceived ease of use and attitude toward the sites, which results in stronger behavioral intentions to use the sites (Singh, Fassott, Chao, & Hoffmann, 2006).

By analogy with product design, Web site design can be described as a specific set of instrumental or technical, economic, social, aesthetic, and symbolic attributes or qualities of a site which have to contribute to its users' satisfaction, which in turn depends on the users' cultural habits and values (Hermeking, 2000). In sum, a culturally well-designed Web site may be defined as communicating the right information at the right place with the right layout in the right manner and in the right time according to the culture of each of its users.

Still many executives who are responsible for corporate communications and Web site design, and many technical Internet experts especially, expect their sites to be used worldwide with the same preferences and in the same way as in their own culture. This is generally not the case, however, since also those who are responsible for Web site design are influenced by specific cultural values, attitudes, communication preferences, and habits of their own. Because of this self-reference criterion, the WWW as a product originating from a low-context culture in sum may still reveal a significant lack of adaptation and appropriateness with respect to many high-context cultures, which may be mirrored by the global usage disparities or the digital divide.

Most so-called global target groups, like young adults, are a myth, since even with identical consumption data (e.g., ownership of products), their consumption patterns (e.g., buying motives, forms of usage) generally do differ depending on country and/or culture (de Mooij, 2004). For example, studies on U.S.-American and South Korean students revealed that both groups use the Internet frequently, with an average usage time of nearly two hours per day. However, the importance of this medium as a source of information among other media differs (i.e., magazines rank first in the United States, TV ranks first in Korea) (Choi & la Ferle, 2004). In addition, these Korean students, in contrast to their U.S. counterparts, might well use the Internet primarily for participation in online communities, but less for e-commerce (Park & Jun, 2003).

Consequently, like all product developers, Web site designers should distinguish between the "internal qualities" of a site (i.e., site attributes perceived and appreciated from their own frame of reference or point of view) and the "external qualities" (i.e., site attributes perceived and appreciated from a foreign user's point of view), which, according to their different cultural values and preferences, may differ enormously.

Basic Design Criteria

Often the general criteria— (1) site quality, which is also equated with usability; (2) establishment of trust; and (3) creation of positive affection during site use—are quoted as the most essential Web site characteristics to which designers should answer.

Regarding the establishment of impersonal, abstract trust on the WWW, Morrison and Firmstone (2000) refer to the general phenomenon of a high trustworthiness of things that are familiar, expected, and conform to habits, in contrast to a low trustworthiness of things that are unfamiliar, strange, and unexpected. This phenomenon can be carried over to the forms of communication via the Web. The more the design of a site conforms to culturally familiar communication styles and cultural habits, the more trust is established. This aspect is equivalent to the "not-invented-here-syndrome" or "country-of-origin" (COO) effect well known in international marketing, which may result in a generally more positive image of products or services from the home country than of those from foreign countries (Usunier, 1993; Knight & Calantone, 2000). Accordingly, even global brands are often marketed and advertised as if they were large local brands (on which most locals trust). Trust is another essential argument in support of culture-specific Web site design, which also may increase general Internet acceptability.

Positive affection depends on a user's experience both of successful site use, which is subject to culture-specific definitions of site usability, and of a culturally appropriate or expected ratio of emotional vs. informational or rational messages communicated. In sum, these very general design criteria are subject to cultural communication styles, but they also depend strongly on each other. Accordingly, some more operational, culture-related design criteria need to be introduced.

Contributions from Advertising Research

In promoting people or organizations and their products, services, or ideas, communication via the Web is generally equivalent to marketing communication like advertising or public relations. In addition, it also combines several basic elements of traditional communication via newspaper, radio, TV, telephone, or direct mail in a specific way (Ju-Pak, 1999). Consequently, both the standardization/adaptation discussion and central criteria of cross-cultural advertising research on culture-specific advertising styles are relevant to Web site design.

A significant body of research on international advertising has shown some differences in preferred advertising styles and creative strategies, in particular between high-context and low-context cultures: in general, indirect and transformational advertising messages creating emotions through pictures and entertainment are more favored in high-context cultures like France or Japan, for example, whereas direct and rational advertising messages providing product information above all play a more important role in low-context cultures such as Germany or many parts of the United States (Mueller, 1996, 2004; de Mooij, 1998).

The following examples of cross-cultural Web site research demonstrate the relevance of these general results from advertising research for the design of international sites.

A Comparison of Content Appeal and Layout

Early evidence for the analogy of Web sites with international advertising is given in Ju-Pak's (1999) analysis of 310 commercial Web sites from the United States, the UK, and South Korea. Her results reveal some remarkable national differences in preferred creative strategies and, in particular, in regard to verbal and visual content appeal. Rational, fact-based appeals (especially visual) are dominant in the U.S. and the UK, whereas emotional, symbolic appeals (both visual and verbal) are dominant in South Korea. The UK shows the highest rate of text-heavy layout, whereas South Korea shows the highest rate of text-limited, visual layout. These differences are equivalent to the preferred advertising styles of low-context cultures (here: UK, U.S.) and high-context cultures (here: South Korea).

This very general relation between content, layout, and the culturally preferred communication style best suited for both advertising and Web site design is demonstrated in Figure 4.

Figure 4. General relation between creative strategy and communication styles

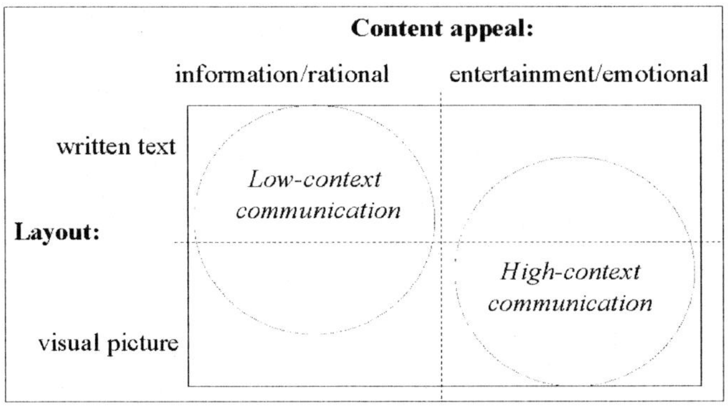

With respect to the creative strategies chosen, messages of both traditional media (print, TV) and new media (Web sites) can generally be adapted in varying degrees to either high-context or low-context communication preferences. Because of its visual and acoustic nature, however, to date TV is the most appropriate medium for the transformational or entertaining communication highly preferred in high-context cultures, whereas print media based on written texts are more appropriate for rational, informative communication preferred in low-context cultures. To date, the nature of the Web overall seems to be more similar to print media; Ju-Pak's (1999) results, however, indicated quite early that culturally adapted creative strategies in Web site design can contribute to making this medium more appropriate to the communication preferences of high-context cultures.

Some additional design characteristics analyzed by Ju-Pak (1999) reveal cultural differences that are not unexpected. The length of each single Web page seems to correspond to copy format, as the UK leads both in text-heavy layout and in shorter pages, whereas South Korea leads both in visual layout and in longer pages. South Korea utilizes much more (high-context type) multimedia presentation (text, sound, and/or video) than the United States and the UK, where presentation more often is based on (low-context type) text only. Finally, according to Ju-Pak's results, the degree of interactivity by number of hyperlinks or clickable items is higher in (high-context) South Korea.

A Comparison of Hard-Sell and Soft-Sell Appeals

For their comparison of 150 Web sites of Japanese multinational companies in Japan, Spain, and the United States, Okasaki and Alonso (2003) utilized a classification of advertising appeals into "soft-sell" and "hard-sell" appeals. Soft-sell appeals are characterized by indirect approaches and a primary emphasis on creating emotions, a mood, or an atmosphere by visuals and symbols; in contrast, hard-sell appeals are characterized by direct approaches highlighting product merits with explicit information, competitive persuasion, and comparative claims. In general, the former advertising appeals prevail in high-context cultures like Japan or South Korea, whereas the latter prevail in low-context cultures like the United States (Mueller, 1996, 2004). Accordingly, the Japanese sites analyzed revealed the highest rate of soft-sell appeals, whereas the U.S. sites revealed the highest rate of hard-sell appeals and vice versa; the Spanish sites revealed a nearly equal rate of both appeals (Okasaki & Alonso, 2003).

The primary emphasis on emotional stimulation vs. rational information has been described similarly by the hierarchy of advertising effects (formerly known as the classic formula AIDA: attention-interest-desire-action), which equally differs by cultural communication preferences. In low-context cultures, most products are advertised with primary emphasis on cognitive, rational information ("learn"), so the hierarchy is "learn-feel-do"; in high-context cultures, most products are advertised with primary emphasis on affective emotions ("feel"), and the hierarchy is "feel-do-learn." This general tendency, however, is weakened by the category of products: advertising for "high-involvement" products like cars or other expensive technical, durable consumer goods with higher perceived risk often puts stronger emphasis on rational information ("learn"), whereas advertising for non-durable, "low-involvement" consumer products like detergents or food often puts stronger emphasis on emotions ("feel") (de Mooij, 1998).

The majority of products included in Okasaki's and Alonso's (2003) Web site research are digital high-tech and high-involvement products. The same holds true for Ju-Pak's (1999) research, which explicitly points out that the national differences in Web sites revealed in her study are not due to differences in products featured. Nevertheless, some further differentiation into both brand and the product type represented is necessary in general, since the familiarity with a certain brand or type of product, its complexity or required know-how, and its relevance for consumer (B2C) or business (B2B) interests may influence the demand for explicit information as well as the time budgeted for a site use, and may therefore influence the creative strategy for Web site design.

Structural Design Criteria

In order to introduce operational Web site design criteria and in order to verify the cultural tendencies discussed above, Web sites of different international companies and brand types in various European countries (Germany, the UK, France, Italy, and Spain), as well as the United States and Japan, have been compared in a long-term, qualitative study along the following eight structural criteria, which proved to be culturally relevant following low- vs. high-context communication preferences (Hermeking, 2004, 2005):

1. Content appeal
2. Layout
3. Length of pages
4. Multimedia presentation
5. Interactivity
6. Structure of content
7. Total volume of Web site
8. Degree of navigation support

Low-Context Web Sites

Web sites representing a strong low-context style are dominated by rational content appeals, text-heavy layout presenting small pictures only, missing multimedia presentation, high interactivity with respect to many internal links, large total site volume, and deeply structured content accompanied by intensive navigation support.

According to Hall and Hall (1990), low-context communication often correlates with a "monochronic" cultural time orientation: in monochronic cultures, time is linear, passing ("time is money"), handled precisely, and divided in an orderly fashion. Thus action chains generally are compartmentalized and sequential; one thing is completed after the other. Correspondingly, low-context-style Web sites that carry a great deal of explicit informational content are detailed, compartmentalized, structured in an orderly way, and also supported by many explicit navigation tools such as sitemaps, index registers, and search engines to facilitate a quick orientation and a time-saving site use. In addition, missing multimedia presentation and a text-heavy layout of the naturally voluminous Web sites reduce download times and correspond to a monochronic time orientation.

High-Context Web Sites

Web sites that are adapted to high-context style reveal more transformational content appeals with text-limited layouts including more colored backgrounds, larger pictures, and a much higher rate of animated illustrations or moving visuals in particular. Multimedia presentation tends to be higher, including sound effects, jingles, or occasional options for downloads of video or radio interviews, for example. Corresponding to the visual-heavy layout, navigation support tends to be less intensive in high-context sites, where symbolic visual elements like pictures or hidden pop-up icons are frequently used as clickable items to lead to further site content; those Web sites consequently appear to be less detailed and voluminous.

Such sites tend to be more appropriate to the communication preferences of high-context cultures, which according to Hall and Hall (1990) are often characterized by a "polychronic" time orientation: in polychronic cultures, time is generally regarded as circular and repeating, subject to social relationships or needs and therefore handled in a flexible, imprecise way. Action chains are structured in a less detailed way and interrupted more often since many things are done simultaneously. Entertaining visuals, animated illustrations, and even real multimedia elements certainly increase download time, but this does not matter so much. Explicit navigation support is rare since neither a strictly ordered route through the less detailed site structure nor very quick orientation is necessary. Implicit symbolic cues, however, support an intuitive navigation, which enhances both entertainment appeals and positive affection in Web site use.

In general, the compared Web sites reveal different degrees of cultural adaptation along these design criteria and correspond more or less to the specific (national-cultural) communication preferences (Hermeking, 2004, 2005). Exceptions to this finding are Japanese Web sites, which, like other East-Asian sites, often present both much text and many visuals. This seems to correspond to the East-Asia-specific, two-fold media preference.

International marketing and advertising research provides many additional criteria that are also relevant for culture-specific Web site design. For example, background colors are generally subject to culturally specific aesthetic and symbolic conceptions and preferences (Usunier, 1993; Madden, Hewett, & Roth, 2000; Aslam, 2006), and generally have higher importance in high-context cultures. These culture-specific symbolic and aesthetic concepts also relate to visual and pictorial presentations, which tend to reveal the highest degree of adaptation. Even strongly standardized sites tend to be different in their visual presentations of people, products, artifacts, nature, and so forth, since many visual scenes, motifs, and images, like in advertising, are implicit representations of specific cultural values (de Mooij, 1998; Mueller, 1996, 2004; Tharp, 2001). Verbal headlines or slogans presented in Web sites may well differ in quite the same respect.

In sum, however, most compared international Web sites—even in high-context cultures—still seem to be somewhat dominated by an overall low-context style along the structural design criteria (Hermeking, 2004, 2005). This obviously reflects the nature of the WWW, which, to date, is generally more similar to print media. The degree of such an overall low-context style standardization vs. an adaptation to high-context preferences depends on a set of some basic conditions beyond local culture, which always must be considered as well.

Findings on Global Web Site Standardization

Generally, a basic differentiation into brand and product type represented in Web sites is necessary which is also supported by the findings on the compared international sites (Hermeking, 2004, 2005). Their different degrees of global standardization vs. local adaptation are demonstrated in Figure 5.

Web sites of global companies of the New Economy that are based primarily on the WWW and e-commerce (e.g., Amazon) tend to be strongly dominated by low-context style and, as such, strongly standardized. In the Old Economy, sites of companies representing industrial goods and primarily addressing business users (B2B) tend to be standardized to a similar degree (e.g., ABB). Web sites that mainly address consumers (B2C) reveal different tendencies: Global brands of durable, high-involvement products (e.g., Sony) are generally characterized by only modest standardization, that is, those sites tend to exhibit a somewhat higher degree of cultural adaptation in high-context cultures. Web sites representing global brands of non-durable, low-involvement products (e.g., Coca-Cola) reveal an even higher degree of such cultural adaptation.

Figure 5. Web site adaptation of different product types

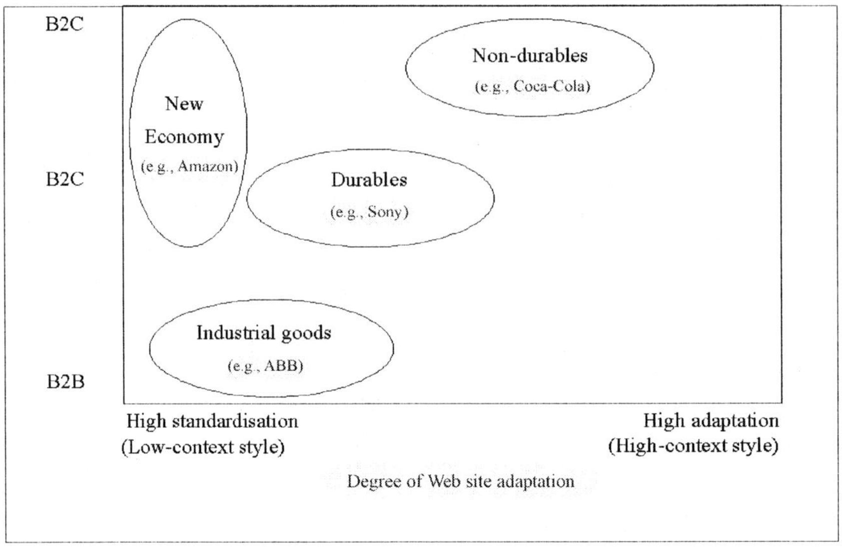

Consumer-related (B2C) sites of the latter kind tend to be most open for adaptations to communication preferences of high-context cultures. Therefore, analysis of non-durable product Web sites tends to be the best starting point to demonstrate cultural adaptation on the WWW.

One explanation might include the large majority of consumers addressed worldwide, who—in contrast to businesspeople—do not belong to the "information elite" of frequent Web users; therefore, they may be less open to being addressed in culturally inappropriate low-context styles. In addition, non-durable consumer goods are often characterizable as low-involvement products, therefore less explicit and rational information may need to be provided.

In addition to such a product-type-related adaptation, local Web sites of global consumer brands occasionally seem to be more standardized worldwide by utilizing the typical communication style of their brands' country of origin. Global sites of a French food brand, for example, would tend to be somewhat more dominated by (French) high-context style, whereas global sites of a German car brand, for example, would tend to represent more (German) low-context style. Such a "country-of-origin" (COO) effect generally reflects cultural values of the represented brand's or product's home country, in addition to or instead of reflected cultural values of the local site's home country. This has also been found in a cross-national analysis of Indian, Chinese, Japanese, and U.S. sites by Singh, Zhao, and Hu (2005). A positive COO effect works only if the images of the product/brand and its country of origin are positively related in the targeted countries (Knight & Calantone, 2000). Given this, national representation through site design can create adequate trust; however, in all other cases, local culture-specific adaptations are more appropriate or even necessary in order to reduce trust-diminishing "not-invented-here" feelings.

These findings represent only a few of the many influential factors on Web site standardization beyond local communication preferences. In each single case, designers must recognize the complex contingencies of influential hard and soft factors, as demonstrated in Figure 6.

The design criteria discussed above represent controllable variables, which are dependent on a country-specific set of environmental conditions known as uncontrollables. Recognizing the

Figure 6. Influential factors on international Web site design

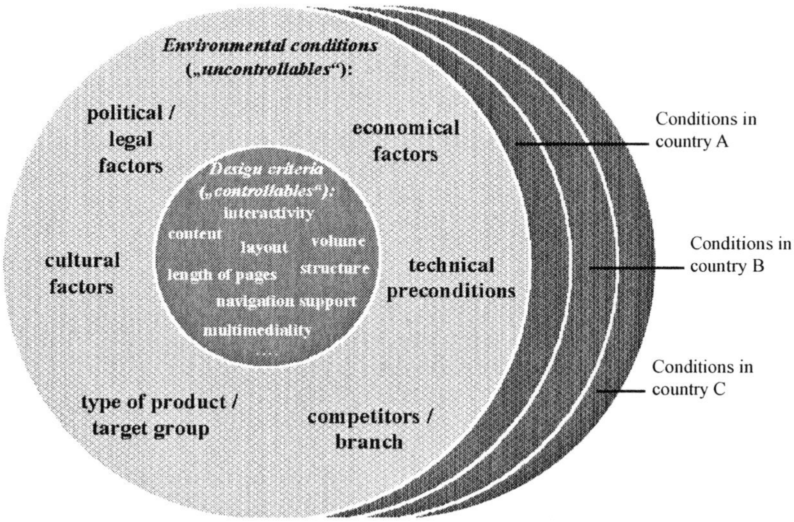

cultural soft factors among diverse and equally important hard factors behind those design criteria may help to create sites that meet cultural communication preferences on a much more sufficient degree.

CONCLUSION

Cross-cultural marketing and advertising research provides some useful insights into the global disparities and differences in Internet usage. These are essentially based on influences of uncertainty avoidance, individualism, and low- vs. high-context communication preferences. According to their conceptual premises, those cultural criteria do well explain general tendencies of unconscious, mentality-related collective Internet usage behavior and Web site design preferences on an aggregated (inter-)national level. They do not well explain, however, any conscious or identity-related usage behavior or preference on an individual or sub-group level.

With respect to the disparities in global Internet consumption, a country's low rate of Internet usage could also be considered as a general indicator of an overly low quota of culturally appropriate Web sites. Like the worldwide disparity in print media consumption, a significant digital divide will be normality unless the design of Web sites—and finally of the Internet in general—completely meets the specific communication preferences of many high-context cultures. According to the structural and other design criteria introduced, such preferred sites may be summed up in a very simple way: "less Print, but more TV style." Since technical conditions (including data transfer) continuously become more efficient, such a culturally appropriate Web site design, in sum, may turn the Internet into a real global medium in the future.

REFERENCES

Anonymous. (2000a). Weltweite Suche nach den Vielsurfern. *Media & Marketing,* (11), 94-96.

Anonymous. (2000b). Internet-Nutzer sind aktiver. *Marketing Journal, 33*(5), 267.

Anonymous. (2005). Medienlandschaft Westeuropa: Fakten und Zahlen aus Marketing, Werbung, Medien. *W&V Compact,* (2), 6-22.

Aslam, M. M. (2006). Are you selling the right colour? A cross-cultural review of colour as a marketing cue. *Journal of Marketing Communications, 12*(1), 15-30.

Choi, S. M., & la Ferle, C. (2004). Convergence across American and Korean young adults: Socialization variables indicate the verdict is still out. *International Journal of Advertising, 23*(4), 479-506.

de Mooij, M. K. (1998). *Global marketing and advertising: Understanding cultural paradoxes.* Thousand Oaks, CA: Sage.

de Mooij, M. K. (2004). *Consumer behavior and culture: Consequences for global marketing and advertising.* Thousand Oaks, CA: Sage.

Gudykunst, W. B. (2001). Anxiety, uncertainty and perceived effectiveness of communication across relationships and cultures. *International Journal of Intercultural Relations, 25*(1), 55-71.

Hall, E. T. (1977). *Beyond culture.* Garden City, NY: Anchor-Doubleday.

Hall, E. T., & Hall, M. R. (1990). *Understanding cultural differences: Germans, French and Americans.* Yarmouth, ME: Intercultural Press.

Hammond, K., Turner, P., & Bain, M. (2000). Internet users versus non-users: Drivers in Internet uptake. *International Journal of Advertising, 19*(5), 665-680.

Hermeking, M. (2000). The cultural influence on international product development. *Global Player: International Management and Business Culture, 7*(5), 8-13.

Hermeking, M. (2004). Cultural influences on Internet diffusion and Web site acceptance: Some findings from cross-cultural marketing research. In F. Sudweeks & C. Ess (Eds.), *Proceedings of the 4th International Conference on Cultural Attitudes Towards Technology and Communication* (pp. 109-122). Murdoch, Western Australia: Murdoch University.

Hermeking, M. (2005). Culture and Internet consumption: Contributions from cross-cultural marketing and advertising research. *Journal of Computer Mediated Communication, 11*(1). Retrieved June 15, 2006, from http://jcmc.indiana.edu/vol11/issue1/hermeking.html

Hofstede, G. (1991). *Cultures and organizations: Software of the mind.* New York: McGraw-Hill.

Hofstede, G. (2001). *Culture's consequences* (2nd ed.). Thousand Oaks, CA: Sage.

Holden, N. (2004). Why marketers need a new concept of culture for the global knowledge economy. *International Marketing Review, 21*(6), 563-572.

Ju-Pak, K. H. (1999). Content dimensions of Web advertising: A cross-cultural comparison. *International Journal of Advertising, 18*(2), 207-232.

Knight, G. A., & Calantone, R. J. (2000) A flexible model of consumer country-of-origin perceptions: A cross-cultural investigation. *International Marketing Review, 17*(2), 127-145.

la Ferle, C., Edwards, S. M., & Yutaka, M. (2002). Internet diffusion in Japan: Cultural considerations. *Journal of Advertising Research, 42*(2), 65-79.

Levitt, T. (1983). The globalization of markets. *Harvard Business Review, 61*(3), 92-102.

Madden, T. J., Hewett, K., & Roth, M. S. (2000). Managing images in different cultures: A cross-national study of colour meanings and preferences. *Journal of International Marketing, 8*(4), 90-107.

Morrison, D. E., & Firmstone, J. (2000). The social function of trust and implications for e-commerce. *International Journal of Advertising, 19*(5), 599-623.

Mueller, B. (1996). *International advertising: Communicating across cultures.* Belmont, CA: Wadsworth.

Mueller, B. (2004). *Dynamics of international advertising: Theoretical and practical perspectives.* New York: Peter Lang.

Müller, S., & Gelbrich, K. (2004). *Interkulturelles Marketing.* München, Germany: Vahlen.

Nielsen NetRatings. (n.d.). Retrieved from http://www.nielsennetratings.com/news.jsp?section=dat_to)

Okazaki, S., & Alonso, J. (2003). Right messages for the right site: On-line creative strategies by Japanese multinational corporations. *Journal of Marketing Communications, 9*(4), 221-239.

Park, C., & Jun, J. K. (2003). A cross-cultural comparison of Internet buying behaviour: Effects of Internet usage, perceived risks, and innovativeness. *International Marketing Review, 20*(5), 534-553.

Rhee, K. Y., & Kim, W. B. (2004). The adoption and use of the Internet in South Korea. *Journal of Computer Mediated Communication, 9*(4). Retrieved June 15, 2006, from http://www.jcmc.indiana.edu/vol9/issue4/rhee.html

Rogers, E., & Hart, B. (1998, November). Edward T. Hall and the origins of the field of intercultural communication. In *Proceedings of the Meeting of the National Communication Association, International and Intercultural Communication Division,* New York.

Schugk, M. (2004). *Interkulturelle Kommunikation: Kulturbedingte Unterschiede in Verkauf und Werbung.* München, Germany: Vahlen.

Singh, N., Zhao, H., & Hu, X. (2005). Analyzing the cultural content of Web sites: A cross-national comparison of China, India, Japan, and U.S. *International Marketing Review, 22*(2), 129-146.

Singh, N., Fassott, G., Chao, M. C., & Hoffmann, J. A. (2006). Understanding international Web site usage: A cross-national study of German, Brazilian, and Taiwanese online consumers. *International Marketing Review, 23*(1), 83-97.

Steenkamp, J. B. E., ter Hofstede, F., & Wedel, M. (1999). A cross-national investigation into the individual and national cultural antecedents of consumer innovativeness. *Journal of Marketing, 63*(2), 55-69.

Terpstra, V., & Russow, L. C. (2000). *International dimensions of marketing.* London: South-Western.

Tharp, M. C. (2001). *Marketing and consumer identity in multicultural America.* Thousand Oaks, CA: Sage.

Usunier, J. C. (1993). *International marketing: A cultural approach.* New York: Prentice-Hall.

Yeniyurt, S., & Townsend, J. D. (2003). Does culture explain acceptance of new products in a country? An empirical investigation. *International Marketing Review, 20*(4), 377-396.

Section IV

Governance and Legal Factors in Global Cyberspace

Chapter XII
Global Governance and the Local Internet

Yasmin Ibrahim
University of Brighton, UK

ABSTRACT

The Internet is often thought of as a unified technology that transcends traditional geopolitical boundaries. For this reason, early discussions of the Internet often advocated the idea that nation states would have either limited or no control over this exclusively electronic domain. The subsequent need for protocols and standardization of the Internet, however, led to the development of formal institutions such as the Internet Corporation for Assigned Names and Numbers (ICANN). It also led more nations to adopt policies designed to promote, control, and mold the Internet culturally and politically within their jurisdictional boundaries. The trajectory of the Internet's history and global governance nevertheless reveals the increasing encroachment of the nation-states in shaping the Internet as a local entity. This chapter outlines the characteristics and intrinsic qualities of the Internet which make it a global entity and how it has been deemed ungovernable in the conventional terms.

INTRODUCTION

The Internet is often conceived as a monolithic global entity that defies the territorial boundaries of nation-states. The earlier discourses prophesied the withering of the nation-state and the loss of sovereignty over this electronic terrain. The subsequent need for protocols and standardization of the Internet saw the emergence of formal institutions such as the Internet Corporation for Assigned Names and Numbers (ICANN) on the one hand, and the rise of nation-states as active players in enacting policies and regulations to promote, control, and mold the Internet culturally and politically within their jurisdictional boundaries, on the other. The governance of the Internet is also mediated by transnational and semi-formal institutions, as well as civil society organizations and interest groups leading the way toward multigovernance. The imposition of territorial boundaries on the 'borderless entity' of the Internet creates tension between the local and global aspects of the Internet, making it both a localized as well as an international entity in terms of its characteristics and conditions of use, and these interactions create issues and challenges in reconciling them. The trajectory of the Internet's history and global governance nevertheless reveals the increasing encroachment of the nation-states in shaping the Internet as a local entity. The enmeshing of the global and local dimensions of the Internet has implications for e-commerce, security, expression of freedom, and privacy, just to name a few, in cyberspace.

This chapter outlines the characteristics and intrinsic qualities of the Internet that make it a global entity and how it has been deemed ungovernable in the conventional terms. It also examines the difficulties of globally orchestrating the Internet and the increasing role of the nation-states in mediating the character, rules of engagement, and culture of the Internet, and assesses how the Internet is becoming defined locally or nationally despite its global presence.

DEFINING CHARACTERISTICS OF THE INTERNET

Early discourses of the Internet often portrayed it as a homogenous virtual platform. However, in reality the electronic frontier has a multitude of usages and is a collection of different multimedia technologies and networks. Additionally, while they are governed by standard computing protocol—commonly termed as transmission control protocol/Internet protocol (TCP/IP)—this does not necessarily mean that the users can access all its domains. Some usages of the Internet such as encrypted person-to-person electronic mail, invited Internet relay chat or videoconferencing, and password—protected file transfer protocol or World Wide Web (WWW) sites may be relatively private; others such as electronic mail-based distribution lists, Usenet groups, and WWW pages are more public in orientation (Bennahum, 1996).

Rheingold (2000, p. 5) describes the online platform as a "conceptual space where words, human relationships, data, wealth and power are manifested by people using computer mediated communications technology" (CMC). It has also been dubbed as the 'virtual world' by many, "for when you're online you're both everywhere and nowhere at the same time" (Crew & Thierer, 2003, p. xv). These discourses described the emergence of new forms of society and identity (Rheingold, 2000) in which the "virtual was often disembodied from the real" (Miller & Slater, 2000, p. 4) or physical spaces of the world. This disembodiment represented a form of escapism from the real society where individuals could invent, deconstruct, and re-invent their identities. As such it symbolized fluidity in terms of identity as well as a form of release from the confines of the real world. This notion of 'disembeddedness' from the physical world then raises questions about the relevance of territorially bound laws and policies to regulate this space.

One of the main characteristics of the Internet is the element of connectivity. The Internet allows for many computers to be connected to the same network, with the only requirement being adherence to the TCP/IP protocols. This brings users from different physical locations to the virtual platform giving a global connectivity and allows communication between people who are spatially dispersed (Slater, 2002, p. 534; Miller & Slater, 2000; Hine, 2000; Rheingold, 2000). Besides this feature of global connectivity, the online environment is defined by the ability to communicate anonymously. These features are seen as empowering while raising question about credibility, as users are not constrained by both time and space. Another feature of the Internet is hypertextuality, which links information stored in different places all over the world through the World Wide Web and allows the provision of links that connect information that is geographically distant to merge and be available in one space.

The interconnected global nature of the Internet is further characterized by a 'de-centeredness' due to its lack of hierarchy in terms of how information is disseminated. This gives it the potential to be a subversive entity pulling it in the direction of anarchy and indifference to state control. This has, undoubtedly, precipitated discourses about the rise of virtual communities leading to the disintegration of the nation-states through these permeable electronic borders. The deconstruction of cultural and political territorial boundaries by the Internet emanate from the notion that the governance of the Internet cannot be confined to nation-states as the Internet is non-territorial. The demarcation of boundaries on this borderless electronic presence then raises concerns about the applicability of territorial jurisdictions and policies on these globally shared spaces. The regulability of the Internet has invariably become an issue of global concern as more and more users join the virtual world. As Johnson and Post (1996) point out, cyberspace radically undermines the relationship between legally significant online phenomena and physical location.

In short, the Internet's methods of establishing communication are not bound territorially; and the routing structure is independent of political jurisdictions, and connection costs are insensitive to distance and political boundaries (Mathiason, Mueller, Klein, & Holitscher, 2004, p. 2; Lloyd, 1997). This has led to a degree of contestation between different players (both at national and international levels) wanting to regulate the Internet. These struggles have taken various manifestations and agendas, and this chapter traces these dialectics between the local and global dimensions of power on the Internet.

THE INTERNET AS A "COLLECTIVE RESOURCE"

The World Wide Web (WWW) began in 1989, the brainchild of Tim Berners-Lee at the European Organization for Nuclear Research (CERN) to address the need for a collaborative tool to help scientists share knowledge; it was offered to the entire Internet community for experimentation and thus the WWW was born (Horton, Taylor, & Hoft, 1996, p. 13). According to Abbate (1997) the most important values expressed in the design of the Internet encompass the network's "informal, decentralized, user-driven development." In addition to this, the U.S. military's role in the development favored values such as survivability, flexibility, and high performance over commercial goals such as low cost, simplicity, or consumer appeal.

The U.S. government, through the National Science Foundation, directly funded the development of the Internet from the 1960s until April 1995. While it is privately operated presently, much of the investments are from public finance, and invariably the development of the Internet infrastructure, access, and literacy issues are linked to the real world of policy making and funding decisions of governments. The Internet, as a result of this, gained its footing as a public network, while the ownership of hardware such as routers, servers, and fiber is almost totally private (Crew & Thierer, 2003, p. xx; Reidenberg, 1997). It can readily be perceived as a collective resource and as a vast information commons which should be amenable to a degree of collective decision making. In this sense, many view cyberspace as a hybrid of the public/private enterprises, and often it is difficult to distinguish whether it should be defined as a private resource of a collective global good.

The uneasy dependence of the Internet between the private and public organizations means that the issues of control and governance becomes highly politicized and, in so far as this relationship is sustained, Internet governance becomes dialectical not just between nation-states but also between private organizations and public bodies. The myriad of political, economic, and social issues accruing from the Internet both locally and globally means that self-regulation alone, or the market forces of demand and supply, cannot regulate the Internet. This means that the traditional notions of jurisdiction in national and international terms encounter difficulties as the non-ownership and the common-property status of the Internet presents challenges for issues of regulation. No one completely owns the Internet, and yet claims to ownership through investments in hardware and software are not entirely dismissible either, and this is one major factor for the rise of various regulatory and legislative disputes, as well as the promotion of the Internet as a common resource that should be safeguarded from the power structures of the real world. The common property status of the Internet consequentially has implications for the governance of the Internet and the ad hoc nature in which governance has occurred.

THE LIBERTARIAN IDEAL OF SELF-GOVERNANCE

Due to the unique features of the Internet environment, the earlier writings on the Internet tended to be overly optimistic and perhaps deterministic, bound up with articulations of liberation and freedom of expression. The Internet was deemed to offer the potential to revolutionize political communication and the nature of public opinion, expression, and formation. Since the medium and message were perceived to be free from distortions that characterize the traditional channels of political communication, the Internet was seen as decentralizing the production of knowledge and democratic discourses through interactivity, thus deterring dominant authorities from subverting the flow of information (Savigny, 2002; London, 1995; Lippert, 1997; Randall, 1988). This decentralization and the possibility of greater access thus suggest an increase in opportunity for citizens to participate in the political process (Savigny, 2002; Cairncross, 1997; Hagel & Armstrong, 1997).

The propensity to anarchy and the plausible loss of sovereignty of nation-states also floated the possibilities of self-governance of the online spaces. According to Barlow (1996a), the co-founder of the Electronic Frontier Foundation, cyberspace represents "the promise of a new social space, global and anti-sovereign, within which anybody, anywhere can express to the rest of humanity whatever he or she believes without fear." Barlow's (1996b) discourses conveyed valiant attempts to keep the Internet free from the hegemony of nation-states, business conglomerates, and institutions. He called for a form of self-governance among the users where the virtual community formed and reformed its own codes of ethics and regulations governing users' behavior. This romantic zeal to keep the virtual world unregulated by corporations and the nation-states prompted a wave of activism against regulations. Undoubtedly, early cyber activists wanting to preserve the libertarian values of the Internet clamored for self-policing rather than organizations and nations imposing rules on the Internet.

While associated with freedom and liberation, the Internet is also often viewed as an instrument of power and surveillance where one's actions and transactions can be monitored and recorded (Buchstein, 1997; Agre, 2002; Lyon, 2001; Gandy, 1993). The new media presenting a 'lethal' threat to autocratic regimes is based on the belief that it is impossible to completely control all the individual activities that can be distributed across borders using these technologies. On the other hand, totalitarian regimes can enhance their central management, surveillance, and control through these new technologies (Burnham, 1983; Gandy, 1993; Mulgan, 1991; Lyon, 1994; Beniger, 1996; Van Dijk, 1999; Mowshowitz, 1994).

While the rationale behind implementing regulation on ICTs may differ from country to country, there are common areas of concern which have made regulation necessary. These include taxation laws for e-commerce, laws pertaining to the security of sensitive data, and prosecution of cyber crime. Even liberal democratic states accept the need to regulate the undesirable areas of the Internet, such as child pornography, racism, and the instigation of violence, terrorism, and hate speech. Most countries have enacted legislation to address some of these issues, thus reinstating the sovereignty of the nation-state and re-enacting national borders on cyberspace previously conceived as unviable (Hughes, 2003).

THE NOTION OF "REGULABILITY"

According to Lessig (1999) 'regulability' means the capacity of a government to regulate behavior within its proper reach, and with reference to the Internet it relates to the ability to regulate the behavior of their citizens on the Internet by respective authorities or governments. Advocates of digital libertarianism are naturally aghast at suggestions of restraining activities and behavior on the Internet, but in reality, every government has enacted various sorts of legislation to regulate aspects of this digital communication. These have ranged from the sophisticated to the crude. For example, Cuba simply outlaws the sale of personal computers to individuals. While the government may see the need to regulate the Internet, the question then hovers over whether it is technically possible, or even effective, given the impact of commercialization which not only leads to greater global connectivity but also to a more complex Internet architecture (Wacker, 2003).

In conventional terms, the Internet was deemed ungovernable due to the sheer expanse of the Web, and its global reach raises two pertinent issues: 'regulability' on the one hand, and the application and reach of regulations on the other. While global in terms of reach and presence, the Internet as a technology in reality is embedded within material and political structures. This means local governments have to be involved with global issues and often policies have to be enacted with regard to the Internet's infrastructure as well as its content.

Local governments must decide on regulating it, or perhaps privatizing it or controlling it indirectly through Internet service providers (ISPs). Local policies, when applied to the global entity of the Internet, may be problematic with regard to the criteria of legitimacy and universality. The legitimacy of one enactment with regard to Internet activities in a particular physical space may be contested or non-applicable in another. Policymakers who believe in the classical liberal tradition of individual rights may seek to ensure that certain rights are acknowledged and protected as the growth of the Internet continues to blur the jurisdictional boundaries between real and virtual spaces (Crew & Thierer 2003, p. xvii; Reidenberg, 1997; Johnson & Post, 1996). Inevitably, separated from territorial jurisdictions, new rules will emerge, in a variety of online spaces, to govern a wide range of new phenomena that have no clear parallel in the non-virtual world (Johnson & Post, 1996, p. 2; Menthe, 1998; Proksch & Schweighofer, 2001; Loundy, 1997; Weinberg, 2000).

The issue of regulability of the Internet is a tenuous one, but that does not imply that it cannot be governed either. Nation-states at a national and local level use various means to control the Internet. According to Lessig (1999), there are four elements that influence the online environment, namely regulations, social norms, the market, and the architecture of the Internet itself. These four constraints directly or indirectly regulate the Internet.

Hence, at the national and local level, there may be different rules of engagement with the Internet for the users. The local environment, in defining the terms of engagement to a large extent, localizes the Web. Miller and Slater (2000), who studied the Internet in Trinidad, argue that the Internet is mediated through the context of use. Thus, social norms and cultural practices, as well as overt rules of a particular context, can influence how users engage with the medium. From a cultural and social perspective, the Internet is localized through the immediate context. Consequently, various online interactions can be embedded in disparate ways in the larger social structures, such as professions and social movements (Friedland, 1996; Miller & Slater, 2000; Slevin, 2000: Wynn & Katz, 1997, Slater 2002).

According to Hine (2000), the Internet can be investigated in two ways: either as a culture in its own right or as a cultural artifact. From the former perspective, researchers observe the forms of communication and identity which are produced and sustained in the online setting as a space in its own right. Analyzing the Internet as a cultural artifact, however, entails studying the Internet as an entity that could be shaped by its offline social context. This involves locating and observing how a means of communication is used within an offline social setting or context. In making a case for studying the Internet as a culture in its own right, many studies have alluded to the lack of contextual social cues in shaping Internet behavior (Kiesler, Siegel, & Mcguire, 1984; Sproull & Kiesler, 1986, 1991). These researchers contend that the text-based or discursive medium of electronic mail tended to strip out social cues, particularly features such as gender, age, race, social status, facial expression, and intonation. In negating the 'reduced social cues' model, there is an acknowledgement that computer-mediated communication (CMC) does not negate but, in effect, reinforces the spatial and hierarchical barriers present in the offline social world (Mantovani, 1994). While the Internet is global, local norms in a physical context can localize the Web.

This view reiterates the notion that technology has specific social effects which are essentially shaped by the context of its use (Hine, 2000). The meanings and perceptions produced in online interactions may themselves be mediated by the offline context, highlighting the fact that the online and offline worlds are enmeshed in complex ways. According to Castells (2002), cultural expressions are increasingly captured in the electronic hypertext of the multimedia system that is at the same time global and local: global in its interaction, local in its sources of emissions and in the destination of its messages. In this sense, the Internet is culturally defined locally, while these local norms can create a cultural hybridity when it interacts with the global Internet.

STANDARDIZING THE INTERNET

The use of this global electronic platform by various users in different parts of the world invariably calls for common rules of engagement and norms. The need for concerted actions and decisions is the origin of global governance of the Internet. There have been difficulties in arriving at a consensus of what global governance can entail. In view of this, the United Nations (UN) Working Group on Internet Governance (WGIG)[1]—set up in November 2004 as a result of the World Summit on the Information Society (WSIS)—agreed to a working definition of Internet governance.[2] Two main interpretations of this concept were discussed. A narrow definition would entail incorporating the administration and management of the technical infrastructure of the Internet, while a broad perspective would envelope political and policy issues such as privacy, security, and the fight against child pornography.

Internet governance, then, involves a form of collective action by governments and/or private sector operators of the networks connected by the Internet to establish agreements about the standards, policies, rules and enforcement, and dispute resolution procedures of global networking activities (Mathiason et al., 2004, p. 3). According to Padovani (2003), the governance concept is generally accepted as a way to describe negotiation practices between public, private, and third sector agents, creating a complex network of interdependence at different levels of authority. Thus, various actors are implicated in the global governance of the Internet, and this new transnational order is characterized by increased networking of individuals, private institutions, commercial enterprises, and various government bodies (Slaughter, 1995).

States constitute one of the main players as individual entities as well as parts of international organization in forming the foundations for global governance. Other players involved in global governance include organizations within the United Nations, the International Telecommunications Union (ITU), the World Intellectual Property Organization (WIPO), the World Trade Organization (WTO), and the United Nations' Educational, Scientific and Cultural Organization (UNESCO). States can form blocs along regional proximity and economic interests, as in the case of the Organization for Economic Cooperation and Development (OECD), which comprises 30 states, as well as the G8, the Council of Europe, and the Asia-Pacific Economic Co-operation (APEC). These regional and economic blocs can convene to develop principles, norms, and rules that bind member states.

OECD member states have taken measures to facilitate e-commerce and to ensure economic activity is not hampered by government intervention of member states. While regional organizations and blocs can make agreements with regard to issues such as civil rights and political liberties on the Internet, these are often non-enforceable and may be left to the member state to negotiate internally. Various agreements made by organizations can also be sector specific and may not necessarily reconcile all other laws that may have been made in that remit. For example, open source and open access laws may clash with privacy and copyright issues. Thus, when disputes arise, there may be no universal recourse to solving issues. While international treaties and agreements are beginning to cluster nation-states on issues such as consumer protection, intellectual property, taxation, and to some extent privacy, these shifts are incremental and their promulgations are often left to the individual states to negotiate and implement (Zittrain, 2003, p. 9).

A large degree of governance can also happen through formal and informal non-state institutions. A number of bodies such as Network Solutions Inc. (NSI), the Internet Assigned Numbers Authority (IANA), and the Internet Engineering Task Force (IETF) have played a vital role in the development of the Internet. The most notable of these organizations is ICANN, which was established during the Clinton administration in 1998 as a non-profit organization under the laws of California to formally report to the U.S. Department of Commerce. The founders agreed upon encouraging private sector leadership of ICANN in order for it to operate in a manner that is representative of global interests and open in its management decision making (Palfrey, 2004, p. 10). ICANN was entrusted with the key responsibilities of managing the domain name system, allocating Internet protocol (IP) address space, assigning protocol parameters, and managing the root server system, the backbone of the Internet.[3] ICANN was conceived to operate on a 'bottom-up' basis, seeking to involve individuals and organizations of other countries and its supporting organizations and to keep all other governments at bay by receiving advice only through a Government Advisory Committee (GAC) (Alfonso, 2005).

The establishment of ICANN was partly attributable to the lack of clarity over who actually owned the root, and also due to the exponential rise in both state and non-state interests in the Internet in the mid-1990s (Klein, 2002, 2003). The rise in the litigation of domain names coincided with discourses about the applicability of trademark and copyright law to the Internet. Additionally, governments were becoming more aware of the importance of electronic commerce, and the need to regulate investment and economic activities accruing from the Internet. After four years of debate and negotiation, ICANN formed the Country-Code Names Supporting Organization (ccNSO) in 2003 through a partnership with the worldwide community of registries to allow local, regional, and global groups to have a greater voice (Singer, 2003). With the domain name system (DNS), individual countries are tending to claim responsibility for administration of the national country-code top-level domains (ccTLDs), however as Mueller (2002) points out, the U.S. rules the root, and hence geopolitics of the physical world affect the cyberpolitics of the Internet.[4] Furthermore, ICANN has generated controversy by its seemingly arbitrary and disproportionate allotment of highly conveyed top-level domain and IP addresses that seem to privilege developed nations over developing ones. Most recently, ICANN sparked controversy by granting a top-level .xxx domain name to an independent company run by a British businessman that will make it available for pornographic Web content. This topic was cited by WGIG members, especially representatives from developing countries, in calling into question ICANN's legitimacy as arbiter of culturally sensitive users (Mclaughlin & Packard, 2005, p. 362).

While there is a need to ascertain that there is no duplication of Internet names and addresses (either domain names or IP numbers), ICANN's dominance has been subject to massive scrutiny and criticism (Kennedy, 2000; Komitis 2003; Alfonso 2005; Mayer-Schoenberger & Ziewitz, 2006). Many critics deem ICANN as undemocratic and unrepresentative, as well as lacking legitimacy due to its close links to the U.S. government (Weinberg, 2000). In this sense, the association between the U.S. government and ICANN seemingly gives the U.S. a global role in regulating the Internet. According to McLaughlin (2003, p. 8), ICANN is responsible for "coordinating a chunk of the Internet plumbing." But many argue that the control of this brings with it an immense degree of power and responsibility. With the Internet as a public resource becoming more transnational in character, the global community is requesting institutional mechanisms and guarantees that allow them to participate meaningfully in crafting not just norms, but concrete and representative governance structures (Nye & Donahue, 2000; O'Siochru, 2003; Padovani & Tuzzi, 2003).

As ICANN is spatially located in America and governed by the laws of California, it constitutes an example of an attempt to deal with a global public good via a form of private global platform (Pal & Teplova, 2003). In 2005, the U.S. National Telecommunications and Information Administration (NITA) announced that it will not relinquish oversight of root server administration to a private or public international body (Mclaughlin & Packard 2005, p. 362). The Bush government's announcement in May 2006 to renew its exclusive contract with ICANN revived debates about the U.S. Commerce department's role in overseeing ICANN's work (Mohammed, 2006). In the months leading up to the November 2005 Tunis meeting of the WSIS, China, India, Brazil, and even the European Union (EU) called for phasing out the department's oversight in favor of a multilateral body that would increase international input.

While ICANN can be representative of a form of global governance, policy making beyond the technical administration of domain names can entail public struggles between numerous other bodies ranging from "the American Bar Association, to French courts, to the International Telecommunications Union (ITU), and the United

Nations" (Crew & Thierer, 2003, p. xvii). Other informal groupings of civil society, including corporations, have worked to develop norms and standards relevant to the Internet; cases in point are the Anti-Spam Technical Alliance (ASTA) and the World Wide Consortium, which deals with application software standards and the development of standards for Internet accessibility for disabled users (Mathiason et al., 2004, p. 12). The strong role of civil society in establishing the Internet is also paving the way for new forms of multilevel governance; this is evident in the growing reference to 'global civil society' in the Internet governance literature as well as political discourse (Kaldor, 2003; Keane, 2003). Themes debated by civil society with regard to Internet governance have included pluralism and access to information, legal issues concerning individual freedoms and privacy, communication rights, and the application of ICTs (Padovani, 2005).

NATIONAL IDENTITY AND DOMAIN NAMES

Domain names can be regarded as a form of national identity, and the need to control and own country-specific domain names can be associated with national sovereignty and pride; hence the ownership of ccTLDs can become both a political issue and one of economic priority for countries (Froomkin, 2004). However, country names in the virtual world are not automatically legally protected, and the trajectory of assigning ccTLDs has at times not mirrored the geographical realities in terms of the location of registrants or the political transformations of territorial boundaries. These anomalies have triggered a degree of negotiation between countries and ICANN via the Government Advisory Committee. The GAC has been used by governments as a platform to voice their jurisdictional issues. GAC from the onset declared the Internet name space a 'public resource', mounting a persistent campaign to redefine the legal delegation procedure and practical relationship between ICANN, governments, and ccTLD administrators, and to make name space references to countries exclusive and grounded in the existing political order (Whittaker, 2002; Rimmer, 2003).

The well-publicized dispute between South Africa and a Seattle-based company, Virtual Countries Inc., over the ownership of *southafrica.com* raised various controversies with regard to the delegation and ownership of ccTLDs and the rights of national governments. South Africa argued that nations have a legal right to their names and to any economic benefits that flow from the use of that name on the Internet, and the taking of their name by the West amounted to an "act of colonial expropriation akin to their taking of indigenous people's cultural property" (Froomkin, 2004, pp. 869-871; Rimmer, 2003).

The delegation of ccTLDs were initially assigned to country representatives in an ad hoc fashion, and the need for more systematic delegation was felt when the demand for ccTLDs substantially increased in the first half of the 1990s, reflecting the rising awareness of the commercial value of ccTLDs, prompting many governments to take part in an international and legal campaign to nationalize, or at least deprivatize, ccTLDs (Froomkin, 2004; Rimmer, 2003; Yu, 2003).

In June 1998, a U.S. government white paper acknowledged the role that national governments have in managing policy for their own ccTLDs (see http://www.icann.org/cctlds); since 2000, ICANN has been working with managers of ccTLDs, and these individual relationships tend to be complex depending on the policies pursued, economics, language, culture, legal environment and relations with governments of different ccTLDs, and the organizations that operate them.

ICANN attempted to harmonize the process of resolving disputes over domain names by creating a specific resolution process, the Uniform Domain Name Dispute Resolution Policy (UDRP), for domain name registrars to abide by. The UDRP is aimed at addressing disputes about domain names, for example cybersquatting, in a way that is more expedient and economical than recourse to international courts, as it involves arbitration and decision making by a small number of non-government arbitral bodies instead of litigation involving lengthy legal processes and multijurisdictional problems. The UDRP, however, is based on the concept of U.S. trademark law and works for disputes among U.S.-based claimants and for domain names registered in the U.S. by non-U.S. parties (Mayer-Schoenberger & Ziewitz, 2006).[5] The U.S. trademark law nevertheless runs into difficulties when disputes involve two-non-U.S. claimants before a non-U.S. registrar. In such circumstances it is very likely that non-U.S. law will apply, and the disputing parties may have claims that differ greatly from those that may arise under U.S. trademark law (Froomkin, 2004; Rimmer, 2003; Mayer-Schoenberger & Ziewitz, 2006). Hence, non-U.S. registrars, as well as managers of ccTLDs, have resisted adherence to the UDRP, especially if following the policy would force them to violate the laws of their home jurisdiction (Mayer-Schoenberger & Ziewitz, 2006).

Around 60% of the URDP disputes have been heard by the World Intellectual Property Organization (WIPO).[6] The others are handled by the CPR Institute for Dispute Resolution, the National Arbitration Forum, and the Asian Domain Name Dispute Resolution Center (ADNRC). Decisions are made on a case-by-case basis rather than using detailed guidelines or an extensive body of precedent. A report by Syracuse University's Convergence Center[7] and Geist's (2001) study of WIPO arbitration suggest that the interpretation of the URDP by WIPO and forum-shopping by litigants have created a system biased towards large trademark holders. Others like the Max Planck Institute study (2002)[8] feel that the UDRP has been operating satisfactorily but needs fine-tuning regarding free speech and rules concerning the burden of proof. Most nations have rules of varying complexity to deal with domain disputes. Increasingly these rules are based on the UDRP with modification to reflect the local circumstances.

ICANN's overarching role has raised serious questions about the intersection between Internet and national sovereignty, as governments maintain they should be final arbiters over their country-code domains (Mclaughlin & Packard, 2005). Governments also feel that ICANN has been slow to establish multilingual domains that would allow their citizens to register domain names in their native languages (Geist, 2006). Hence it is viewed by many in the international community as the province of a small technocratic elite with ties to the U.S. Department of Commerce (Mclaughlin & Packard, 2005).

Today, national governments recognize ccTLDs "as a component of their sovereignty and a vital national interest." While ICANN through its auxiliary bodies has established guidelines for resolution of disputes primarily based on the process of arbitration, not all governments are interested in working with ICANN, and some might prefer to act alone or to abandon ICANN for a more favorable international forum such as the ITU. Some governments have sought to use national legislation to regulate local ccTLD managers (Yu, 2003; Geist, 2002); a case in point is the government of South Africa which recently introduced legislation to reclaim control of the .za name space from the incumbent ccTLD managers. Over the years the control of the DNS and ccTLD delegations will likely continue and will include ICANN, IANA, ccTLD Managers, national governments, GAC, ITU, and WIPO, and will also feature other actors such as individual ccTLD managers, intellectual property right holders, ISPs, and major telecommunications and information technology companies (Yu, 2003, p. 8).

The ICANN system provides a means of determining how to co-exist in at least some parts of cyberspace in particular for anyone who wishes to obtain and hold a domain name. Those who acknowledge ICANN's authority are agreeing to join into a social contract and submit to it (Palfrey, 2004, p. 28). Nevertheless, there are persistent concerns that this emerging system of Internet governance lacks fundamental mechanisms to prevent government overreach and to safeguard liberties (Rundle, 2005, p. 2; Mclaughlin & Packard, 2005).

THE TENSIONS BETWEEN LOCAL AND GLOBAL GOVERNANCE

The borderlessness of the Internet and the existence of borders in real physical spaces present various problems when laws or norms are violated. The diffusion in sources of control and ownership also means that there is a degree of constant contestation between the global and local contexts, which raise issues of accountability and responsibility. Thus, nation-states may adopt different approaches with regard to issues such as e-commerce, gambling, libel, censorship, or privacy laws. The conflict in the virtual and real space arises where there are problems transferring territorial laws into cyberspace. This has been one of the main overarching concerns of the Internet, and this is where the global entity succumbs to local rule-making and social norms. The Internet has seen the emergence of an array of regulations and policies which have brought forth the difficulties in application, not just at the local and global level, but in the United States; for example, there are problems reconciling the laws between different states, and between state and federal levels in terms of gambling, privacy, and copyright. Nevertheless, despite the difficulties in applicability, the Internet is becoming a heavily regulated network that competes with different demands from federal, state, and international governments as well as other special interest bodies and transnational entities.

This raises the argument that with Internet governance, states have to now adjust to a system of sharing powers with international organizations, citizens' groups of the civil society, especially non-governmental organizations, and transnational organizations (Mathiason et al., 2004; Proksch, 2001). The power re-alignment to non-state organizations does not necessarily imply the loss of power for the state as territorial regulations remain a central component of Internet governance (Goldsmith, 1998; Slaughter, 1997). According to Johnson and Post (1996, p. 2):

Physical spaces are of primary importance in determining legal rights and responsibilities; all laws are prima facie territorial and hence there is a general correspondence between borders drawn in physical space (between nation-states or other political entities) and borders in 'law space.'

There is also an acceptance of legitimacy where persons within a geographically defined border are the ultimate source of law-making authority for activities in that border.

As mentioned earlier, content on the Internet is an issue that affects every nation-state. Some states adopt an approach where they believe that they have a right and a responsibility to directly intercede and apply technical control on the content that is available to the citizens. In Saudi Arabia and Singapore, for example, all ISPs must go through a proxy server where authorities can block access to sites which they object to, such as pornography or material that can be culturally or religiously sensitive. Internet cafes in China must hold records of sites visited; this is mainly to prevent access to sites containing pornography, gambling, and those that "harm national unification, sovereignty, and territorial integrity" (Lum, 2006). In Vietnam, the Interior ministry is vested with the power to monitor content, including e-mail communication; this means that Internet users are legally liable for all communication occurring on the personal space. Other nations playing an active role in monitoring content include Algeria, Yemen, Bahrain, United Arab Emirates, North Korea, Iran, and the Maldives.

In other countries, a primary piece of statutory enactment can be the main determinant of what may be acceptable in the country. Australia embraces this statutory approach through the Broadcasting Services Amendment (Online Services) Act 1999, which regulates Internet content. A similar statutory enactment in Singapore dubbed the 'Code' automatically subsumes all users to comply with the code of conduct, particularly with regard to matters of race, religion, and politics. Additionally, political Web sites are required to further register with the authorities. The Singapore government uses this key piece of legislation along with other measures to control cyberspace.[9]

In the U.S., all electronic communication is regulated by the Federal Communications Commission, and in line with the guarantees of Freedom of Speech enshrined in the First Amendment, it imposes minimal content regulations (Reidenberg, 2005; Zittrain, 2003). In view of this, it has regulations of a broad nature that even without amendments, the Internet can be mediated. The U.S., in adopting a constitutional approach to regulating the Internet, seeks to reconcile the laws and the spirit of its constitution. The First Amendment to the U.S. Constitution, for example, provides protection for freedom of speech greater than is provided, or thought appropriate, by many other countries (Thierer & Crews, 2003).

Nevertheless, the non-geographic character of cyberspace makes it difficult to apply territorially based rules to activities online. As such, local sovereigns cannot control online actions whose physical location is irrelevant or cannot even be established (Tantuico, 2001, p. 15). In view of this, territorial sovereigns can simply seek to extend their jurisdiction and to amend their laws as necessary to attempt to govern all actions in cyberspace that have substantial impacts on their own citizenry.

This raises the difficulties of extraterritorial enforcement, and for governments which object to freedom of speech on the Internet, this can be a thorny issue. Myanmar (Burma), for example, objects to political criticism on the Internet and requires licensing of computers. The military junta in Myanmar severely restricts online activities and only allows a limited package of approved Web sites, referred to as the Intranet. But while it can enforce this legislation within its territorial boundaries, it cannot enforce such practices on countries which may host Web sites critical of the government in Myanmar (Krebs, 2001). The government also licenses cyber cafes in the country. On the other hand, in South Korea the high number of broadband subscribers attracts a high level of citizen and civic group participation (Kyu, 2003).

The Internet has been seen as a precious conduit for the worldwide export of democratic ideas by libertarians, but the reach of such ideas and the appropriation of the Internet for democratic empowerment can be limited by the context of use.

LOCALIZING THE NET

The development of infrastructure to support the Internet and the content that is available on the Web are two issues which concern nation-states. State governments are increasingly starting to control the global Internet through intermediaries such as Internet service providers, who are often required to have a license; further, the award of this license may also depend on certain conditions being fulfilled (Graham, 1999, p. 108). Additionally, these conditions could stipulate what materials may or may not be made available on the Internet to local residents. Evidently, judicially isolated nations with harsh censorship laws that may not be popular in the international platform opt for solutions on their home ground which conform to their local customs, and these solutions often involve controlling intermediaries and the use of technologies such as proxy servers and filtering devices. While filtering software is not perfect, it is getting more sophisticated and sensitive to the context. In China and Saudi Arabia, comprehensive policies are in place to ensure filtering requirements of the state and authorities concerned.[10] Thailand also uses proxy servers to block objectionable material, and Thai ISPs receive official guidance from the Ministry of Information and Communication Technology (Bangkok Post, 2003). States can use filters and packet interceptors, as well as hacker tools such as viruses and worms, to enforce decisions and in the process establish electronic borders that prevent offending material from entering the state like the firewall established by China (Reidenberg, 2005).

Multinational corporations aspiring to cut lucrative deals with repressive regimes have altered the character of the Internet locally to win deals with local authorities. Those looking to do business in China have complied with local policies by signing a Public Pledge on Self-Discipline for China Internet Industry, whereby they agree to refrain from producing, posting, or disseminating pernicious information that may jeopardize state security and disrupt social stability, contravene laws and regulations, and spread superstition and obscenity (Zittrain, 2003, p. 8). On the global stage, IT companies that have complied with Chinese policies have come under pressure from human rights groups, as in the recent case of Google. Along side Google and Yahoo!, other IT giants including Microsoft and Cisco have been taken to task by the members of Congress for their business practices in China (Zeller, 2006). Weblog entries on some parts of Microsoft's MSN site in China using words such as 'freedom', 'democracy,' and 'demonstration' have reportedly been blocked (BBC News, 2006). Human rights activists have also criticized Yahoo! for collaborating with the Chinese government in gathering evidence which had led to the imprisonment of dissidents (McCullagh, 2006). According to Reporters without Borders, several cyber dissidents and journalists are in prison in China over Internet postings criticizing Chinese authorities (Reporters Sans Frontieres, 2002).

Besides controlling the ISP through licensing, they can also be regulated through laws relating to libel and copyright. The UK Defamation Act of 1996, for example, extends existing legislation to the Internet by treating ISPs as 'publishers', thus giving them the same legal responsibilities and liabilities as publishers of print. An industry can also self-regulate itself through voluntary initiatives, as in the case of Britain, where there is no written constitution and government has shown no wish to legislate. The ISP industry in Britain instead established the Internet Watch Foundation (IWF) in 1996 which operates through a 'notice and take down' procedure. The industry self-regulation, many argue, may prompt the industry to be overtly cautious in excluding content that the ISPs may deem as offensive to a local community. Nevertheless, IWF in the UK has been very careful to act only where there is clear evidence of criminal activity, almost invariably involving child pornography. While IWF has worked effectively in the case of child pornography, its initiatives to root out illegal or offending speech has been seen by civil liberty groups as curbing freedom of expression and making users conform to politically correct speech.

LIBEL AND OFFENSE

The expanse of the Web means the content of the electronic platform can range from the intellectually stimulating to the downright offensive and non-credible. The Internet, by combining the traits of traditional communication media, poses problems for censorship because it becomes difficult to classify it and decide who regulates it and how. Due to the convergence of traditionally distinct forms of communication (i.e., print, wire, radio, television, and interpersonal communication) on the Internet, it is difficult to categorize the Internet under existing legal classification schemes. When countries find material harmful or offensive, there may be problems in enforcing a claim. Salient features of the Internet, such as relative anonymity, decentralized distribution, multiple points of access, lack of ties to geography, and the absence of a simple system to identify content, as well as the tools of encryption protocol, make it arduous to control speech in cyberspace. While the structure and design of the Internet can affect user behavior and actions online, other elements such as social norms can also constrain behavior in indirect and less discernible ways. These social norms, often embedded in the culture of a physical space, can mediate Internet use. In Japan, suicide pacts in chat rooms were raising concerns about the use of technology in this manner (Harding, 2004).

Content on the Internet can also break religious taboos and cause enough offense for it to be deemed blasphemous. In 2002, the Vatican wanted a crackdown on the Internet's 'radical libertarianism', and the Rome police force acted by closing down five Web sites that contained blasphemous material about Catholicism and the Madonna. In November 1999, Amazon.com announced that it would stop selling Adolf Hitler's book *Mein Kampf* to customers in Germany in view of the historical sensitivity to it.

The benchmark Yahoo! case involving the Nazi paraphernalia and the objection to the sale of these in French courts is a case in point. When Yahoo! faced an order from a French court threatening damage unless Yahoo! took measures to preclude French citizens from viewing online auctions of Nazi memorabilia, it obtained a declaratory judgment from an American court indicating that any finding of damages would not be enforced in the United States (Reidenberg, 2005). The French courts asserted their rights to demand that Yahoo! cease trading such material, but only after a professional panel could assess that such a ban could apply without affecting the rights of other users in different locations. This was made possible through a 'geocoding' technology which identifies users through the countries of use. While technology made the application of the French court ruling possible, such technologies are only 70% effective. Similarly, in September 2002, a federal court ordered a Web site in Australia to remove material which casts doubts on the occurrence of the Holocaust (Gomez, 2004). This reinforces the degree of customization and tailoring that is occurring on the Internet due to local, cultural, and political sensitivities, for example, Google, which offers country- and language-specific variants, and apparently obeys the informal requests of officials from Germany to eliminate potentially illegal sites from its Google.com counterpart at Google.de.

Physical location and territorial laws are also crucial in the case of libel on the Internet as it can be consequential where the claim is being fought or heard. Due to the fact that content on the Internet can be potentially available globally, this means that the rules with regard to libel and prosecution can vary from state to state, and hence encourage 'forum shopping' where legal experts may shop for the most favorable geographical location for the libel case to be heard. In the *Dow Jones* and Gutnick case, for example, the publisher argued for the case to be heard in America instead of Australia, but the Australian court ruled in 2002 that in this particular case the location of the download was more consequential than the upload, as its dissemination in the local community had the potential to damage the reputation of the defendant. One main criticism of the defamation laws and the Internet laws is that the former were created mainly for the print media, and their application to cyberspace raises new problems, such as whether there is a publication only when the defamatory statement is posted, or each time it is accessed and read by a reader.

Nevertheless, the transference of the defamation laws to cyberspace presents various problems, including the onerous task of deciding which jurisdiction should apply (i.e., the location of the upload or download). According to the Center for Democracy and Technology, the main problem of jurisdiction over the online world is that "no single law defines what goes on the Internet. Online activities that are socially acceptable and legally protected in one locale may not be in another, and each country's rules and norms are globally unique" (as cited in Tuccille, 2001). In the United States, the First Amendment to the Constitution guarantees the right of freedom of religion and freedom of expression to all Americans, even those whose opinions may be deemed reprehensible to a community, with the Supreme Court having affirmed that the government could not regulate Internet content any more than the traditional areas of expression (Whittaker, 2002, p. 111). Slander is the defamation of a subject by word of mouth, while libel deals with defamation recorded and published to a public audience. The jurisdiction to deal with libel and slander may differ in different countries with varying standards, for example in the UK a distinction is made between libel and fair comment. In the U.S., whether the subject is judged as a private or public figure can also come to play a part, as lower standards apply in the former (Whittaker, 2002, p. 113).

PRIVACY AND NATIONAL SECURITY

The Internet as a global network caters to a myriad of activities that combine both private and public forms of communication. The conflict between the private and public hybrid character of the Internet means that states have to decide on issues concerning the public and private nature of the Internet which can have implications for national security and privacy. While there has been much debate about the north-south divide, there has also been disagreement between the western industrialized countries about matters of privacy protection, copyright, and free speech. There has been some debate as to whether data (people's electronic profiles) are tradable or entail something akin to inalienable personhood (Rundle, 2005, p. 4). The EU has argued for a personhood conception of personally identifiable data, while the United States has pushed for the notion that individuals may contract away rights to this data (Rundle, 2005, p. 4). In the contemporary political landscape, with the rising awareness of electronic surveillance by nation states, new initiatives such as Carnivore in the U.S. and the Regulation of Investigatory Powers Act (RIP) in the UK have extended governmental and commercial rights to spy on citizens and employees.

Since the terrorists attacks of September 11, 2001, various states have adopted a number of enactments in terms of national security and the Internet giving authorities more powers to access data that occur on the Internet. The U.S. government, for example, since September 11 has revised its policy enabling authorities to usurp power to wiretap and to allow the FBI to monitor the Web browsing behavior of individuals suspected of being involved in terrorist activity. The U.S. government has also halted tightening of privacy laws or initiatives in the private-sector context, specifically in the online environment. Hence, global events can affect a global network such as the Internet, but with local and global repercussions for users and business communities. The OECD and Council of Europe developed rules in the early 1980s with regard to the handling of personal data. However, the OECD principles are non-binding and the signatories of the COE convention are European only. OECD members have continued to craft guidelines to respond to technological developments, and this includes the drafting of the Guidelines for Consumer Protection in the Context of Electronic Commerce, reflecting a more recent trend to shift some responsibility for security to computer users themselves (Rundle, 2005, p. 5).

The Indonesian government, following the Bali bombings, passed anti-terrorism regulations in October 2002, increasing the powers of police and allowing for detention without trial (Gomez, 2004). Under this new ruling, the security forces can intercept an individual's e-mails and tap people's phone lines (Luwarso, 2003). The government in Pakistan in 2002 required Internet café owners to keep detailed records of usage under the banner of fighting terrorism. This also justified the monitoring of Internet use and the blocking of several Web sites, including an Al-Qaeda Web site and other pages that provide 'anti-Islamic' information.[11]

Sovereign authority can prevail on the Internet with regard to privacy issues; for example, in the United States, the Children's Online Privacy Protection Act applies various laws from the American Statute to any Web site that collects personal information from children regardless of its place of origin. Similarly, the European Directory on data privacy applies substantive European law to any organization that uses means within the EU to collect personal data (Reidenberg, 2005).

In 1998, the EU implemented the Privacy Directive applicable to both online and off-line platforms. In a general sense, the EU provides a stronger protection of personal information from access by private-sector companies than it does from government, while the reverse was the case in the United States until September 11. Citizens nevertheless are calling for stronger enforcement and compliance of privacy laws in Europe as well as in America. In May 2003, the Council of Europe's 'Declaration on Freedom of Communication on the Internet', which was adopted by the committee of ministers, specified a series of norms to protect free flow of information and provide for privacy in terms of communication and anonymity for senders and receivers. According to the Electronic Privacy Information Center (EPIC), the top 100 e-commerce sites on the Web fail to live up to all the "fair information practices of privacy protection." EPIC is calling upon lawmakers to produce legislation to enforce these practices. Besides the U.S., other countries such as Australia, Britain, Canada, Denmark, France, Germany, India, Singapore, and Sweden have also passed new anti-terrorism laws which may threaten the privacy of the individual user. Policymaking on the Internet again reinforces the discordance between enactments and enforcements.

Often governments grapple with the Internet's fast-changing technical environment and the technological challenges that it can pose through traditional forms of governance, and this can often be inadequate. For example, encryption can be seen as a right by regular users of e-mail, but it may be viewed as a threat by certain governments. The issue of encryption of information transmitted over the Internet, which may be a necessity for confidential business transactions, clashes with government interests in national security in countries such as the United States. Citing concerns over terrorism, drug trafficking, and money laundering, the United States forbids exporting of certain kinds of encryption technology for Internet use outside the U.S. Vietnam and Myanmar also cite national security as the main reason for restricting the diffusion of the Internet in the respective countries. In Vietnam, Web sites of foreign news sites and those belonging to human rights organizations are blocked by the government. It also forbids the use of the Internet for political opposition against national sovereignty and security and violations of morality of the law. In January 2000, the Chinese government passed a rule that any information conveyed via the Internet must first be viewed and approved by security forces. The government has also demanded that corporations and individuals report the specifics about encryption devices they possess.[12]

According to Lyon (2003, p. 9) surveillance entails "focused attention on behaviors and trends of persons and of populations with a view of managing, controlling, protecting or influencing them." Online surveillance can be done by both governments and corporations, and the degree of surveillance can be influenced by both global phenomena as well as local political, social, and economic considerations. In communities where the ethics of privacy and civil liberties are not valued or vocalized or actively defended through civil society and lobby groups, there may be fewer safeguards in place to defend the rights of Internet uses. While national security issues can override privacy issues in many countries on a global basis, the degree of variation is again determined by the local context and level of perceived threat to the national or local environment.

DEVIANCE AND THE INTERNET

The Internet, while a platform for information dissemination and commercial activity, can harbor behavior which may be illegal or deemed as criminal in most jurisdictions, and these can include copyright infringements, theft, credit card fraud, financial scams, money laundering, hacking, industrial espionage, cyber terrorism, gambling, defamation, and cyber stalking. In the U.S., the state of Pennsylvania passed a law allowing the state attorney to call a Web page to the attention of the local judge. If the judge finds probable cause that child pornography exists on the page, the Attorney General can demand that any Internet provider with Pennsylvania customers makes sure that the page is not visible to those customers. At times, states could appeal to the global ethics of the world community to fight vice. A case in point is the Bangkok-based organization called End Child Prostitution in Asian Tourism, which has urged governments to take action against those who use the Internet for promoting child prostitution and child pornography. The Council of Europe has also attempted to harmonize national laws dealing with several different types of crime. In a WTO case in which Antigua and Barbuda challenged U.S. restrictions on the supply of gambling and betting services via the Internet, for its defense the United States relied in part on Article XIV(a) of the WTO's General Agreement on Trade Services (GATS). This article permits countries to go against their market-access commitments if it is deemed as taking measures "necessary to protect public morals or to maintain public order" (Rundle, 2005, p. 8).

Anti-pornography measures are complicated by the varying definitions of what constitutes pornography. Hong Kong, for example, introduced an Internet content rating system in 2001 which classifies gay and lesbian Web sites as "harmful media." India has also declared the electronic publication of pornography an offense under the Technology Act 2000. The First Amendment provides constitutional protection and may therefore make the U.S. a haven for those wishing to spread hate speech on the Internet, and similar issues are raised by the recognition that Internet pornography receives constitutional protection within the United States and data privacy is a fundamental right outside the U.S. (Reidenberg, 2005, p. 1959).

Besides pornography, hacking is another activity that authorities and business corporations are grappling with which can endanger national security and privacy, leading to criminal activities. Many countries are enacting rules to guard against unauthorized access and entry into computer networks. These include India's Information Technology Bill 1999 as well as Japan initiatives in January 2000 to boost computer security after embarrassing raids on government Web sites. Since 1999, nations of the Association of South-East Asian Nations (ASEAN) have also agreed to intensify their efforts to fight Internet crime by setting up a framework to share information (Net-Security, 2002). The United Nations has also held annual conferences highlighting the problems law enforcement officials face when dealing with organized international crime networks that have access to global technologies such as the Internet. In July 2003, APEC held a conference on strengthening international law enforcement cooperation to deal with cyber crime.[13] Additionally, in an initiative under the International Civil Aviation Organization (2003), 188 governments have pledged to follow common standards to embed biometric information (e.g., photos, fingerprints, or digitized eye scans) in travel documents (Rundle, 2005, p. 13). While there has been consensus to a large degree from the world community with regard to child pornography and racist communication over the Internet, there are issues with enforcing and monitoring activities on the Internet on a global basis.[14]

THE ISSUE OF DIGITIAL DIVIDE

According to the Computer Industry Almanac Study, the worldwide number of Internet users surpassed 1 billion in 2005, up from 45 million in 1995 and 420 million in 2000, and 2 billion is expected to be reached in 2011.[15] Much of the current and future Internet user growth is coming from populous countries such as China, India, Brazil, Russia, and Indonesia while the U.S. continues to lead with nearly 200 million Internet users at year-end 2005. The two most populous countries, China and India, are now in second and fourth place in terms of Internet users. In Latin America and the Caribbean, in comparison, on average less than 6% of the population have regular access (if any access at all) to the Internet; they are usually those who reside in the main urban centers, and in their majority belong to a social strata which can afford to pay for a telephone line and access services besides owning or having free access to a computer in their home or office (Alfonso, 2005, p. 22).

The term *third world* came into currency with the crumbling of imperialism in the decade proceeding World War II to denote the geographical divide between the rich and poor nations. In the contemporary global landscape, the predominance of Europe and America in the real world is seen to be sustained in cyberspace. Often the Internet is narrated as a space dominated by the west where there is a 'threat' posed by the occurrence of foreign intellectual production and cultural values (Bosah, 2000; Gunkel, 2000) in terms of both content and consumption. This hegemony of foreign or Western presence is deemed as undermining local values and knowledge, and often elicits concern that this to some degree be matched by the expression and communication of local content that is relevant to indigenous contexts (Batchelor, 2002, p. 4). Linguistically, the vernacular of the West (i.e., English and European languages) largely dominates the Internet (Gunkel, 2000) intensifying the problems of access associated with Internet illiteracy.

The development of the Internet is often seen as integral to the notion of globalization, and while parts of the Internet have global aspirations, there are gaps between reality and these aspirations (Whittaker, 2002, p. 7). Affluent countries and regions such as America, Western Europe, and parts of eastern Asia lead the way in terms of Internet penetration usage, however the spread of computer-mediated communications is evidently uneven, with less than 0.1 of the population having access to the Internet in many countries (Whittaker, 2002). The two key issues that played a part in forming the idea of a World Summit on the Information Society (WSIS) at the start of the new millennium were growing inequalities and technological gaps, and the emerging challenge to reorder a world system in which multilateralism no longer seems to be a recognized principle and institutional legitimacy is a contested matter (O'Brien et al., 2000).

While the Internet can theoretically wire the globe, the issue of access, infrastructure, and literacy can also be defined by the spatial location. Most government initiatives in information technology are geared towards increasing access by making networked computers widely available in schools and colleges, and often this is augmented by commercial forces (Graham, 1999, p. 87). These can include initiatives such as the

European Commission's 'eEurope' to transform Europe into a computer savvy society. Many states already subscribe to the development of a network or information society. Malaysia's 'Multimedia Super Corridor', for example, reflects its endeavor to attract investments in this sector while transforming the country into an information society. In contrast, in July 2001, the Taliban regime in Afghanistan banned the use of the Internet by its citizens, and while this was overturned with the swearing in of the new interim president, Hamid Karazi, the problem of access was mainly due to poor infrastructure.[16] In developing countries and transitional economies, there is often a conflict between liberal economic policies targeted at the IT industry and strict Internet content regulations to curb materials on the Web. The issue of digital divide means that not all countries can have equal access to the Internet, and invariably there will be discrepancies between literacy and penetration rates not just between local and global spaces, but between rural and urban spaces within bounded territories.

The existing data on computer usages, while limited, nevertheless reveal that the majority of users tend to be from affluent countries and have professional backgrounds (Shahin, 1999). The rate of penetration, access, and usage in the different countries will be dependent on policies and corporate planning, and this will involve the role of governments, institutions, and corporations in the real world. The Regional Conference of Africa held in Bamako in May 2002 insisted in its conclusions on the need for Africa to address the economic, technological, and political aspects of the use of African languages on the Internet, including production and maintenance of Web sites in African languages, training of African data-processing specialists, and so forth (Alzouma, 2005). The Internet operates within a set of social conditions that give it meaning and value. The images of the Internet in Africa, for example, are constructed through a discourse which reflects the ideology of developmentalism and the interests of particular groups of actors. Therefore we should look behind the screen, where the real actors with real interests engage in 'socially anchored practices' and these instances of global culture flow combine a localized sense of belonging (Uimonen, 2003, p. 146).

CONCLUSION: CONTESTATIONS BETWEEN LOCAL AND GLOBAL

The issues, with regard to the Internet, nation-states, and territorial-based regulations, capture the complexities and dialectics, in terms of enacting overarching global and regional frameworks of governance. It nevertheless does not overshadow the role of nation-states (and as such the physical locations) in influencing user culture and terms of engagement with regard to new information and communication technologies. Nation-states are in fact leaving their mark on cyberspace by negotiating their rules while reconciling the need for a certain degree of standardization and cooperation with global entities and communities. This colonization of the Internet by nation-states is by no means an all-encompassing phenomenon (i.e., it may not affect whole Internet populations per se), but it resurrects boundaries and invisible barriers in the virtual spheres. It puts down rules of engagement for its citizens, thus infusing a unique user culture with regard to the Internet within the bounded physical spaces of the nation-state. The localization of the global Internet is inevitable as this global entity is tamed and adapted to the political, economic, and social conditions in the physical terrains of appropriation.

REFERENCES

Abbate, J. (1997). *Inventing the Internet.* Cambridge, MA: MIT Press

Alfonso, C. (2005, July). *Internet governance: A review in the content of WSIS process.* Research Paper. Montevideo, Uruguay: Instituto del Tercer Mundo.

Alzouma, G. (2005). Myths of digital technology in Africa: Leapfrogging development? *Global Media and Communications, 1*(3), 339-356.

Agre, P. E. (2002). Real-time politics: The Internet and the political process. *The Information Society, 18,* 311-331.

Bangkok Post. (2003, July 3). *Government forces ISPs to block 'inappropriate' Web sites.* Retrieved January 9, 2006, from http://www.thaivisa.com/index.php?514&backPID=58&tt_news=434

Barlow, J. P. (1996a, January 15). Thinking locally, acting globally. *Cyber Rights Electronic List.*

Barlow, J. P. (1996b, January 8). Declaration of Independence of cyberspace. *Cyber Rights Electronic List.*

BBC News. (2006, June 14). *Microsoft censors Chinese blogs.* Retrieved July 1, 2006, from http://newsvote.bbc.co.uk/mpapps/pagetools/print/news.bbc.co.cuk/1/hi/technology/40887

Beniger, J. R. (1996). *The control revolution.* Cambridge, MA: Harvard University Press.

Bennahum, D. S. (1996, March 2). Call for comments around the world. *MEME.*

Boyle, J. (1997). *Foucault in cyberspace: Surveillance, sovereignty and hard-wired censors.* Retrieved November 6, 2001, from http://www.law.duke.edu/boylesite/foucault.htm

Bosh, E. (Ed.). (2000). *Cyberimperialism? Global relations in the new electronic frontier.* Westport, CT: Praeger.

Buchstein, H. (1997). Bytes that bite: The Internet and deliberative democracy. *Constellations, 4*(2), 248-263.

Burnham, D. (1983). *The rise of the computer state.* New York: Pantheon Books.

Cairncross, F. (1997). *The death of distance.* London: Orion.

Castells, M. (2002). Introduction. In *Internet in everyday life.* Oxford: Blackwell.

Froomkin, M. (2004). When we say us, we mean it. *Houston Law Review, 41*(3), 839-884.

Friedland, L. A. (1996). Electronic democracy and the new citizenship. *Media, Culture and Society, 18,* 185-212.

Gandy, O. (1993). *The panoptic sort.* Boulder, CO: Westview.

Goldsmith, J. L. (1998). The Internet and the abiding significance of territorial sovereignty. *Indiana JGLS, 5,* 475-492

Geist, M. (2001, August). *Fair.com?: An examination of the allegations of systemic unfairness in the ICANN UDRP.* Retrieved June 9, 2006, from http://aix1.uottawa.ca/ngeist/~geistudrp.pdf

Geist, M. (2002) Fair.com?: An examination of the allegations of systemic unfairness in the ICANN UDRP. *Brooklyn Journal of Law, 27*(3), 903-938.

Geist, M. (2005, July 7). *Unease over how the Net is run.* Retrieved July 11, 2006, from http:newsovte.bbc.co.uk/mpapps/pagetools/print/news.bbc.co.uk/1/hi/technology/48716

Gomez, J. (2004, July). Dumbing down democracy: Trends in Internet regulation, surveillance and control in Asia. *Human Rights, 1.*

Graham, G. (1999). *The Internet: A philosophical inquiry.* London: Routledge.

Hagel, J., & Armstrong, A. G. (1997). *Net gain: Expanding markets through virtual communities.* Boston: Harvard Business School Press.

Harding, A. (2004, December 7). *Japan's Internet suicide clubs.* Retrieved December 7, 2005, from http://news.bbc.co.uk/2/hi/programmes/newsnight/4071805.stm

Horton, W., Taylor, L., & Hoft, N. (1996). *The Web page design cookbook.* New York: John Wiley & Sons.

Hine, C. (2000). *Virtual ethnography.* London: Sage.

Hughes, C. R. (2003). Fighting the smokeless war. In C. R. Hughes & G. Wacker (Eds.), *China and the Internet.* London: RoutledgeCurzon.

Johnson, D.R., & Post, G.P. (1996). Law and borders—the rise of law in cyberspace. *Standford Law Review, 48.*

Kaldor, M. (2003). *Global civil society; an answer to war.* Cambridge, UK: Polity Press.

Keane, J. (2003). *Global civil society?* Cambridge: Cambridge University Press.

Kiesler, S., Siegel, J., & Mcquire, T. (1984). Social psychological aspects of computer-mediated communication. *American Psychologist, 39,* 1123-1134.

Kennedy, G. (2000). New codes and protocols for cyberspace: Current issues in Internet governance. *CTLR, 6*(8), 223-229.

Klein, H. (2002). ICANN and Internet governance: Leveraging technical coordination to make global policy. *The Information Society, 18*(3), 193-207.

Klein, H. (2003). Technology, contracts and the Internet: Private governance for global communications. In S. Braman (Ed.), *The emergent global information policy regime.* New York: Palgrave.

Komaitis, K. (2003). ICANN: Guilty as charged? *The Journal of Information, Law and Technology,* (1). Retrieved February 3, 2006, from http://elj.warwich.ac.uk/jilt/03-1komaitis.html

Krebs, V. (2001). The impact of the Internet on Myanmar. *First Monday, 6*(5). Retrieved January 3, 2006, from http://www.firstmonday.org/issues/issue6_5/krebs/index.html

Kyu, H. Y. (2003). The Internet and democracy in Asia. In I Banerjee (Ed.), *Rhetoric and reality: The Internet and challenge for democracy in Asia.* Singapore: Eastern Universities Press.

Lessig, L. (1999). *Code and other laws of the cyberspace.* New York: Basic Books.

Loader, B. D. (Ed.). (1997). The governance of cyberspace. Politics, technology and global restructuring. In *The governance of cyberspace.* London: Routledge.

Lloyd, I. (1997). *Information technology law.* London: Butterworths.

Lippert, P. J. (1997). Internet: The new agora. *Interpersonal Computing and Technology, 5*(3-4), 48-51.

London, S. (1995). Teledemocracy vs. deliberative democracy; a look at two models of public talk. *Interpersonal Computing and Technology, 3*(2), 33-55.

Loundy. D. (1997, October). *Local government, global headaches.* Retrieved February 2, 2006, from http://www.govtech.net/magazine/stroy.print.php?id=95441

Lum, T. (2006, February 10). Internet development and information control in People's Republic China. *CRS Report for Congress.* Retrieved March 5, 2006, from http://www.fas.org/sgp/crs/row/RL33167.pdf

Luwarso, L. (2004). Manufacturing control: New legislation threatens democratic gains in Indonesia. In S. Gan, J. Gomez, & U. Johanne (Eds.), *Asian cyberactivism: Freedom of expression and media censorship.* Bangkok: Friedrich Naumann Foundation.

Lyon, D. (2001). Surveillance after September 11. *Sociological Research Online, 6*(3).

Lyon, D. (2003). Cyberspace, surveillance and social control. In K. C. Ho, R. Kluver, & K. Yang (Eds.), *Asia.com: Asia encounters the Internet.* London: RoutledgeCurzon.

Mantonvani, G. (1994). *New communications environments: From everyday to virtual.* London: Taylor and Francis.

Mathiason, J., Mueller, M., Klein, H., & Holitscher, M. (2004). *Internet governance: The state of play. The Internet Governance Project.* Retrieved January 3, 2006, from www.internetgovernance.org

Mayer-Schoenberger, V., & Ziewitz. M. (2006). Jefferson re-buffed—the United States and future of Internet governance. *KSG Faculty Working Papers, 6*(18). Retrieved June 6, 2006, from http://ksgnotes1.harvard.edu/research/wpaper.nsf/rwp/RWP06-018

McCullagh, D. (February 16). *Politicians lash out at tech firms over China.* Retrieved July 1, 2006, from http://news.com.com/2102_1028_3-6039834.html

Mclaughlin, A. (2003, December 17). The virtues of deliberative policymaking: A response to public participation in ICANN. *Berkman Center Research Paper.* Retrieved January 3, 2006, from http://cyber.law.harvard.edu/mclauhglin/mclaughlin-reponse-public participation.html

Mclaughlin, A., & Packard, V. (2005). What is bottom-up about global governance? *Global Media and Communications, 1*(3), 357-373.

Menthe, D. (1998). Jurisdiction in cyberspace: A theory of international spaces. *Michigan Technology Law Review, 4*(3). Retrieved February 5, 2006, from http://www.mttlr.org/volfour/menthe.html

Miller, D., & Slater, D. (2000). *Internet: An ethnographic approach.* Oxford/New York: Berg.

Mohammed, A. (2006, February 5). U.S. renews contract for oversight of Internet. *Washington Post.* Retrieved July 11, 2006, from http://www.washingtonpost.com/wpdyn/content/article/2006/05/19AR2006051901684

Mowshowitz, A. (1994). Virtual organization: A vision of management in the information age. *The Information Society, 10,* 267-288.

Mulgan, G. (1991). *Communication and control: Networks and the new economies of Communication.* New York: Guilford Press.

Net-Security. (2002, November 27). *E-cop.net endorsed by the e-ASEAN task force.* Retrieved January 4, 2006, from http://www.net-security.org/secworld.php?id=1140

Nye, J., & Donahue, J. (2000). *Governance in a globalizing world.* Washington, DC: Brookings Institute Press.

O'Siochru, S. (2003). *Global governance of information and communication technologies: Implications for global civil society networking.* Report written for the Social Science Research Council. Retrieved June 7, 2006, from www.ssrc.org

Padovani, C. (2003). *Notes on global governance analysis and democracy: From output (strategies and impact) to outcome (are more participatory processes thinkable?) A theoretical approach (looking at WSIS).* Memo prepared for the Social Science Research Council. Retrieved June 3, 2006, from www.ssrc.org/itc

Padovani, C., & Tuzzi, A. (2003, July). Changing modes of participation and communication in an international political environment: Looking at the World Summit on the Information Society as Communicative Process. *Proceedings of the Political Communication Section of the IPSA Congress,* Durban.

Pal, L., & Teplova, T. (2003, May). Domain games: Global governance and the Internet. In *Proceedings of the Annual Meeting of the Canadian Political Science Association,* Halifax.

Palfrey, J. (2004, January). The end of the experiment: How ICANN's foray into global Internet democracy failed. *Public Law Research Paper Series, 93.* Retrieved June 9, 2006, from http://ssrn.com/abstract=487644

Proksch, W., & Schweghofer, E. (2001, April). Internet governance and territoriality nationalisation of cyberspace. In *Proceedings of the 16th BILETA Annual Conference,* Edinburgh, Scotland. Retrieved January 28, 2006, from http://www.bileta.ac.uk/01papers/proksch.html

Rundle, M. (2005). Beyond Internet governance: The emerging international framework for governing the networked world. *The Berkman Center for Internet & Society Research Publication Series,* (16). Retrieved June 8, 2006, from http://cyber.law.harvard.edu/publications

Randall, V. (1988). *Democracy and the media.* London: Frank Cass.

Reidenberg, J. (1997). Governing networks and rule-making in cyberspace. In B. Kahin & C. Nesson (Eds.), *Borders in cyberspace: Information policy and the global information infrastructure.* Cambridge, MA: MIT Press.

Reidenberg, J. (2005). Technology and Internet jurisdiction. *University of Pennsylvania Law Review, 153,* 1951-1974.

Reporters Sans Frontieres. (2002). *Torture, arbitrary detention and self-censorship, four months later: Consequences of the state of emergency and the fight against 'Maoist terrorism' attacks on freedom of press.* Retrieved February 22, 2006, from http://www.rsf.fr/article.php3?id_article=7245

Rimmer, M. (2003). Virtual countries: Internet domain names and geographical terms. *Media International Australia Incorporating Culture and Policy, 106,* 124-136.

Rheingold, H. (2000). *The virtual community.* New York: Harper Collins.

Savigny, H. (2002). Public opinion, political commitment and the Internet. *Politics, 22*(1), 1-8.

Slater, D. (2002). Social relationships and identity on/off-line. In L. Lievrouw & S. Livingstone (Eds.), *Handbook of new media: Social shaping and consequences of ICTs.* London: Sage.

Shahin, J. B. (1999, October). Regulation, governance and the Internet: The quest for a global ethic? In *Proceedings of Ethicomp.* Retrieved January 3, 2006, from http://www.ccsr.cse.dmu.ac.uk/conferences/ethicomp/ethicomp99/abstracts/shahin.html

Sharrock, L. (2001). The future of domain name dispute resolution: Crafting practical international legal solutions from within the UDRP framework. *Duke Law Journal, 51*(2), 817-849.

Singer, M. (2003, June 27). *ICANN makes peace, forms ccNSO.* Retrieved July 11, 2006, from http://www.internetnews.com/infra/print/php/2229111

Slaughter, A. M. (1995). International law in a world of liberal states. *EJIL, 7,* 503-538.

Slaughter, A. M. (1997). The real new world order. *Foreign Affairs, 76*(5), 183-538.

Slevin, J. (2000). *The Internet and society.* Cambridge, UK: Polity Press.

Sproull, L., & Kiesler, S. (1986). Reducing social context cues: Electronic mail in organizational communication. *Management Science, 32*(11), 1492-1512.

Sproull, L., & Kiesler, S. (1991). *Connections: New ways of working in the networked organization.* Cambridge, MA: MIT Press.

Tantuico, D. (2001). Governance in cyberspace. *Inter Media International Institution of Communications, 29*(5/6), 14-16.

Thierer, A., & Crews, C. W. (2003). Introduction. In A. Thierer & C.W. Crews (Eds.), *Who rules the Internet.* Washington, DC: Cato Institute.

Tuccille, J. D. (2001, August). *The thwarting of Internet censorship.* Retrieved January 3, 2006, from http://www.free-market.info/main0107/messages/028962908.html

Uimonen, P. (2003). Networking as a form of life: The transnational movement of Internet pioneers. In G. Garsten & H. Wuff (Eds.), *New technologies at work: People, screen and social virtuality.* Oxford: Berg.

Van Dijk, J. (1999). *The network society.* London: Sage.

Wacker, G. (2003). The Internet and censorship in China. In R. Hughes & G. Wacker (Eds.), *China and the Internet.* London: RoutledgeCurzon.

Weinberg, J. (2000). ICANN and the problem of legitimacy. *Duke Law Journal.* Retrieved June 9, 2006, from http://www.law.duke.edu/journals/dlj/articles/dlj50p187.htm

Whittaker, J. (2002). *The Internet: The basics.* London: Routledge.

Wynn, E., & Katz, J. E. (1997). Hyperbole over cyberspace: Self-presentation and social boundaries in Internet homepages and discourse. *The Information Society, 13*(4), 297-327.

Yu, P. K. (2003). The never ending ccTLD story. *Cardozo Law School Public Research Paper, 65.* Retrieved June 7, 2006, from http://ssrn.com/abstract=388980

Zeller, T. (2006, February 15). House member criticizes Internet companies for practices in China. *The New York Times.* Retrieved July 1, 2006, from http://www.nytimes.com/2006/02/15/technology/15end-internet.html?ei=5088&en+id2d

Zittrain, J. (2003). Be careful what you ask for: Reconciling a global Internet and local law. In A. Thierer & C. W. Crews (Eds.), *Who rules the Internet.* Washington DC: Cato Institute.

ENDNOTES

1 The UN created the Working Group on Internet Governance (WGIG) composed of 40 members from several countries and constituencies (governments, business, academics, organized civil society).

2 See WSIS 2003. Retrieved February 2, 2006, from http://www.itu.int/wsis/

3 The specific set of functions assigned to ICANN by the U.S. Department of Commerce's "Memorandum of Understanding" gave it the authority to set policy for, and to manage the allocation and assignment of, Internet protocol addresses; add new names to the top level of the Internet domain name hierarchy; and maintain responsibility for root servers that distribute authoritative information about the content of the top level of domain name space (Mueller, 2002).

4 At the heart of the Internet are 13 so-called root servers, 10 of which are in the United States. These maintain records allowing a domain name such as *wsw.org* to be translated into an IP address pointing to a specific computer on which a Web site is held. The root servers perform a critical role in routing traffic that ends in a top-level domain such as country-specific code. The U.S. has made it clear that it intends to retain control of the Internet's root servers indefinitely.

5 U.S. trademark law and the related UDRP process may not be familiar to such registrants, but at least they have arguably voluntarily availed themselves to U.S. jurisdiction and legal principles.

6 WIPO is a specialized agency of the United Nations which administers 21 international treaties regarding intellectual property protection. It also plays a significant role in the development of Internet domain policy in particular by preparing reports and recommendations based upon submission of members. In response to requests from national governments, WIPO initiated international consultations to address outstanding issues relating to intellectual property and Internet domain names.

7 The Syracuse University Convergence Center's Study on WIPO is available at http://dcc.syr.edu/report.htm

8 The Max-Planck Institute Report is available at https://www.zar.uni-karlsruhe.de/admin/get_data.php?resID=95

9 Other measures include controlling via ISPs and through ideology that stresses 'Asian values' as being deferential to the government and authorities for the welfare of the nations, and where civil liberties and human rights are not seen as priorities or even detrimental to the development of the nation.

10 Refer to "Internet Censorship: Law and Policy Around the World" on the Electronic Frontiers Web site. Retrieved December 4, 2005, from http://www.efa.org.au/Issues/Censor/cens3.html

11 RSF, June 18, 2003; Index on Censorship, August 6, 2002.

12 See "Cryptography and Liberty 2000: An International Survey of Encryption Policy." Report by EPIC, Washington, DC. Retrieved November 4, 2005, from http://www2.epic.org/reports/crypto2000/countries.html

13 APEC, July 25, 2003.

14 Sixty-seven out of 190 states have become parties to the "Optional Protocol to the Convention on the Rights of the Child" and in the Durban Declaration of the World Conference Against Racism, Racial Discrimination, Xenophobia, and Related Tolerance which took place in 2002, which indicates a general consensus on the norm that extreme racist content should be prevented.

15 Computer Industry Almanac Study 2005. Retrieved February 3, 2006, from http://www.c-i-a.com/pr0106.htm

16 "Internet Under Surveillance." RSF Report 2003.

Chapter XIII
Web Sites and the Law:
An Avenue for Localization

Kendall Kelly
Texas State University, USA

Jennifer Jones
University of the Incarnate Word, USA

ABSTRACT

By adhering to cultural expectations, a Web site can build ethos with an intercultural audience. Those who create successful intercultural Web sites and Web site content must understand the laws and legal systems of the target country to create Web sites that the target audience will view as trustworthy. Laws create certain expectations in audiences, and when a seller violates such laws or even appears to violate such laws, it can create a loss of consumer confidence. This chapter examines certain legal and cultural issues that will instantly brand a Web site as unreliable, because part of developing a successful international Internet business is developing an ethos that potential customers can trust. The ideas and opinions presented here should be viewed as a means for raising awareness of diverse legal cultures and reduce the need to make costly changes as a result of legal advice given outside of an appropriate cultural context.

INTRODUCTION

When La Ligue Contre Le Racisme Et l'Antisemitisme sued Yahoo!, Inc. in a French court, it became clear to everyone that Web sites, even when not intended, were international and that companies with Web sites were subject to legal action in almost any country where they could be viewed. In fact, even though Yahoo! had not specifically localized its site for the French or France, it was using software to identify users who accessed the Web site from French ISPs and then pushed French advertisements to those users. The TGI, the Tribunal de Grande Instance, ruled that by targeting French users, Yahoo! had established a basis for jurisdiction in their court.

Yahoo! pursued further appeals but still decided to ban Nazi and other anti-Semitic materials from its Web site (Reinsch, 2005, pp. 179-180). As Hestermeyer (2006) notes, "Where the other country [in this case France] turns out to be an important market the company will often comply with the judgment whether it can be enforced or not" (p. 275). Yahoo! understands that a country's legal system and its laws are an intrinsic part of its culture. By adhering to such cultural expectations, a Web site can build ethos with an intercultural target audience. Those who design and create successful intercultural Web sites and Web site content must understand the laws and legal systems of the target country, so that they can build Web sites that the target audience will view as trustworthy.

Laws create certain expectations in audiences, and when a seller violates such laws or even appears to violate such laws, it can create a loss of consumer confidence. E-businesses, already perceived by consumers as somewhat risky, can ill afford such a loss. "Along with the potential advantages of electronic commerce come uncertainties and risks. The very attributes that make electronic commerce advantageous also create risks that undercut consumer confidence" (Alboukrek, 2003, p. 433). Consequently, consumers may perceive a Web site as illegitimate, even if it is not actually illegal, because the Web site does not adequately accommodate cultural sensitivities—for example, by displaying banned material or advertising in a manner considered to be in poor taste. If intercultural Web sites operate differently than the Web sites the target audience regularly views, the audience might not recognize the specific ways in which the site deviates from familiar sites originating in their geographic area, but they unconsciously recognize its difference. And while some difference is refreshing, these kinds of differences add an air of immorality and illegitimacy to a Web site that most e-businesses would prefer to avoid. Understanding the basic legal system of the target country even when one might not understand specific laws will help Web developers create more trustworthy and, therefore, more successful Web sites. The manner in which cultures traditionally reason is particularly important when addressing cross-boundary issues related to Web site development. For example, in addressing jurisdictional issues for cyberspace, August (2002) writes:

Judges, legislators, and treaty draftsmen are using analogy—the tried-and-true tool of legal reasoning—to modify existing rules to fit this new paradigm. Reasoning by analogy can have its problems, however, especially when differences in context are not taken into account. Nevertheless, analogy is the reasoning tool of choice for creating new law, and existing analogous rules have to be considered in any description of cyber-jurisdiction as it currently stands. (p. 531)

This chapter touches on some legal and cultural issues that will instantly brand a Web site as unreliable, because part of developing a successful international Internet business is developing an ethos that potential customers can trust. When a Web site offers products that may be banned as Yahoo! did, or defames the character of a company as Dow Jones was accused of doing, or uses deep-linking or frames in a manner that is inconsistent with other Web sites that customers view, or even relies on disclaimers when the audience is unfamiliar with disclaimers, the Web site undercuts its reliability and loses customers. By recognizing that legal issues represent a localization front even as graphics, content, and rhetoric do, Web site designers will create better localized and ultimately more effective Web sites for intercultural audiences.

The purpose of this chapter is *not* to offer techniques for avoiding litigation associated with international Web sites or to address specific issues of international cyberlaw. Before making a localized Web site available to the public in the target country, prudent businesses will seek qualified counsel in the target country. The advice in this chapter should raise awareness of diverse legal cultures and reduce the need to make costly changes as a result of legal advice given outside of an appropriate cultural context. Although an increased awareness of cultural legal context could potentially save businesses money, it is not a substitute for legal counsel.

BACKGROUND

To create a global Web site, developers start with internationalized source content: marketing material, product specifications, documentation, and so forth. This internationalized content allows translators and localizers to modify the Web site content easily for their target audience. This process is referred to as localization. Early localizers and intercultural communication experts relied on the theories of Hall (1976) and Hofstede (1991) to inform localization efforts until localization theorists directly applied their theories to Web site design (St.Amant, 1999; Singh & Periera, 2005). In addition to the most obvious aspect of localization, translation, localizers consider rhetoric and content (St.Amant, 2003) as well as graphical pitfalls that await the unaware international Web developer (Locke, 2003; Kostelnick, 1995; Quiye, 2000; Forslund, 1996). And yet other articles give developers practical advice on how to streamline the globalization process (Jones, 2002). Several books also cover the cultural, rhetorical, graphic, and practical barriers to creating global Web sites (Yunker, 2003; Esselink, 2000; Singh & Pereira, 2005).

The private sector is beginning to respond to the research of localization experts. A 2005 study by Singh, Kumar, and Baack (2005) demonstrated that international Web sites created by American companies did have some similarities with their local counterparts, but that they were not identical. Legal issues are the source of some of these differences. Driskill (1996) notes that legal issues can be problematic for intercultural collaboration "especially in the writing of warnings" (p. 27). Schadbach (1998) warns that although legal scholars recognize that law must deal with "rules that occur everywhere and often cross national borders" and that this necessitates translation, ultimately, meaningful translation is impossible (p. 336). Ulijn (1996) points out that mere compliance with European product liability laws that require translation into the consumer's native language may be insufficient; "differences in culture may cause documents to be interpreted in a way the U.S. supplier did not intend" (p. 70). Language and cultural constraints "create the potential for textual disharmony—the possibility of having several interpretations for a single word or phrase" (Alboukrek, 2003, p. 436). Obviously, technical communicators cannot rely merely on translation to create legal, safe, and ethical Web sites.

Unfortunately, the literature on Web site localization does not recognize legal issues as a localization issue, but exclusively a legal one. Several articles and Web sites discuss the problems associated with Web site design to avoid litigation (Reinsch, 2005; Hestermeyer, 2006; Ott, 2006; Bolin & Tysver, 2006) Anders Permeus, a Swedish attorney specializing in intellectual property and computer and technology transactions, offers a wonderful article on Web site design in Sweden (2001). While all of these articles present ways to avoid litigation, none recognizes that ignoring these legal issues can create a gap both in credibility and comprehension within a culture. Certainly, avoiding litigation is important for e-businesses; engendering consumer confidence so that customers will buy the site's products is also important. Since a people's legal system is an intrinsic part of their culture, a site that does not appear to abide by the local laws will not attract many customers even if it does avoid litigation. Consequently, legal issues must be a factor in localization, not just during the legal review at the end, but during the localization process, just as design, layout, graphics, and rhetoric are.

So far, no one has considered the legal issues that apply to the Web site localization process in part because looming litigation is a complex and pressing issue left exclusively to the realm of international attorneys. The perception is that the international legal landscape is too complex and varied for Web site developers to even consider. In *Beyond Borders,* Yunker (2003) admonishes Web site developers that "Every country has its own unique regulations ... you need to rely on legal experts in each local market to help you play by the rules" (Yunker, 2003, p. 189). Problems arise with legal localization because much of the localization source material comes from the United States and much of the guides to legal issues governing the Internet have a decidedly American bent. August (2002), in a footnote, recognizes the limiting effect of legal culture and tradition in addressing problematic cross-boundary issues, and describes the American Bar Association's report, "Achieving Legal and Business Order in Cyberspace: A Report on Global Jurisdiction Issues Created by the Internet, 55 Bus. Law. 1801" (2000), as "limited by its decidely American perspective" (p. 550).

While this article reiterates the advice that companies should have local legal experts review their Web sites, it also seeks to provide strategies to accommodate target countries' legal systems, not in an attempt to avoid litigation, but rather to create more credible and hence more successful Web sites. By explaining the two basic legal systems, common and civil law, that most countries base their legal systems upon, Web site developers may more easily create localized Web sites that target audiences will both understand and believe.

COMMON LAW AND CIVIL LAW

Most of the world's population—99.58%—is subject to legal systems influenced by civil law or common law traditions. The civil law tradition and hybrids thereof account for 59.03% of global population, whereas the common law tradition and hybrids thereof account for 34.81% of the global population. In terms of global political division, the civil law tradition and its hybrids account for 44.77% of political entities, whereas the common law tradition and its hybrids only account for 27.91% of political entities (University of Ottawa, 2006). Therefore, for companies in common law jurisdictions—particularly those in most of the United States given its historical lack of familiarity with civil law traditions—an awareness of diverse legal traditions is necessary for meaningful global business decisions in localizing content for Web sites (Gruning, 2004, p. 437) .

Although a definitive history of civil law and common law systems, as well as explanations of similarities and differences between the two systems, is outside the scope of this chapter, Schadbach (1998) provides concise definitions useful for raising a company's awareness of different legal systems to inform its business practices:

The legal systems of the so-called 'civil law' countries were originally based on Roman law. These legal systems are characterized by a primary law-making approach of comprehensive codifications. The common law family, on the other hand, is an organically grown body of case law supplemented by relatively unextensive statutes with a narrower scope ... another crucial difference between common law and civil law systems lies in the perception of the role of the law in each system. A civil law system's code generally states new law and abolishes the previous law in the same area whereas the statutes in common law systems generally supplement or codify preexisting case law. (p. 332)

Moreover, Stein states that reasoning applied by legal thinkers across civil and common law systems varies because civil law is based on the premise that "there can be only one right answer to any legal problem" whereas common law is "open-ended in that new extensions to existing rules can be revealed at any time by the courts" (Stein, 1992, p. 1594). In addition, Koch (2003) describes the "overarching goal" of civil law systems as "legal certainty," yet in a common law system, legal certainty is impossible to achieve (p. 33). Finally, Curran (2001) describes common law as "defined in terms of past judicial decisions. The resulting methodology is such that the common law perpetually is in flux" (p. 75).

To illustrate the possible disconnection between the civil and common law traditions, Stein (1992) provides the following insightful example:

The common-law lawyer is accustomed to opening his case with a discussion of the facts and only later to come to the relevant rules of law. He finds it difficult to invert his normal mental processes, even when the issue turns on the interpretation of a legislative text. An English barrister, arguing a case on the application of the Value Added Tax before the Court of the European Communities at Luxembourg, where the civil-Law ethos prevails, began by reciting the circumstances that had given rise to the case. The judges interrupted him, asking what rules were applicable and what was the issue. How can you understand the problem, he replied, until I have explained what the case is about? (p. 1599)

Consequently, civil law countries strive to create relative legal certitude with the codes and regulations, while common law countries view such artifacts as a basis for judicial decision making. Consequently, civil law countries generally provide greater legal certainty for businesses, and that certainty creates options for localizers.

Disclaimers and Common Law / Civil Law

On his comedy audio CD entitled "Dangerous," Bill Hicks (1990) highlights the attitude that many consumers in common law countries have about legal disclaimers. He points out that every pack of cigarettes comes with a different surgeon's general warning; he always asks for the "low-birth weight" packs, while avoiding the "may cause lung cancer" packs—just to be safe. He gets big laughs because the audience recognizes that legal disclaimers rarely represent the actual outcomes consumers experience, but merely warn of possible risks. Consumers in common law countries and the United States in particular take legal disclaimers for granted and understand that the dire outcomes described in the disclaimer may not come to pass or, at least, not any time soon. Consumers in civil law countries, who are not as inundated with legal disclaimers as consumers in common law countries, may not accept disclaimers in this manner. They may not view disclaimers as unlikely or at least not immediate outcomes to guard against, but rather as statements of probable outcomes. At the very least, legal disclaimers will undercut the perceptions of reliability in civil law countries, because competing Web sites do not contain such disclaimers.

Imagine Web sites for two similar products, servers for example. One Web site touts the speed and reliability of the server, while the other Web site combines descriptions of speed and reliability with statements like "misuse of equipment may result in data corruption or loss." Obviously the former Web site is far more persuasive. Consequently, Web designers should not automatically insert single-sourced disclaimers, but rather contract legal counsel from the target country to determine if disclaimers are required in civil law countries, because unnecessary disclaimers could easily destroy the credibility the Web designer has tried so hard to build. In other words, legal disclaimers serve merely to underscore a product's weaknesses while not always providing adequate legal protections. Furthermore, even in common law countries where legal disclaimers do provide considerable legal protections, they are unlikely to offer such protections when translated from another language.

In other words, what a case is about, or legal language means, turns not on an objective legal definition imbued with cross-cultural meaning, but rather finds meaning amid the cultural and systemic context of applicable law. Therefore, mere translation of legal language may not properly contextualize words in a legally meaningful manner. Instead, the "socio-economic, politic-cultural, and systemic differences of the respective jurisdictions" must be considered (Schadbach, 1998, p. 419).

Valcke (2006) posits that participants within each legal system will provide legal interpretations distinct from those provided by outside observers: "Legal materials, be they judicial decisions, statutes, or whatever else qualifies as 'legal materials' in any given legal system, indeed may be interpreted very differently depending upon whether the interpreter is, or not, immersed in the local legal culture" (p. 133). As a result, a company seeking legal advice related to cross-border and cross-cultural transactions would likely benefit from engaging counsel from within the local legal system. Just as rhetorical and cultural contexts can render translations meaningless even when the audience may understand the meaning of the words themselves, translations of legal language can also be rendered meaningless. Hall (1976) explains: "The problem lies not in the linguistic code in the context, which carries varying proportions of the meaning. Without context, the code is incomplete since it encompasses only part of the message" (p. 75).

Consequently, technical communicators must be aware of the effect context has on all language, but especially legal language and legal disclaimers. Civil law and common law systems create a distinctive context for readers. For example, as Koch (2003, p. 33) explains, civil law systems attempt to achieve legal certainty. The codes themselves attempt to establish a company's liability in a particular circumstance. The existence of disclaimers may not afford e-businesses in civil law countries the legal protections that such disclaimers provide in common law countries. For example, "a forum selection clause," which is typically enforceable in the United States, is not enforceable in the European Union. Thus, a Web site developer's inclusion of this clause would be rendered meaningless.

Forum selection clauses are enforceable worldwide for almost all kinds of transactions. The exception is for consumer contracts. In the European Union, the clauses are unenforceable; while in the United States and elsewhere, they are enforceable. These differing approaches are reflected in the Draft Hague Convention on Jurisdiction and Foreign Judgments in Civil and Commercial Matters, where the negotiating parties have yet to agree to a uniform treatment; and it looks as if they will include both approaches in the final convention. (August, 2002, p. 573)

Consequently, including such a clause on one's Web site affords companies targeting EU companies no legal protection in business-to-consumer transactions, but they do create a culturally inappropriate Web site.

Companies in common law countries use legal disclaimers to manage the dynamic nature of the legal system. Readers in these countries anticipate disclaimers, understand the litigious context that fosters them, and read them accordingly. Further, consumers in common law countries may be more inclined to use products inappropriately without the legal disclaimers that they expect and, of course, their exclusion can expose companies to substantial liability. Disclaimers are only one area in which the common and civil law systems have ramifications.

PROTOTYPING IN REVERSE: A MODEL FOR LEGAL LOCALIZATION

In his 2005 article, "A Prototype Theory Approach to International Web Site Analysis and Design," Kirk St.Amant suggests a method for analyzing Web sites for design, layout, and graphics, so that designers can quickly localize their Web sites. St.Amant explains that "the key to using prototype theory in International Web site analysis and design has to do with concepts of recognizability and acceptability/credibility" (p. 80). He suggests that designers examine a range of Web sites from the target country in the same line of business that one wishes to start. After viewing a variety of Web sites, the designer develops a range of acceptability that his or her Web site must meet. For example, when examining Web sites from the target country, a designer may develop a prototype based on the number and style of graphics and photographs, the content of photographs and graphics—that is, people or places or something abstract, the presence and location of navigation bars, the density of text and hyperlinks, the typefaces used, and so forth. Eventually the designer can develop a prototype of preferred Web design for the target country and design his or her Web site appropriately.

Prototyping for legal localization works in reverse. When examining target Web sites for legal issues, designers should not try to develop a legal prototype, but rather consider their internationalized source content as the prototype and look for other examples. For instance, when considering a marketing strategy or a critique or description of a competitor or his products, look for similar occurrences on Web sites in the target country. If the Web designer can find similar marketing strategies on several other Web sites in the target country, such marketing strategies are clearly credible and effective for the target audience. If the potential global marketer cannot find similar products and marketing strategies for similar, reputable businesses in the target country, it is highly unlikely that the Web site designer has discovered an untapped market or extremely novel marketing strategy for the target market, rather the sale of such products and the use of such marketing strategies is probably banned in that particular country.

Obviously this technique has some drawbacks such as what if the target country really is an untapped market? Or the country has very few Web sites originating from it? In such instances, designers should consult similar lines of business and similar countries. Furthermore, earlier and more frequent consultations with e-commerce attorneys from the target country are probably in order. Businesses should not risk expenditures on developing marketing for products that have a high likelihood of failure.

Reverse prototyping for legal issues should prove to be a success in civil law countries, since the civil law system strives for clarity in legislation. Consequently, reverse prototyping is likely to yield Web sites with credibility. However, only a qualified attorney can provide adequate protection from litigation claims. Reverse prototyping can work for common law countries as well, but it is less likely to be successful, because common law legislation does not create guidelines that are as detailed as civil law legislation. In the following section, we will explain how reverse Web site prototyping can work in both civil and common law countries regarding marketing and defamation.

Marketing

Most people who travel abroad have a momentary disorientation or shock when confronted with advertising strategies that are not practiced in their native country. This author never ceases to be vaguely shocked and amazed at the ability of European marketers to employ semi-clad women to hawk products and even more amazed at the effectiveness of such marketing. As Janet

Jackson discovered, the FCC is not very tolerant of bare-breasted women in the United States, so many Americans traveling abroad are vaguely disoriented when such images are used to sell products. Local laws influence the development of consumer marketing tastes and a consumers' culture influences local laws. Consequently, because laws governing marketing strategies and regulations vary greatly from country to country, they can really brand your Web site as foreign and unreliable.

Advertising to children is an example of marketing strategies that can vary from country to country. In Sweden it is illegal to direct advertising at under-12s. In Greece there is a ban on toy advertisements between 7 a.m. and 10 p.m. And Italy has a ban on the use of under-14s in advertisements (although this law is currently under review). In the UK, the government's recent White Paper on Health demanded voluntary action from the advertising industry on the advertising of food (Oliver, 2005, p. 19). Web sites clearly marketing to children will probably seem unreliable to parents in those countries and are unlikely to encourage such consumers to purchase products from the vendors they represent. Marketing strategies that are legal and effective in the United States such as empirical head-to-head demonstrations of efficacy, like the paper towel commercial that illustrates how much more absorbent brand A is than brand B, could easily be considered bad taste in other countries. Some of the marketing staples that may be acceptable in some countries, but unacceptable in others are testimonials, nudity, any representations of people, and certain claims of effectiveness.

Civil law countries lay out clear guidelines that govern acceptable and unacceptable marketing strategies, and the Web sites in those countries reflect those guidelines. For example, the EU lists 24 specific claims, including phrasing that producers may make about their foods (Murphy, 2006, p. 17). Because all food producers may only use these 24 claims, Web sites for food sold in EU countries have clearly identifiable trends. Consequently, reverse prototyping can be quite effective for vetting marketing strategies. Before using any marketing strategy from playing a jingle to showing pictures of the product, ensure that such a marketing tactic is common on other Web sites in the target country.

Marketing regulations are not so clear cut in common law countries, since the subtleties of case law generally govern them. One could still rely on reverse prototyping, but should be more conservative when adopting marketing strategies from other Web sites. A Web site designer should find numerous instances of the strategy in question before adopting that strategy.

Defamation

One country's free speech is another country's defamation. In 2002 an Australian business man, Mr. Gutnick, sued *Barrons* for reporting on the possible improprieties of religious charities that had traded the publicly owned securities of several of his companies. Likewise, Harrods sued Dow Jones for an article about Harrods entitled "The Enron of Britain." Additionally Mr. Bangoura sued the *Washington Post* for articles criticizing his leadership of a UN program (Goel, 2005, p. 16). In the United States, such criticisms do not approach defamation, but they could be the source of litigation and even large judgments in countries like Australia and England. Consequently, before publishing any materials critical of any person, product, or company on a Web site targeted for another country, consider the content carefully for signs of defamation. Reverse prototyping will work well in civil law countries in this regard. Web site designers should look carefully for similar kinds of criticisms before putting them on their Web sites. Once again, as demonstrated with marketing strategies, removing possibly slanderous materials from the localized Web site will create a more credible Web site for the target audience, but one should consult local legal counsel to ensure that Web materials do not contain any possibly slanderous materials.

In countries, particularly common law countries, with stringent defamation laws that allow for broad interpretation, the prudent business person should avoid any language or comment that could possibly be construed as defamation, not only to avoid possible litigation, but because consumers will view such language as outside the mainstream and from a generally unreliable source. It is notable that Goel (2005, p. 17) recommends blocking Web site access for people from "high-risk jurisdictions" such as Australia and England, because the threat of litigation is too great.

DEEP LINKING, FRAME-LINKING, AND TRADEMARKING

By understanding civil and common law as well as prototyping, Web designers can create credible Web sites that should require little modifications after legal review. The new and changing nature of the technology makes conclusive decisions regarding Web site design impossible, because they simply are not a matter of settled law and consumers will not recognize a standard. Deep linking and framing are examples of this. Deep linking is the practice of linking to someone's Web site below or deeper than the homepage. Such linking can allow audiences to bypass registration and advertising on the homepage. Of course, the technology does exist to require users to register and view advertising regardless of the point at which they link to the site, and several courts have noted that. Other courts have interpreted such practices as a copyright violation as I note below.

Frame-linking, or just framing, occurs when one Web site provides a link that opens someone else's Web site within the original Web site's frame. Because such a technique frequently makes the content of the secondary Web site appear as if it originated from the original Web site, some courts have identified this technique as a copyright violation as explained below.

Finally, because the use of trademarks by anyone other than the trademark holder is illegal in many countries, using trademarks as the clickable text that links one Web site to the trademark holder is illegal (Permeus, 2001). Such a practice is common and legal in other countries.

Unfortunately, even in countries that have passed legislation to address this issue, sufficient case law has not clarified the issue. For example, Lubens (2003) notes that "although copyright and unfair competition claims against deep linking practices have been largely unsuccessful in the United States, the 1996 European Community Directive on the Legal Protection of Databases ('the Directive') has been successfully deployed in Germany and Denmark to make deep linking illegal" (p. 447). However in a 2006 ruling, the Danish Maritime and Commercial Court ruled that such linking was legal (Ott, 2006). Permeus (2001) explains that the Supreme Court in Sweden held that deep linking was not illegal, but that procedural issues prevented this ruling from setting precedent in the manner hoped for (pp. 23-24).

Framing is even more difficult to understand. Permeus (2001) notes that while framing is not specifically illegal in Sweden, the difficulty in differentiating the original Web site from the secondary Web site creates a copyright issue (p. 24). A recent Austrian decision found that framing was not illegal while, CNN, Reuters, Dow Jones, the *Times Mirror*, and the *Washington Post* reached a settlement with TotalNews banning framing (Ott, 2006). And one can find numerous framed Web sites within the U.S.

The unsettled nature of Internet laws makes it difficult for e-businesses to avoid litigation, but easy to create Web sites that look and feel like their homegrown counterparts. Because the case law is not definitive, no Web site structure will brand a Web site as unknowledgeable and unreliable, although avoiding practices like deep linking, frame-linking, and trademarking might help e-businesses avoid lawsuits.

LOCAL LEGALIZATION CHECKLIST

To recap, when localizing Web sites, consider these legal issues:

- **Common or civil law:** Does the target country use a common law or civil law system? (Consult the *CIA World Factbook* for clarification.)
  - If civil law, exclude disclaimers completely and allow legal counsel from the target country to review the Web site at the end of the process and add disclaimers as appropriate within the legal cultural context.
  - If common law, contract with legal counsel from the target country early to write disclaimers as appropriate within the legal cultural context and add them to your single-source content.
- **Marketing:** Identify Web sites for similar lines of business in the target country.
  - Do the market strategies from your source Web site match?
  - If they do not, what marketing strategies should you use?
- **Defamation:**
  - If civil law, do criticisms match Web sites in the target country?
  - If common law, exclude any potentially defamatory language.
  - Always have local legal counsel review a Web site before posting online.
- **Linking:**
  - Avoid deep linking, frame-linking, or use of trademarking.
  - If you must employ these strategies, contract with legal counsel from the target country to ensure their legality in the target country before you create a Web site that relies on them.

CONCLUSION

Legal localization presents another front for Web developers in their localization efforts. Just as culturally inappropriate graphics, layout, design, and rhetoric can confuse and repel target audiences, so can culturally inappropriate legal gaffes. In part because the global legal climate regarding e-commerce has focused primarily on jurisdiction issues and because legal materials are generally difficult to understand, Web developers have avoided localizing for these issues. Instead they have delayed consideration of legal issues until late in the localization process just as they delay legal review of the source materials until late in the development process. However, Web developers unconsciously account for legal issues in the content they develop in their home countries because they have internalized the legal issues. While content developed by native Web developers may have some legal problems, such content generally adheres to the legal standards of their home countries. Such is not the case when localizing content for other countries. While native localizers will correct some of the legal issues, without the direction of the supervising Web developer, they may not consciously recognize such issues. Hopefully this chapter has demonstrated the importance of legal issues as a localization front to gain legitimacy and provided some strategies for approaching this front.

We have attempted to provide a legal lens, a framework, for viewing the international legal landscape that will help Web developers approach its shifting nature. By understanding the differences between the common and civil law systems, Web developers can build a foundation for examining and creating Web sites that engender confidence in consumers, and ultimately contribute to successful e-businesses and broad-ranging acceptance of e-commerce in the consuming public.

REFERENCES

Alboukrek, K. (2003). Adapting to a new world of e-commerce: The need for uniform consumer protection. *The George Washington International Law Review, 35,* 425-460.

August, R. (2002). International cyber-jurisdiction: A comparative analysis. *American Business Law Journal, 39,* 531-562.

Barber, W., & Badre, A. (1998). *Culturability: The merging of culture and usability.* Retrieved March 29, 2006, from http://www.research.att.com/conf/hfWeb/

CIA. (2006). *The world factbook.* Retrieved from http://www.cia.gov/cia/publications/factbook/

Curran, V.G. (2001). Romantic common law, enlightened civil law: Legal uniformity and the homogenization of the European Union. *Columbia Journal of European Law, 7,* 63-138.

Driskill, L. (1996). Collaborating across national and cultural borders. In D. Andrews (Ed.), *International dimensions of technical communication* (pp. 23-44). Arlington, VA: STC.

Esselink, B. (2000). *A practical guide to localization.* Philadelphia: John Benjamins.

Forslund, C. (1996). Analyzing pictorial messages across cultures. In D. Andrews (Ed.), *International dimensions of technical communication* (pp. 45-58). Arlington, VA: STC.

Gray, O. (2005). Responsible advertising in Europe. *Young Consumers, 3,* 19-22.

Goel, M. (2005). Internet defamation: Where must a defendant defend? *Intellectual Property and Technology Law Journal, 17,* 15-18.

Gruning, D. (2004). Bayou state bijuralism: Common law and civil law in Louisiana. *University of Detroit Mercy Law Review, 81,* 437-461.

Hall, E. T. (1976). *Beyond culture.* New York: Doubleday.

Herrington, T. (2003). *A legal primer for the digital age.* New York: Longman.

Hestermeyer, H. (2006). Personal jurisdiction for Internet torts: Towards an international solution? *Northwestern Journal of International Law and Business, 26,* 267-288.

Hicks, B. (1990). Smoking. On *Dangerous* [CD]. Salem, MA: Rykodisc.

Hofstede, G. (1991). *Cultures and organizations: Software of the mind.* New York: McGraw-Hill.

Jones, A. (2002). *Localization of single-source content.* Retrieved from http://www.stc.org/confproceed/2002/PDFs/STC49-00043.pdf

Koch, C. (2003). Envisioning a global legal culture. *Michigan Journal of International Law, 25,* 1-66.

Kostelnick, C. (1995). Cultural adaptation and information design: Two contrasting views. *IEEE Transactions on Professional Communication, 38,* 182-196.

Locke, N. A. (2003). Graphic design with the world in mind. *Intercom, 50*(5), 5-7.

Lubens, R. (2003). Survey of developments in European database protection. *Berkeley Technology Law Journal, 18,* 446-472.

Murphy, C. (2006, May 24). In no uncertain terms. *Marketing,* 17.

Ott, S. (2006). *Links and law.* Retrieved August 11, 2006, from http://www.linksandlaw.com/

Permeus, A. (2001). Applying Swedish law to Web sites. *Computer and Internet Lawyer, 18*(7), 23-28.

Quiye, W. (2000). A cross-cultural comparison of the use of graphics in scientific and technical communication. *Technical Communication, 47*(4), 553-560.

Reinsch, R. (2005). E-commerce: Managing the legal risks. *Managerial Law, 47*(1/2), 167-196.

Singh, N., & Pereira, A. (2005). *The culturally customized Web site.* Burlington, MA: Elsevier.

Singh, N., & Boughton, P.D. (2005). Measuring Web site globalization: A cross-sectional country and industry level analysis. *Journal of Web Site Promotion, 1*(3), 3-20.

Singh, N., Kumar, V., & Baack, D. (2005). Adaptation of cultural content: Evidence from B2C e-commerce firms. *European Journal of Marketing, 39*(1/2), 71-86.

Schadbach, K. (1998). The benefits of comparative law: A continental European view. *Boston University International Law Journal, 16,* 331-422.

St.Amant, K. (2003). Designing Web sites for international audiences. *Intercom, 50*(5), 15-18.

St.Amant, K. (2005). When culture and rhetoric contrast: Examining English as the international language of technical communication. *IEEE Transactions on Professional Communication, 42,* 297-300.

St.Amant, K. (2005). A prototype theory approach to international Web site analysis and design. *Technical Communication Quarterly, 14,* 73-91.

Stein, P. (1992). Roman law, common law, and civil law. *Tulane Law Review, 66,* 1591-1598.

Tysver, T., & Bolin, B. (2006). *Bitlaw: A resource on technology law.* Retrieved August 10, 2006, from http://www.bitlaw.com/internet/Webpage.html

Ulijn, J. (1996). Translating the culture of technical documents: Some experimental evidence. In D. Andrews (Ed.), *International dimensions of technical communication* (pp. 69-86). Arlington, VA: STC.

University of Ottawa. (n.d.). *World legal systems.* Retrieved April 8, 2006, from http://www.droitcivil.uottawa.ca/world-legal-systems/eng-systeme.html

Valcke, C. (2006). Comparative history and the internal view of French, German, and English private law. *The Canadian Journal of Law and Jurisprudence, 19,* 133-160.

Yunker, J. (2003). *Beyond borders: Web globalization strategies.* Boston: New Riders.

Chapter XIV
A Frame Analysis of Privacy Regulation in an International Arena:
Understanding Interdependency in a Globalized World

Vinita Agarwal
Purdue University, USA

Charles J. Stewart
Purdue University, USA

ABSTRACT

The conceptualization of privacy has been the subject of much debate for more than a century in scholarship ranging from social philosophy and sociology to law, finance, and medicine. Legal issues concerning the right to privacy, moreover, are part of a complex web of state and national laws. This chapter examines the international legal issues resulting from the European Union's 1998 Data Protection and Privacy Directive and its effects on online interactions in a global context. The authors believe this focus will help individuals interacting in international cyberspace understand and adapt to varying cultural and national perspectives of privacy in such contexts.

INTRODUCTION

While the authors realize that legal issues concerning the right to privacy are part of a complex web of state and national laws and court decisions, and that strict constructionalists reject the notion that privacy can be found in the emanations of guarantees embodied in the penumbras of constitutional amendments, we have focused this chapter explicitly on the international legal issues resulting from the European Union's 1998 Data Protection and Privacy Directive. We

Levin and other conservative dissenters recognize, as Justice Hugo Black did in his dissenting opinion in 1965, that "there are, of course, guarantees in certain specific constitutional provisions which are designed in part to protect privacy at certain times and places with respect to certain activities" (Levin, 2005, p. 58). Levin concludes his discussion of the controversy over a general right to privacy by stating:

So the right to privacy means everything and nothing. It has no constitutional basis and no tangible form. But what is clear is that the Supreme Court, by usurping the legislature's authority to set social policy, has seized from the people the power to make determinations. (p. 60)

The confusion and conflict over the right to privacy in the U.S., particularly as it relates to electronic data and transmissions, entered a new realm in 1998 with the passage of the European Union's Data Protection and Privacy Directive (referred to hereafter as the Directive). It reiterated the status of privacy as a fundamental human right in the European Union. Joel Reidenberg, professor of law at Fordham University, states that the Directive embodies the approach of European democracies towards the "state as the necessary player to frame the social community in which individuals develop [and citizen autonomy therefore] depends on a backdrop of legal rights. Law, thus, enshrines prophylactic protection through comprehensive rights and responsibilities" (Hearing..., 2001).

The Directive is an encompassing legislation that regulates access to an individual's personal data in the commercial domain. It follows the precedent set by the privacy principles contained in the Organization for Economic Cooperation and Development (OECD) in 1980. However, when the EU began to enforce the Data Protection and Privacy Directive, it came to the forefront of political deliberations because of its huge potential impact on trans-Atlantic commerce and data transfers involved in virtually all commercial transactions. A national or regional principle, the right to privacy, became a global concern.

The Subcommittee on Commerce, Trade, and Consumer Protection's hearings that commenced on March 8, 2001, were the first of their kind with the goal of raising privacy-related issues of significant importance to increase knowledge in the information-based economy. According to subcommittee Chairman Cliff Stearns, its agenda was to deliberate explicitly on the trade ramifications of the EU privacy directive and, by emphasizing the urgency of examining this issue carefully, to "make our [U.S.] wishes known in more definitized ways." In his introduction, Stearns stated that the subcommittee was created to hear how the EU had chosen to balance its commercial and consumer privacy interests, and to set the ground for deliberating the implications of international regulations such as the privacy directive.

These hearings were especially important because, with increasing cross-border commercial transactions and financial flows, regulations within one nation or region such as the EU significantly affect the economy or commerce of another. As ranking member of the subcommittee Edolphus Towns observed: "Compliance with the EU is not optional." As an object of analysis, the hearings, according to Joel Reidenberg, emphasized the urgent need to "understand the international dimensions of the global information marketplace," especially since culturally and contextually grounded notions such as privacy intersect with regulations that cross national borders.

PURPOSE, BACKGROUND, AND METHOD

In this age of globalization and e-commence, it is essential that we understand how vital principles embodying the notion of privacy are perceived and practiced in different cultures and traditions both to enhance the flow of data and commerce and to avoid philosophical, political, and legal conflicts. Misunderstandings may impede the growth of e-commerce globally. A witness at the hearings said that "the EU privacy directive is probably the most important law by which the EU is writing the rules of cyberspace." The purpose of this study, therefore, is to discover, through a framing analysis of the subcommittee hearings on the EU Directive, how Americans and Europeans perceive privacy, and how these perceptions may lead to disagreements and potential problems in commerce between the EU and the U.S. Before delving into the framing analysis of the hearings, it is important to understand the situational factors that led to the first-ever hearings on privacy.

Background

The European Union's comprehensive privacy legislation, the Data Protection Directive, dealing with the protection of the fundamental human right to privacy with regard to the processing of personal data, became effective on October 25, 1998. It was enacted to take into account the development of a "frontier-free international market" within the overall context of an "information society" and the increasing cross-frontier flows of personal data between the member states of the EU (http://europa.eu.int/comm./justice_home/fji/privacy/overview). The Directive attempts to harmonize regulation within its member states, as well as to ensure protection of personal data of its citizens when exported to other countries.

Accordingly, transfer of personal data of EU citizens could take place only to non-EU countries that adhered to the EU's standard for an "adequate" level of privacy protection (Swire & Litan, 1999). The key term "adequate" had to be defined for each case. Since U.S. privacy regulations were perceived to be "inadequate" by the Directive's stipulations, the Directive suggested a potential interruption in the flow of data transfer that could have major consequences for e-commerce and other forms of trade and financial transactions between the EU and the U.S.

To resolve this potential stalemate, the U.S. negotiated through a series of meetings the provisions of an "adequate" level of privacy protection of consumer data in the U.S. as embodied in a set of principles called the "safe harbor" (http://www.export.gov/safeharbor/USPrinciplesJune2000.htm). These principles are notice, choice, onward transfer, security, data integrity, access, and enforcement. On May 31, 2000, the European Union member states unanimously approved the proposed safe harbor principles, covering most industry sectors (with the notable exclusion of the financial sector) and almost all forms of personal data. The safe harbor principles went into effect on November 1, 2000, and attempted to balance the privacy principles relating to individual control of personal information with the legitimate concerns of businesses in commercial transactions involving transfer of individual data.

The safe harbor negotiations exemplify a common dilemma that faces most international regulation in a globalized world: the attempt to negotiate a balance between freedom of information vital for trade, commerce, and business with the protection of that information within and without national borders. While the U.S. and EU share common privacy principles, the U.S. takes a fundamentally different approach to assuring privacy. Traditionally, the U.S. favors an industry-driven self-regulatory approach to ensure a balance in meeting business needs with the privacy needs of consumers. It follows a sectorial approach that relies on a mix of legislation, regulation, and self-regulation. The EU has demonstrated a regulation-heavy approach consistent with its belief that privacy is a fundamental human right. Given these differences, many U.S. organizations expressed uncertainty about the impact of the EU-required "adequacy standard" on personal data transfers from the EU to the U.S. (www.export.gov/safeharbor).

Speaking on behalf of the U.S. government in a letter to the EU Parliamentary Workshop on the Privacy Directive, Robert La Russa, then acting undersecretary for the International Trade Administration, underscored the critical importance of the passage of the safe harbor in noting that it was essential for the private sector to facilitate the management of routine information exchange (http://oecd/org/document). However, despite the hectic pace of the negotiations and the political uproar caused by the Directive, approximately 225 American companies signed the safe harbor accord by August 2002 (http://europa.eu.int/comm/justice_home/fsi/privacydocs/lawreport).

To define the parameters of the global privacy debate, the Subcommittee on Commerce, Trade, and Consumer Protection of the Committee on Energy and Commerce House of Representatives, 107th Congress, chaired by Cliff Stearns (Florida), held a series of three hearings on privacy on March 8, 2001. The second of these, called "The EU Data Protection Directive: Implications for the U.S. Privacy Debate," considered the testimony of several key negotiators of the safe harbor, notably David Aaron, who had been the chief negotiator of the treaty as the U.S. ambassador to the EU. The purpose of the hearings was for the U.S. to learn more about the European approach to information privacy and to consider the impact of the European Data Protection Directive on international commerce in general and e-commerce in particular.

In his introductory remarks, Stearns highlighted the subcommittee's intent to "see how the EU Data Protection Directive can inform the United States' own approach to information privacy." While the deliberations of the subcommittee hearing were intended as a means to achieve an understanding of privacy with regard to transfer of personal data in commercial transactions, they referred to and exemplified the culturally and historically situated nature of privacy. Quoting Ambassador Aaron, W. J. "Billy" Tauzin (Louisiana) stated, "While we and the Europeans share many basic values, the European Union directive comes from a different legal tradition and historical experience. The safe harbor principles are reflective of those European traditions and experiences, and as such at times don't harmonize well with our American legal tradition and historical experiences.

By including testimonies from a variety of perspectives, the subcommittee hearing presented an insightful deliberation into the cultural, historical, and nationally grounded parameters of the privacy debate instigated by the EU Directive.

Method

William Gamson (1985) conceptualizes Irving Goffman's notion of frames as "frameworks for understanding available in our society for making sense of events and to analyze the special vulnerabilities to which these frames are subject" (pp. 615-617). Through different strategies and narrative forms, frames may generate "theories about what exists, what happens, and what matters" (Gitlin, 1980, p. 6). Research has demonstrated the "robust influence" of frames "in shaping social perception and policy" (Gandy & Kopp et al., 1997 pp. 162-164; Iyengar, 1991). Frames or the process of framing, as "principles of organization" or "a central organizing idea or story line," may affect deliberations in a variety of ways and enable us to gain important insights into the subcommittee's hearings on privacy because they may suggest "what the controversy is about, the essence of the issue" and outline the boundaries of the debate while defining the issue (Gamson & Modigliani, 1987, p. 143). Frames enable us to understand and interpret complex information and to determine how to react to complex situations such as privacy (Lau & Schlesinger, 2000; Rein & Schõn, 1993,

1994). For example, we may use frames to orient witnesses and participants to "locate, perceive, identify, and label" deliberations in their terms (Gamson & Modigliani, 1987, pp. 615-617).

A narrative form may embody rhetorical devices that enable us to control the agenda and structure understandings about a subject (Goffman, 1974; Fillmore, 1975; Cappella & Jamieson, 1997). For example, we may use a *frame extension* strategy to broaden the appeal of a frame to a greater audience. On the other hand, we may use a *counter-framing* strategy to define negatively "an idea in terms of what it is not—an act of distinguishing one thing from another" (Davies, 2002, p. 271). In the case of the subcommittee hearings, frame analyses of elite discourse may be useful in revealing what the speaker sees as relevant to the topic at hand, or what witnesses perceive as relevant to an issue (Gamson & Modigliani, 1987; Gitlin, 1980).

The frames employed influence the choices people make between essentially equivalent positions or options through subtle alterations in the presentation of judgment and choice problems (Graber, 1984; Iyengar, 1991; see also Kahneman & Tversky, 1997, 1990, 1984). The inherent ambiguity of frames may determine these influences while serving as a bridge between cognition—the process of knowing and judgment—and culture. Goffman writes that "Would-be challengers face the problem of overcoming a definition of the situation that they themselves may take as part of the natural order" (1961). However, as Goffman elaborates, this ambiguity also embodies the fundamental fragility of frames in use and their vulnerability to reframing acts.

In examining the varying interpretations of the Directive, our analysis draws on the dynamic, shifting nature of frames. Frames embody within themselves the potential for appropriation and the possibility for ambiguous points to invert stable meanings (Benford & Snow, 2000). Taking the conceptual notion of "information privacy" as a starting point, our analysis of the flexible, situated nature of the frames used in the subcommittee hearings by representatives from different national, industrial, political, and cultural perspectives reveals a set of overlapping, intersecting, and occasionally opposing frames. It examines the potential of these overlapping and intersecting frames as points of transformation and utilizes the concepts of *frame extension* and *counter-framing* (Snow, Rochford, Worden, & Benford, 1986) to construct points of arguments for and reveal points of ambiguity that may suggest emerging recommendations.

Our analysis presents and elaborates on four key frames and their sub-frames articulated in the subcommittee hearings to reveal differences in the conceptualization and interpretation of information privacy between the EU and the U.S. Examining the hearings as a dialog that reveals the narrative construction of privacy through the use of implicit culturally and historically grounded organizational principles, the four key frames exemplify the four key definitions in charge of the situation (Goffman, 1961). These frames are: (1) trust, (2) sovereignty, (3) democracy, and (4) interdependency. The discussion section will note the points of intersection and overlap among these four frames to reveal recommendations that may reconcile the differences between the EU and U.S. frames, and enhance the free flow of information and commerce.

ANALYSIS OF KEY FRAMES

Trust Through Speed and Globalization: Technology and Privacy

Privacy, understood in its classical interpretation as the "right to be left alone" (Warren & Brandeis, 1890), is framed in the hearings on information privacy through the lens of trust and consumer confidence. However, almost a century after Warren and Brandeis (1890) wrote their famous definition, this is a right that has complex and often highly contradictory implications for the globalized economy. This frame elaborates the challenges confronting witnesses in the subcommittee's deliberations in employing the conceptual notion of privacy for negotiating the delicate balance between increasing freedom of information and the corresponding need for control, protection, and trust. The fundamental difference between the U.S. and the EU in this frame embodies these tensions.

Gamson (1985) suggests that we analyze frames to see which particular definition "is in charge of the situation," thus revealing the central organizing principle underlying this representation. The critical frame of trust is emphasized through two different perspectives at the hearings (p. 14). An examination of Representative Towns' (New York) statement reveals several key sub-frames within the overarching theme of technological implications on information privacy in business transactions: change, trust, and equality. By articulating the need for privacy within a counter-frame of freedom of information, Towns emphasizes the differing central logic of e-commerce within the U.S. and the EU. While there is an ever greater need for businesses to share information across national boundaries, the corresponding risk to privacy has also increased. He presents as evidence a survey by AT Kearney management consultants (Towns, Congressional Testimony) that reported: "Invasive information requests are blamed for 52% of the sales that fall apart, followed by reluctance to enter credit card data (46%)."

Towns claims that "while privacy can limit business marketing opportunities, the effect of privacy policies on consumer confidence is a far more important fact in the future success of e-commerce." Consumer confidence is bolstered through greater privacy, and greater trust is related to greater control of personal data. The key tensions of trust, privacy protections, and consumer confidence structuring Towns' testimony are driven by well-constructed best practices within business itself. Together they construct a narrative that employs privacy policies to increase business by engendering confidence in protection of information.

By utilizing the lens of globalization as the central organizing principle in his narrative, Towns reveals a fundamental paradox of the last century: as technological innovation accelerates change in different spheres, it highlights the need to integrate trust alongside an increasing perception of risk. This frame incorporates the counterframes of risk and trust to negotiate change. It reframes privacy policies as a business tool for building confidence and trust through counterposing freedom of information with increasing protection and control. The central theme of e-commerce recalls the notion of business and its role as an agent of change.

By contrast, Stefano Rodota, chairman of the European Union Data Protection Working Group, adopts the narrative of globalization and change to counter-frame Towns' frames through the lens of the runaway "speed" of e-commerce. This implicitly locates the narrative focus of the frame firmly within the EU perspective, emphasizing the need for greater control by claiming that the Directive "aims at protecting fundamental rights and freedoms, although this objective is intertwined with the free movement of services."

The narrative of "free movement" structures the need for greater protections as a means to control speed characterized by e-commerce. The Directive states its central concerns as minimizing risk and maximizing protection while enabling EU citizens to maximize the benefits from the free movement of goods (see also http://europa.eu.int/comm/justice_home/fsj/privacy/overview/index_en.htm). Such an understanding is representative of privacy scholarship that weighs common good as constitutive of civil society and permitting individuals to engage in their endeavors.

Rodota's testimony directs attention to the role of governmental legislation and regulatory bodies in guarding trust. He addresses this position in his statement by bridging the apparent dichotomy between "the domain of law" and self-regulation as mutually constitutive:

It is an instance of misrepresentation to simplify the picture by making Europe the domain of law and the U.S. the domain of self-regulation. Indeed, it is exactly the framework provided by European Directives and national laws which is making it possible to develop self-regulatory codes and contract models on a large scale.

Rodota seeks to reconcile two counter-frames within a single overarching frame of regulation wherein the Directive is a choice that "makes it possible to develop self-regulatory codes." Framing is a process that, by emphasizing selective aspects of an issue, "promotes a particular problem definition, causal interpretation, moral evaluation, and/or treatment evaluation for the item described" (Entman, 1993, p. 52). Rodota's frame strategically counters the notion of business as a positive driver of change by framing the freedom of self-regulation as dependent upon the framework of laws and regulation symbolized by the Directive.

Rodota's testimony states that while "privacy was invented in the U.S. and has long been considered to be typical of American society, [Europe] is now the region of the world where maybe personal data are most protected." By drawing on the vulnerability of the trust frame, especially with regard to the "speed of e-commerce," Rodota's testimony creates a paradoxical shift in the dialog towards a necessity for greater regulatory frameworks to protect that trust. Recalling the ethos of the "American society" and, implicitly, its related ethos of self-regulation, Rodota directs the need for protection of this ethos under the aegis of the EU Directive by extending the need for regulation to include trust and to counter the notion of "a full-speed world," as well as self-regulation and indeed the protection of the American ethos.

Rodota extends his testimonial counter-frame of speed to refer to the cultural shift in privacy. Here again, privacy is used as constitutive of the fabric of civil society (Cohen, 2000), and the need to build trust to maintain this fabric is counterposed against the larger harm to the overall fabric of American society. Trust in information flow is vital to the existence of the information-driven economy, and regulations concerning the collection, use, and disclosure of such information forms an integral part of all international and national data privacy principles from the OECD adopted on September 23, 1980, the Code of Fair Information Practices, and more recently, the Directive. His frame attempts to reconcile and refute the two major concerns of U.S. representatives: that the Directive is somewhat irreconcilable with underlying U.S. ethos and that the U.S. system of self-regulation provides greater privacy protection.

Rodota also incorporates the notion of "frameworks" to counteract the notion of speed in e-commerce. His testimony states emphatically: "We cannot accept the full-speed world in the data protection field, more and more one of the most important and critical matters in the globalized world." Regulation and guidelines function as a framework within the domain of law to contain this speed of globalization. A close examination of this statement reveals that by using the rhetorical device of situating "a full-speed world" in direct opposition to "data protection" in the globalized world, Rodota reconciles the framework of regulations as a reassuring solution that contains speed within both an implicitly slower, but more protective, "domain of law."

Situating the frame of trust and globalization within the narrative of speed and change, Rodota presents the Directive as a form of "control," regulated within a domain of law, which will build trust and protect consumers. The narrative of speed in his testimony recalls the ambiguous notion of change alongside the insecurities underlying globalization: risk, trust, and control.

The Directive is structured in Rodota's counter-frame of speed as a solution that addresses all three concerns.

According to the U.S. frame, it is in the interest of private enterprise to ensure trust in e-commerce, so private enterprise rather than government is the best guardian of consumer privacy. An important underlying theme in the EU Directive, and a critical issue in the dialogue, has been the ambiguous term "control" and what it means in the context of trust. To representatives of the EU, it means legislation, government, and a regulatory body that would assure adequate privacy protection. As Reidenberg states in his testimony, in Europe "citizens trust government more than the private sector with personal information" and have a basic right to "information self-determination, [which] puts citizens in control of the collection and use of personal information." Ironically, this framing runs counter to the historical mistrust in the U.S. of government involvement and regulatory agencies.

American historian Gary Wills writes that in the U.S., "Government is accepted as, at best, a necessary evil, one we must put up with while resenting the necessity" (Wills, 1999, p. 15). He adds that, "We are pious toward our history in order to be cynical toward our government. We keep summoning the founders to testify against what they founded. Our very liberty depends so heavily on distrust of government that the government itself, we are constantly told, was constructed to instill that fear" (p. 16). Traditionally, free enterprise and capitalism have been revered in U.S. society while government and regulation were feared and distrusted.

While Rodota's strategy extends the counter-frame of speed to incorporate the protection of American ethos and trust privileged, in Towns' testimony within the Directive itself, it does not successfully reconcile the corresponding distrust of government, or the faith in best practices within e-commerce within the frame. However, while the EU emphases are unlikely to allay all the concerns about risk and trust in the U.S., the frame analyses reveal the ambiguity of control as an implicit point of potential transformation in the dialogue.

Harmonizing Sovereignty: Identity and Difference

The notion of information privacy is a good example of how both sides in the privacy controversy have used frame extension strategically in their arguments to appropriate the concept of privacy. As a representative of the U.S., Towns states that, "no consumer, no matter where they live, is due any less than the highest privacy protection a company provides to any other consumer," incorporating the notion of equality within the discourse of privacy protection. In alluding to "no matter where they live," his statement implies the notion of geographically situated sovereignty within the overall privacy protection provided by a company. This frame directs audience perception to a not-so-subtle value scale that evaluates the "amount of privacy" enjoyed by consumers (described in the testimony as geographically located citizens) and suggests that U.S. consumers may be treated as "second class citizens" on the global scene. Joel Reidenberg echoes this sentiment when he states that, "for U.S. citizens, the directive [h]ighlights that American citizens are becoming second-class citizens in the privacy world, the global level" when their privacy is regulated by foreign entities such as the EU. Through the adjacent placement of consumers with second-class citizens and equality with protections, Towns' testimony employs a complex but interesting rhetorical device of appropriating the narratives of sovereignty and equality within the discourse of privacy.

When EU representatives emphasized the frame of sovereignty, along with related attempts to reframe the notion of privacy with the relative power of a nation-state to negotiate superior privileges for its citizens, they equated non-compliance with the Directive as a form of U.S. government negligence toward its citizens. Towns counterframes this perception by stating that, "while we and the Europeans share many basic values, the EU directive comes from a different legal tradition and historical experience [and] as such at times don't harmonize well with our American legal tradition and historical experiences." He reframes the relationship of privacy with national power to highlight traditional and legal differences between the two continents, contextualizing these in the differing *cultural traditions* the two perspectives stem from rather than in differing perceptions of relative *national power.*

The reactions of subcommittee member Steve Buyer, representative from Indiana, to the EU representatives further reveals the historical and political tensions between the U.S. and the EU that make reconciling the two perspectives difficult. He barely contains his disdain for the EU's claims of power, cultural superiority, and protection of privacy:

I am troubled because, as I watch the European Union sort of try to come together, which in world history is amazing. Because you mocked us at the creation of our country, as we were called the Great American Experiment. Perhaps we can now look back across the ocean and sort of mock you back and say, 'Well, let us see if it can succeed.' And then, I find myself here in Congress, and say, 'Well I do agree in a quest for economic harmony.' That is what we are trying to do as each of us, as sovereign nations, seek to protect our own identity, and how we choose to recognize rights and govern.

By emphasizing the historical relativism of the two contexts, the Grand American Experiment and the relatively recent coming together of the European Union, Buyer's comments organize the principle of *economic harmony* through the principle of *political harmony* embedded within an overall historical context. While this framing appears to remove privacy from the dialog altogether and reframe the issue as one of two political entities seeking to protect their sovereign identities within the imperatives of economic harmonization, its implicit historical orientation embeds a temporal seniority, so to speak, of the Grand American Experiment within the narrative.

This is important because as Buyer weaves in a cultural critique of the Grand American Experiment compared to the European Union member states and suggests that the EU must prove itself before claiming to lead, he subsumes the dialogic space of the discourse within the proven success of the Grand American Experiment. Even as regulation becomes an inherent determinant of national identity and sovereignty, economic and political harmony seem to be reconciled within the historical narrative of the Grand American Experiment in whose footsteps the newly defined European Union has only recently begun to follow.

In his response, Rodota reframes the issue by incorporating privacy as a human right and emphasizes its tradition in the U.S.:

I emphasized at the beginning of my statement that there is an important evolution in the European Union, giving an important protection to personal data because they are considered a very important part of the fundamental human rights. And if we are living in the information society, information about individuals becomes more and more important for respecting the individual's rights. There is not an idea to impose a model to the world or to defend a cultural identity. Europe accepted the modern idea of privacy protection coming from the United States ... [And] the idea of privacy was not directly connected with economic at first. We must have a balance. This is our attitude, and I think we can have a fruitful dialogue on these points.

Rodota's testimony identifies several themes of interest in an analysis of the original discourse of information privacy. Rodota downplays the frame of sovereignty that was inserting a new controversy into the original discourse of privacy. Rodota's reclaiming of the original frame is important because it emphasizes, virtually celebrates, the American legal and cultural legacy of privacy, and consequently positions the Directive as an open document that invites a dialog on extending the long-standing American legacy of privacy to the domain of economic information regarding consumers.

In making an explicit connection between the information age and individual information as a form of fundamental individual right, Rodota counter-frames Buyer's framing of the American experiment's historical antecedents with the need to revisit the notion of privacy. In a sense, it removes privacy from a privileged historical position and, by transposing it to the information age, invites American representatives to enter into a dialogue on the issue in a new age. Implicit in the narrative of a new age, of change in the "idea of privacy," of terms such as "evolution in the European Union," of accepting the "modern idea of privacy protection coming from the United States," is the need to revisit the discourse of privacy to meet the needs of "the information age," rather than the European Union proving its eligibility to regulate privacy because of the historical legacy of the Grand American Experiment, as Buyer's testimonial frame seemed to suggest.

Statements from other EU representatives reinforce Rodota's counter-frame. David Smith, the UK assistant information commissioner, describes the capacity of global information flow of individual information which points to the need for an effective international regulatory body to protect consumers globally from increasing risk. He states that such a body "clearly would be an ideal, an international framework which we could all sign up to, which will provide privacy effectively, and what is now [a] global market, where it is difficult to apply some of the nationally based regulatory requirements." This universal conception of a legal framework would eliminate the need for national legislation and the dilemma of national sovereignty while guaranteeing a maximum of privacy protection.

Further, such a framework removes the contextual notion of historical legacy and cultural sovereignty from the privacy discourse while emphasizing the requirements of "effectiveness," to meet the need of "global markets," an "international framework," that is also "ideal." An examination of these terms suggests that "business" and "markets," which implicitly relate to "global" and "international" (as also noted in the trust frame), recall the notion of privacy as a modern and progressive notion, and symbolically frame the Directive as an ideal, modern, progressive, and international document that protects the interests of "business" and "markets" on a global level while guaranteeing or protecting a fundamental human right—privacy.

The functional differences between the two perspectives in this frame may be summarized in relation to their opposing overall goals. The EU's rhetorical strategies identity the EU as a modern, international power in the information age as a counter-frame to Buyer's and Towns' testimony, recalling the symbolic legacy of the Grand American Experiment and its historically privileged position as compared to the newly emergent identity of the European Union. Rodota emphatically states, "there is not an idea [to] defend a cultural identity," but talks instead in terms of "information," "balance," "dialogue," and "individual rights." This contrasts with his earlier framing of "individuals" in geographically situated terms, symbolizing their identity as "citizens" of nations more than neutral, globally situated "individuals."

Rodota's testimony offers a universal structure through which to situate the Directive within the frame of sovereignty while incorporating productive business practices. It speaks to a shared framework of similar values and core principles that is "directly connected with a strong commitment to make effective fundamental human rights in this very sensitive human area." In doing so, he recalls underlying universal core values and commitments shared by being "legally bound by the same core principles." Rodota speaks to the need for a "transitional period" in which cooperation would be important and "more and more privacy protection is considered a value to be offered with goods and services."

With the increasing breadth and range of global business relying on technology, different cultural traditions are increasingly placed in unavoidably close contact with one another, and in this frame, Rodota's testimony employs the openness of "dialogue" and "balance" as a counter-frame to the culturally situated, historically grounded narrative of the American representatives. The deliberations on the Directive illustrate the challenges to harmonizing historically different legacies and understandings of culturally grounded concepts such as privacy. His statement attempts to avoid the issues of "imposing a model," of individual government or agency involvement in privacy, and the thorny matter of sacrificing national sovereignty to a foreign (EU) and international regulatory body. The result, he promised, would be greater privacy, not limited or endangered privacy.

Democratic Tradition: National vs. Self-Regulation

In its implicit embodiment of a normative conceptualization of different freedoms related to individual identity and autonomy, the notion of privacy is an integral part of the western liberal democratic tradition. Embedded within the Directive tying compliance with its privacy protection guidelines is the symbolic critique of national vs. free market regulation. This frame begins with an examination of the testimony of Towns, who stated:

Because the U.S. has no comprehensive national privacy policy, much less one that is comparable to the EU Directive, the EU has decided that all American firms lack adequate privacy protections for personal data. The privacy provisions of the recently enacted financial modernization legislation do not, according to the EU and many others, provide adequate privacy protection. The U.S. firms, therefore, are in a bind.

Towns is referring to the EU's perception of the recently enacted Gramm-Leach-Bliley Act for financial modernization in the U.S., and his sentiments echo what many scholars on privacy have noted about the implicit conceptualization of privacy in the American legal system (Reidenberg, 2002). Its challenge, therefore, harkens to several culturally situated, historical differences within the two political contexts. These are revealed as points of ambiguity within which the conflicting narratives are framed. The most significant, perhaps, is the word democracy.

The U.S. counter-frame to the organization of privacy in terms of national regulation is revealed in Representative John Shimkus' (New York) testimony that reframes the Directive in historical terms, highlighting the relative strength of the U.S. democracy. He stated:

[But] I want to put on record my concern, after hearing the decision rendered by the European Court of Justice earlier this week, that allows the European Union to lawfully suppress political criticism of institutions and of leading figures. In this country, in the history of our country, we have basically had some distrust of national government, symbolically, in the creation of the Bill of Rights to our Constitution over 200 years ago.

In so doing, the first one being the First Amendment, freedom of speech, the implication is here is that probably our strongest allies and democratic countries may not have that faith and trust in the freedom of expression, of political expression.

Shimkus concluded, "This decision is very disturbing, one that could have major implications on the privacy issue and an impact on future business relations between the U.S. and EU countries." The democracy frame that Shimkus offers in his testimony refers, first, to the notion of privacy as inherently connected to the historical principle of individual freedoms of a truly democratic nation (Reiman, 2004). By framing privacy as equivalent to the freedom of speech embodied in the Bill of Rights, Shimkus identifies privacy as a fundamental right in the U.S. It situates privacy within the U.S. Constitution as equivalent to the right of expression offered by the First Amendment even though it is not explicitly mentioned in the Bill of Rights. Unlike many strict constructionalists, Shimkus and others have historically considered privacy to be embodied in the First, Fourth, and Fifth amendments.

Significantly, this sociolegal frame supports Towns' earlier testimony that defended the privacy protections offered by the recently enacted Gramm-Leach-Bliley Act in the financial domain with the historical legacy—"creation of the Bill of Rights to our Constitution over 200 years ago"—of American democratic values. As a rhetorical device, the democracy frame enables Shimkus to label privacy as a deeply rooted historical American value, equivalent to freedom of speech, and tie it simultaneously to the historical distrust of the national government and historical trust in the individual. It is democracy as embodied in the sacred Bill of Rights and allows Shimkus to structure the frame of communication through the principle of democracy.

Shimkus claims that the democracy of the EU does not compare favorably with the U.S. democracy, for instance, in its protection of the freedom of political expression of its citizens and, by implication, the privacy protections of its citizens. If the EU is weak on protecting freedom of expression, it must be weak on privacy protection.

Understanding these interactions through the overarching frame of informational privacy in global trade requires an appreciation of the tension between commercial and consumer interests. While Reidenberg's (1992) testimony before the subcommittee describes privacy as "implicat[ing] core democratic values and market," critiques of privacy protection in the U.S. in the 1990s had noted the lack of overall privacy principles in the collection, transmission, and sharing of individual information in business transactions.

Shimkus structures privacy in the U.S. as a fundamental right of all citizens in terms that emphasize the democratic rights and establish credible protections of these citizens. Simultaneously, he critiques the lack of freedom of political expression in the EU by citing a news event, not directly related to the hearings on information privacy, which questions the credibility of the EU's democratic protections. In doing so, however, his testimony employing the democratic frame employs the device of "narrative fidelity"—the use of his audience's beliefs and values—essential to make this acceptable to the U.S. position. In essence, it is a co-creation with his audience (Stewart, Smith, & Denton, 2001, p. 250).

On the EU side, David Smith of the Office of the UK Information Commissioner organizes the frame of self-regulation and national legislation through the principle of best practice that protects privacy while encouraging global commercial trade. He states that, "we promote good practice which goes wider than simply complying with the law, and it covers conduct which is consistent with those requirements" (i.e., conduct consistent with requirements that promote best practices). Smith's testimony complemented Rodota's frame of control through guidelines and extended it by stating that "we encourage self-regulation, not necessarily instead of statutory regulation but together with it [because] the more that data protection flows out of good business practice than is seen as a simple additional regulatory burden, the more satisfactory and the more effective it will be." Rather than framing the issue as one of competing democracies with different views of assuring privacy, Smith appears to reach out to the U.S. representatives by emphasizing self-

(individual) regulation rather than regulation through government control and legislation. This framing would make the sides appear to be closer in principle than otherwise thought.

Although this frame reveals the roots of privacy in the western liberal democratic tradition, its foundations appear to involve more than just the laws themselves (the Gramm-Leach-Bliley law or court determined Constitutional bases of privacy in the U.S.). It is a critique of the credibility and overarching goal of those laws. Smith notes, for example, that it is more important for data protection to flow eventually out of good business practices rather than national regulation and laws. The EU representatives argue that both statutes and good business practices are necessary to protect the fundamental human right to privacy.

Freedoms and Restrictions of Interdependency: Contingent Notions of Power

An interesting dilemma raised by the Directive is the ambiguous interpretation of another fundamental western democratic hallmark, *freedom*. In their deliberations on the freedom of individual information, one thing all participants at the hearing agreed on was that freedom from one national perspective was not freedom from another national perspective. W. J. "Billy" Tauzin's (Chairman, Committee on Energy and Commerce) statement to the subcommittee articulates this ambiguity in its characterization of the Directive as an attempt to impose something like privacy on the entire world with the premise that one size fits all. He claims that the enactment of the Directive creates a "new privacy regime" meant not to encourage consumer protection but to impede and restrict consumer freedom. The word "regime" carries all of the connotations of a conniving, authoritarian dictatorship counter to the notion of freedom.

Tauzin also structures the notion of privacy in economic terms of a cost-benefit analysis when he states, "the costs of [compliance with the Directive] would be in the multi-billions, and ... all costs will be passed on to consumers." Given that it would encourage a spate of lawsuits in the U.S., it could "cripple the beneficial exchange of information that is a cornerstone of American business practices today." Neither the consumer nor business would enjoy current freedoms under the Directive. Interestingly, the benefit of freedom must be balanced with the cost to industry and the corresponding costs to the consumers. Tauzin's question can be stated in two ways: Are consumers willing to pay the price for privacy? or Do they value the freedom of information enough to pay the price demanded for its protection?

Frame extension is the process by which the meaning of a frame is broadened to appeal to a larger audience (Davies, 2002). By including the notion of information flow and cost analysis with the traditional model of American business, Tauzin extends the privacy directive frame as a restrictive "new privacy regime" through which the EU is trying to impose its "will on the U.S." Imposition equals loss of consumer and business freedom.

This frame extension, while drawing on economic imperatives of information flow and cost-benefit analysis, recalls embedded nationalistic sentiments in references to the traditional principles of American business and the political imposition of the EU's will on the U.S. without due process of dialogue or negotiation. Unlike the EU frame of information privacy in which consumer interests are stated as more protected than commercial interests, commercial interests are equally privileged to consumer interests in this narrative as a cornerstone of the American democratic ethos.

Tauzin also substantiates the need for protection of consumers and business alike with a critique of the EU legal system in which relative consumer disadvantage is portrayed through an inefficient legal system. Tauzin claims:

Compliance and enforcement of the Privacy Directive has, at best, been spotty for the European nations. In fact, a number of nations have not even bothered to enact the required implementing legislation. This lax attitude is something that Americans are not used to. We do not build elaborate restrictions with a wink and a nod so they can be ignored. Given this, we need to know whether enforcement of the Privacy Directive on U.S. companies represent[s] a double standard when compared to enforcement on European firms.

Tauzin's testimony raises questions about equal protection of privacy among different nations and whether American citizen-consumers and businesses would experience a reduction of freedom when doing business with EU member nations. Would EU consumers be allowed greater privacy protections and, therefore, greater freedoms than Americans under the Directive? Or would both suffer? In short, the Directive may produce unfair and unequal results, not privacy rights for all but only for some.

This sociolegal frame draws its critique by legitimating the frame of freedom of information flow in the U.S. as protected by the rigorous enforcement of the rights of all citizens. Tauzin's cost-benefit analysis of the freedom frame, then, compares the cost of this rigorous legal enforcement with the Directive by emphasizing the relative laxity and inefficiency of the EU legal system and the lack of enforcement of the protections it allegedly offers its citizens. It counter-frames the EU Directive as attempting to impose a "new privacy regime" or "a de-facto privacy standard on the world" without due cognizance that "one size does not fit all."

The frame of freedom of information draws on the core American belief in rigorous enforcement and compliance with lawsuits. The unique system of checks and balances offered by each government representative in the global context today is underscored by the fact that no one system of legal or regulatory compliance can be enforced effectively on all. The frame of interdependency brings out the tension between freedom and restriction. Tauzin testifies:

The U.S. economy is interdependent on the rest of the world. The Internet and other electronic networks expand the ability of businesses to reach new or untapped markets worldwide. These technologies fundamentally shrink the size of the globe. Policies affecting electronic commerce made by the world's largest trading bloc—the European Union—will have an impact on the U.S. It also has an impact on how the U.S. Congress will approach the debate over privacy.

While technology has "shrunk" the globe, allowing business to reach "new or untapped markets worldwide," it has also revealed the challenges and potential conflicts of interdependency. While interdependency engenders a corresponding need for harmonization, resolving the challenge involves clarifying the underlying universal principles of democracy and freedom upon which national entities draw. The nature of ambiguity ensures that, while speaking from within a common western liberal democratic tradition, national entities will interpret these essential words and concepts from subtly different frames and ideologies.

Frames, according to Gitlin (1980), are "principles of selection and presentation composed of little tacit theories about what exists, what happens, and what matters" (p. 6). The testimonies reveal that the freedom of information frame is structured on several interrelated assumptions and beliefs regarding technology. It assumes that technology is a determinant influencing the freedom of information flow in the global economy.

It further corroborates this assumption with a shrinking global market, and increased economic and sociolegal interdependencies. Presented in this way, the perception of interdependency counter-frames the need for harmonization of national regulations.

Even though there is no denying the technological logic of freedom of information, the restrictions it imposes on fundamental values such as privacy pose a real ideological, if not economic, challenge to globalization. Its ideological dimensions are revealed in the contradiction between agreements on the importance of continued freedom of information flow on the one hand, and the urgent need for agreement on issues such as best practices and best ideological bases for regulation on the other hand. The issue becomes complex because, as the frames reveal, regulation by itself is not value-free. It embodies the cultural, national, historical, and political preferences and choices of different nation-states. Finally, it is not merely privacy that is contextual and contingent upon the changing economic environment, but more importantly, regulations that govern the interpretation of privacy that are contingent upon the sociopolitical or economic negotiating power of a nation or entity such as the EU.

Furthermore, examination of the frame of interdependency reveals the tension inherent in different cultural interpretations of freedom, in the manner in which these freedoms are translated into rights and duties, and into protection of consumers and business interests. This frame also reveals a significant aspect of challenges faced in reconciling regulations on an international scale: the culturally embodied nature of terms such as democracy and interdependency, freedoms and restrictions, and the negotiation of these terms within different historical, cultural, and political contexts.

Tauzin's testimony elaborated the different historical trajectories of the two contexts in identifying differences of attitudes toward the Directive:

The U.S. and the EU member states approach the issue of privacy from different perspectives. Europeans are instilled with the belief that privacy is a fundamental human right. There are a number of reasons for this belief, including the vast and traumatic experiences of the Nazi regime during the 1940s. Another reason for this perspective is the simple fact that many EU countries are relatively new democracies. It was not long ago that Kings and Queens ruled throughout Europe. In the U.S., we take a different approach towards privacy as we have fundamental protections to free expression provided in the U.S. Constitution, including the First Amendment. By and large, we also rely heavily on the private sector to protect consumer privacy.

By situating the need for harmonization within the different historical lenses through which privacy is interpreted in the EU and the U.S., Tauzin reframes privacy through three contingent sub-frames that differentiate the two entities: recent traumatic experiences with the Nazi regime, newness of the European transition from monarchies to democracy (not mentioning the important transition from Soviet domination to independence by the newest EU members), and greater responsibility of the role of private enterprise in the U.S. Each of these sub-frames is an extension of the larger interdependency frame that encompasses the relative maturity and overall stability of the U.S. democracy. If privacy is fundamentally a democratic principle embodied in the freedom of expression and information exchange, then through the process of frame extension, the U.S. is better situated historically, democratically, and institutionally to protect this fundamental consumer right.

DISCUSSION: TRANSFORMING AMBIGUITIES

Privacy defies reductive and unitary definitions both within and across cultures, disciplines, and societal applications. Integrating technological advances with the constantly evolving, contextual nature of privacy has historically challenged scholars, policymakers, and social philosophers alike. In his classic exposition on privacy and freedom, Westin (1967) defined privacy as "the claim of individuals, groups, or institutions to determine for themselves when, how, and to what extent information about them is communicated to others" (p. 1). Although Westin's definition describes privacy in informational terms, Warren and Brandeis (1890) had earlier defined it as the "right to be left alone." The frame analysis of the rhetorical devices embodied in the deliberations on the privacy Directive reveals its many facets, and highlights the fact that the intimate moral and philosophical principles defined by privacy have evolved in a dialogic relationship with the social, cultural, economic, and technological fabric of society. The frames structure the narrative form of privacy such that its expression in the public and private sphere reveals the values, principles, and morals privileged by a culture at a given time.

To understand the different constructs referenced by the four privacy frames identified in this analysis, a matrix of the privacy frames and the sub-frames situated within their discursive spaces is constructed in Table 1. It identifies the frames, sub-frames, main constructs, and points of ambiguities, and locates the recommendations embodied in the transformative points of ambiguities within the topical frames.

The first frame is a discursive embodiment of the concept of trust in the subcommittee deliberations on information privacy. It reveals that consumer trust in information transfer is expli-

Table 1. Privacy frames and key recommendations

TRUST FRAME • Frameworks ↔ Flexibility • *Create role of "**cultural negotiator**": Address cultural and political dimensions of privacy principles by employing cultural experts in privacy: identify points where an opportunity to integrate flexibility of interpretation through dialogue is built in*	**SOVEREIGNTY FRAME** • *Individual business and educational coalitions establish formal or informal "**regulatory circles**"* • *Individuals designated to work with comparable collaborators in other institutions would operate as circles to involve senior management directly to educate and inform them of historically defined privacy principles*
DEMOCRACY FRAME • *Establishing **codes of conduct** that embody democratic freedoms* • *Integrate best practices in business with democratically defined codes of conduct—as in collection, transmission, and sharing of individual information in business transactions*	**INTERDEPENDENCY FRAME** • Freedom ↔ Interdependency • ***Definition of privacy** should keep flexibility in interpretation, and keep freedoms and protections of consumer central to all such interpretations; Interdependency in globalization balanced with other, more nuanced cultural interdependencies*

cated through the sub-frames of globalization, change, and speed. While the U.S. perspective frames trust as a dynamically changing frame in a positive dialectic with globalization, Rodota articulates the EU perspective by highlighting the need to maintain eroding trust through the speed of changes because of globalization.

The trust frame elaborates on the harms caused to the information privacy interest through increasing risk in information flow and is counter-posed in opposition to risk. Accompanying this risk/trust dialectic are the concerns related to the need to control speed through regulatory frameworks. Freedom of information is counter-posed with increasing protection and control. These constructs embody the dynamic tensions in the trust frame. The business-driven, self-regulatory American paradigm is a counter-frame to the regulation-heavy government approach of the European Union.

While Rodota emphasizes the importance of the Directive as a framework to control speed and guide change, the U.S. perspective emphasizes freedom and control as emergent through business. Isolating the construct of frameworks helps to articulate a point of ambiguity symbolized by the trust frame. It expresses the challenge of regulating change while protecting freedom of information and minimizing the risk to providing individual information. *Our first recommendation is to focus deliberations on flexibility to allow interplay between the Directive and self-regulation, and build consumer trust in transactions as key enablers of freedom of information in constructing frameworks.*

The sovereignty frame is based on the sub-frame of harmonization of regulation across national borders to facilitate information flow. Harmonization discourse involves concerns of equality of national perspectives and, relatedly, relative perceptions of national power. Interwoven within the discourse of national identity and power are the historical contexts and experiences of individual nations. Harmonization in the sovereignty frame implies a negotiation of the history, values, and identities of nations. Underlying such negotiations is the embodiment of national power within the privacy discourse, with individual nations identifying citizens as "second-class citizens" or not, and connecting with notions of national power.

To that extent, harmonization of current regulations is a harmonization of dialogue on the respective historical experiences of the two contexts, with the goal of re-establishing shared values. The points of transformation in this dialogue rest on the dialectic of harmonization and power to negotiate national values. Privacy in this frame is explicated through the lens of national values as embedded within the sovereignty of nations. *Our second recommendation indicated by this frame is to harmonize definitions of privacy in different national contexts and reconcile historical narratives with current negotiations.*

The democracy frame is built on the sub-frames of relative freedom of information and cultural beliefs. In this discursive frame, the historical maturity of democracy is connected to the need for regulation. This need in turn leads to the establishment of best practices and codes of conduct that are required for its enforcement. The point of ambiguity rests on the agreement of a code of conduct for best practices that reconciles the notion of freedoms and beliefs in the agency protecting those freedoms. Privacy is understood in this frame as a ground of democracy, constructed through the instruments of best practices and codes of conduct. The codes of conduct and their enforcement are the means through which information privacy is defined in the democracy frame. *Our third recommendation is to establish best practices that express the freedoms defined in a common understanding of democracy.*

The fourth frame is interdependency based on the notion of the culturally contingent nature of freedoms. This frame and the sub-frames referenced through it negotiate the delicate balance of the right amount of privacy as opposed to too much privacy (leading to accusations of a "privacy regime"). Such discourse highlights the culture-specific nature of privacy regulations. The notion of control embodies the point of ambiguity inherent in this frame. While privacy has been defined as a fundamental human right in Europe, its regulation in a different culture may lead to perceptions of other fundamental freedoms being stifled. Thus, what may be a guardian of the basic freedoms of democracy may be perceived as the regime that stifles these freedoms. *Our fourth recommendation is to create a definition of privacy that achieves agreement on enforcement while protecting control of personal information and thereby achieving a mutual consensus on the information privacy discourse in international regulations.*

The four frames of trust, sovereignty, democracy, and interdependency lead to four different ways of defining information privacy in the subcommittee deliberations on commerce and trade. Although the frames themselves describe the fundamental perceptions of privacy called forth in the concerns expressed by witness testimonies, explicating the sub-frames and the constructs referenced in the discourse fleshes out definitions to achieve a more contextual and richer understanding of the underlying principles and constructs of privacy at stake. Isolating the points of ambiguities identifies how these constructs can be aligned in the negotiations by emphasizing and including all points of concern.

Such deliberations emphasize the situated, emergent, and evolving characteristic of privacy inherent in the many ways it has been described, from Westin's (1967) "claim," to Warren and Brandeis' (1890) "right," to those who consider privacy a "value" or "principle." While privacy is considered conceptually as a fundamental grounding notion for the core western liberal institution of democracy and freedom, it expresses different narrative forms within and across national cultures. The frame analysis reveals how each culture, nation, group, or individual claims a different frame through which to interpret privacy.

Information privacy draws attention to one aspect of privacy in contemporary society. The subcommittee deliberations emphasize the importance of the issue in international regulations as greater quantities of individual information cross national borders daily. With increasing financial flows and interrelated economic dependencies, the four frames emphasize the need to understand the interwoven narratives of history, national identity, power, culture, and freedoms in establishing the key frames through which to interpret information privacy.

The challenge of integrating the four frames of trust, sovereignty, democracy, and interdependency rests on reconciling best practices and their codes through national values and their mutual enforcement. Given the increasing interdependencies of contemporary regulatory environments and the interrelated nature of global finance, national values, and national histories, fundamental democratic values such as privacy need to be relocated continually in the dialectic over national values and culturally located best practices within business to achieve harmonization of regulations and embody the freedoms they so critically define in the so-called information age.

PRACTICAL RECOMMENDATIONS FOR THOSE IN BUSINESS AND EDUCATION

Our frame analysis of privacy regulation in the international arena leads to several practical recommendations. Let us begin with five general recommendations. First, U.S. business and educational institutions can neither go it alone in an increasingly globalized world nor simply impose their values, notions, and traditions on other nations or entities such as the EU. We are too interdependent upon other nations and must respect them as we want to be respected if e-commerce and exchange of information are to flourish in this global village.

Second, we must be aware of and appreciate how different cultures view the notion of privacy. The EU, for instance, sees privacy as a fundamental human right that is *guaranteed* and must be protected, while the U.S. tends to see privacy as a limited right that is *granted* or *given* by the Constitution, the Supreme Court, or Congress, and may be expanded or limited from time to time. We must choose language carefully that recognizes these fundamental differences when creating privacy guidelines and addressing privacy issues with other nations.

Third, we must acknowledge that the EU and other nations place high trust in governments to ensure that the right of privacy is maintained while we in the U.S. mistrust the government and place high trust in business and educational institutions to ensure our rights, including privacy. Merely assuring the EU, for instance, that our business has privacy policies that meet "industry standards" or are "some of the best" will not be persuasive or allay fears in global transmissions of information and financial data. We must establish mutual trust through tangible and credible guarantees of privacy.

Fourth, we must be aware of important principles in the EU Privacy Directive and Safe Harbor Agreement, and how our business and educational institutions can meet them. The seven principles of the Safe Harbor Agreement, for example, are notice, choice, onward transfer, security, data integrity, access, and enforcement. Although, as U.S. attorney Jonathan M. Winer stated at the hearings quoted throughout this chapter, there has been confusion and perhaps extensive violation of the Directive and Safe Harbor Agreement throughout the U.S. and the EU, these seven principles provide a sound foundation for any policy aimed at protecting the individual's right to privacy.

Fifth, we must keep up to date on changes that affect privacy policies. While the Supreme Court, for instance, has greatly impacted the conceptualization and guarantees of the right to privacy in the U.S. during the past century, it is undergoing fundamental changes in membership that may alter past decisions and defer future decisions to the legislative branch. President George W. Bush's recent court appointments have clearly tilted the majority toward strict constructionalists who may choose to overturn controversial decisions such as *Roe v. Wade* and defer to Congress and the states. While this eventuality might appear to be more in line with the European tradition of legislative protections of the right to privacy, it seems unlikely that conservative state legislatures or Congress will move toward recognizing privacy as a fundamental human right that is not recognized as such in the Constitution.

Our frame analysis also suggests specific recommendations for business and educational contexts that can be fruitful steps in harmonizing differences in perspectives between the U.S. and the EU. The integration of these recommendations in regulation, policy, and transactions would enable business and educational institutions to meet the evolving challenges of globalization in the information age.

The first recommendation is to incorporate policy frameworks that counter-balance the imperatives of addressing risk while simultaneously engendering trust. This may be done by focusing deliberations on open points of dialog where flexibility of interpretation can be exploited and built upon. Businesses, for instance, can move in this direction by focusing attention on incorporating "speed bumps" into regulatory frameworks. The notion of speed bumps stems from the EU's concern with the speed of globalization and addresses the need to harness the positive dimensions of speed brought out by U.S. representatives at the hearings.

Speed bumps are points in the framework where periodically scheduled negotiations are built in. These negotiations would provide critical articulation and evaluation of existing understandings to bring about necessary changes and improvements. Therefore, the first recommendation with the goal of furthering such negotiations would be the creation of the post of *cultural negotiator* of privacy. Education and business could benefit by acknowledging and addressing the cultural and political dimensions of privacy principles by employing cultural experts in privacy.

The second recommendation is to harmonize historical narratives of privacy in different national contexts with current negotiations. Although an international body might seem most effective, we recommend that individual business and educational coalitions establish formal or informal *regulatory circles*. Individuals designated to work with comparable collaborators in other institutions would operate as circles to involve senior management directly to educate and inform them of historically defined privacy principles.

The primary responsibility of such individuals might be to reconcile national legislation and resolve the dilemma of national sovereignty while guaranteeing maximum privacy protection for consumers. The regulatory circles could extend or complement the role of cultural negotiators. They would serve primarily an educational role geared toward increasing awareness of the roles national values and historical and cultural traditions play in defining privacy in international transactions.

The third recommendation is to establish *codes of conduct* that embody key democratic freedoms such as free flow of information. In a technologically defined interdependent world, codes of conduct that embody the democratic foundations of privacy necessarily involve the negotiation of continuously evolving information rights of individual consumers. This recommendation can be translated into practice by business and educational institutions initiating processes that integrate best practices with democratically defined codes of conduct that specifically address the imperatives of cross-border transfer of individual information. The central tenets, for example, could involve the central privacy principles of information collection, storage, and dissemination in transactions.

The fourth recommendation is to *create a definition of privacy that balances consumer freedom with cultural interdependencies.* Such a definition of privacy achieves agreement on enforcement while protecting control of personal information, thereby achieving a consensus on information privacy discourse in international regulations. It would frame privacy within the dynamic co-evolution of freedom and interdependency in the global village. This can be translated into practical recommendations within business and educational transactions by incorporating flexibility in interpretation and translation of "freedom."

CONCLUSION

In essence, freedoms and protections must be central in all interpretations. A practical consideration in the application of this recommendation is to incorporate consumer freedom when weighing cost-benefit advantages of privacy alongside privacy protections by integrating the element of choice. All applications, as suggested in the Directive and other privacy policies, must balance the notion of interdependency in the global village with other cultural interdependencies. Implementation of all four specific recommendations must involve cultural negotiations.

REFERENCES

Allen, A. L. (2004). Privacy in American law. In B. Rössler (Ed.), *Privacies: Philosophical evaluations*. Stanford, CA: Stanford University Press.

Benford, R. D., & Snow, D. A. (2000). Framing processes and social movements: An overview and assessment. *Annual Review of Sociology, 26,* 611-639.

Cappella, J., & Jamieson, K. (1997). *Spiral of cynicism*. New York: Harper Collins.

Cohen, J. E. (2000). Examined lives: Informational privacy and the subject as object. *Stanford Law Review, 52*(1373), 1397-1398.

Davies, S. (2002). The paradox of progressive education: A frame analysis. *Sociology of Education, 75*(4), 269-286.

Eden, J. M. (2005). When Big Brother privatizes: Commercial surveillance, the Privacy Act of 1974, and the future of RFID. *Duke Law and Technology Review, 20.*

Entman, R. M. (1993). Framing: Toward clarification of a fractured paradigm. *Journal of Communication, 41,* 51-58.

Etzioni, A. (1999). *The limits of privacy.* New York: Perseus Books.

Fillmore, C.J. (1975). An alternative to checklist theories of meaning. In Cathy Cogen et al. (Ed.), *Proceedings of the First Annual Meeting of the Berkeley Linguistics Society*, pp. 123-131. Berkeley, CA: Berkeley Linguistics Society.

Gamson, W. A. (1985). Goffman's legacy of political sociology. *Theory and Society, 14*(5), 605-622.

Gamson, W.A., & Modigliani, A. (1987). The changing culture of affirmative action. *Research in Political Sociology, 3,* 158-182.

Gandy, O.H. Kopp, K., Hands, T., Frazer, K., & Phillips, D. (1997). Race and risk: Factors affecting the framing of stories about inequality, discrimination, and just plain bad luck. *The Public Opinion Quarterly, 61*(1), 158-182.

Gavison, R. (1980). Privacy and the limits of law. *Yale Law Journal, 89,* 421-455.

Gindin, S. E. (1997). Lost and found in cyberspace: Informational privacy in the age of the Internet. *San Diego Law Review, 34*(1153), 1188-1189.

Gitlin, T. (1980). *The whole world is watching.* Berkeley, CA: University of California Press.

Goffman, E. (1961). *Encounters: Two studies in the sociology of interaction.* Indianaplois: Bobbs-Merrill.

Goffman, E. (1974). *Frame analysis: An essay on the organization of experience.* New York: Harper and Row.

Graber, D.A. (1984). *Processing the news: How people tame the information tide.* New York: Longman.

Hearing. (2001). *Hearing before the Subcommittee on Commerce, Trade and Consumer Protection of the Committee on Energy and Commerce, House of Representatives, One Hundred Seventh Congress.* Washington, DC: U.S. Government Printing Office.

Levin, M. R. (2005). *Men in black: How the Supreme Court is destroying America.*Washington, DC: Regnery Publishing.

Jasperson, A. E., Shah, D. V., Watts, M., Faber, R. J., & Fan, D. P. (1998). Framing and the public agenda: Media effects on the importance of the federal budget deficit. *Political Communication, 15,* 205-224.

Kahneman, D., & Tversky, A. (1984). Choices, values, and frames. *American Psychologist, 39,* 341-350.

McClurg, A. J. (2003). A thousand words are worth a picture: A privacy tort response to consumer data profiling. *Northwestern University Law Review*, 98, 63-65.

Prosser, W. L. (1960). Privacy. *California Law Review, 48,* 383-384.

Reidenburg, J. R. (1992). Privacy in the information economy: A fortress or frontier for individual rights. *Federal Commission Law Journal, 44,* 195-208.

Reiman, J. (2004). Driving to the Panopticon: A philosophical exploration of the risks to privacy posed by the information technology of the future. In B. Rössler (Ed.), *Privacies: Philosophical evaluations.* Stanford, CA: Stanford University Press.

Rein, M., & Schõn, D. (1993). Reframing policy discourse. In F. Fischer and J. Forester (Eds.), *The argumentive turn in policy analysis and planning.* Durham, N.C.: Duke University Press.

Rotenburg, M. (2001). Fair information practices and the architecture of privacy. *Stanford Technology and Law Review, 1,* 89.

Rule, J. B. (1974). *Private lives and public surveillance: Social control in the computer age.* Schocken Books.

Schlesinger, M., & Lau, R.R. (2000). The meaning and measure of policy metaphors. *American Political Science Review, 94*(3), 611-626.

Snow, D. A., Rochford, E. B. Jr., Worden, S. K., & Benford, R. D. (1986). Frame analysis processes, micromobilization, and movement participation. *American Sociological Review, 51,* 464-481.

Solove, D. J. (2001). Privacy and power: Computer databases and metaphors for information privacy. *Stanford Law Review, 53,* 1393-1401.

Solove, D. J. (2002). Access and aggregation: Public records, privacy and the constitution. *Minnesota Law Review, 86,* 1137.

Solove, D. J. (2006). A taxonomy of privacy. *University of Pennsylvania Law Review, 154,* 477.

Stewart, C. J., Smith, C. A., & Denton, R. E. Jr. (2001). *Persuasion and social movements.* Prospects Heights, IL: Waveland Press.

Swire, P.P. & Litan, R.E. (1998). *None of your business: World data flows, electronic commerce, and the European Privacy Directive.* Washington, DC: Brookings Institution Press.

Warren, S. D., & Brandeis, L. D. (1890). The right to privacy. *Harvard Law Review, 4,* 193.

Westin, A. (1967). *Privacy and freedom.* New York: Atheneum.

Wills, G. (1999). *A necessary evil.* New York: Simon & Schuster.

Section V

International Aspects of Online Education

Chapter XV
Cyberspace and Online Education:
The Influences of Global Cyberculture on International Students

Chun-Min Wang
University of Georgia, USA

Thomas C. Reeves
University of Georgia, USA

ABSTRACT

Friedman (2005) argues that the rapid growth of Internet technology is fundamentally changing human life in modern societies. This Internet revolution is also having an impact on higher education in the United States. More than half of all institutions of higher education in the U.S. now offer entirely online or blended courses. The issue becomes more complex when the learners are international students who come from extremely different cultural backgrounds. This chapter explores the nature of cyberculture, and describes how cyberculture affects the online learning experiences of international students. Through the exploration of cyberculture and online education, this chapter reveals what increased global online access is beginning to mean for participants in online education.

INTRODUCTION

Friedman (2005) argues that the rapid growth of Internet technology is fundamentally changing human life in modern societies. Companies in wealthy nations such as Japan and the USA are sending their call centers, document processing operations, computer programming, and other forms of knowledge work to countries where labor is much cheaper such as China and India. People with health problems such as cancer or AIDS seek advice and comfort online from other people facing the same illnesses as well as expert guidance from health professionals. The economic benefits and efficiencies of Internet technology have motivated people to become increasingly engaged in a more wired society (Tapscott, 1998). With the advantages of the Internet, people are creating virtual communities in cyberspace without geographical boundaries (Rheingold, 1993a). These virtual communities, like other human societies, have their own characteristics, and have even come to have their own cultures, which are usually referred to as cybercultures (Bell, 2001).

This Internet revolution is also having an impact on higher education in the United States. More than half of all institutions of higher education in the U.S. now offer entirely online or blended courses (Sloan-C, 2004). Some institutions offer entire degree programs online for a global market (Pittinsky, 2003). One interesting question raised by these developments is how cultural differences are accommodated (or ignored) in online learning and teaching environments. The question becomes more complex when the learners are international students who come from extremely different cultural backgrounds. How does the cyberculture that evolves in cyberspace interact with international students' cultural backgrounds? Is cyberculture a more important factor than students' local (national or indigenous) culture to explain success or failure in international online education?

This chapter explores the nature of cyberculture and describes how cyberculture affects the online learning experiences of international students. Through the exploration of cyberculture and online education, this chapter reveals what increased global online access is becoming to mean for participants in online education. In addition, this chapter identifies future research directions that need to be taken to support instructors and instructional designers faced with designing effective online courses for international students.

BACKGROUND

To begin, a definition of online education is required to clarify the scope of this area. Online education in this chapter is defined as *one form of learning and teaching which uses computer-mediated communication via the Internet to achieve the learning objectives within an educational organization.* This chapter does not distinguish the computer-mediated communication by time (synchronous or asynchronous) or tools (such as online forums or interactive Web sites). Rather, this chapter adopts a general and broad perspective of online education for purposes of investigating cyberculture and exploring the cultural issues that affect international students. Popular and emerging terms such as e-learning, online learning, Web-based learning, and Internet-based learning are included in our conception of online education. It should also be noted that people often view online education and distance education as synonymous (Moore & Anderson, 2003).

While rapidly evolving Internet technology has helped higher education proponents of online education break away from the brick-and-mortar campus and reach a wider range of learners around the world, the challenge of accommodating learners from different cultural backgrounds in online learning environments has not been addressed sufficiently (Collis, 1999; Gunawardena, Wilson, & Nolla, 2003; McLoughlin & Oliver, 2000; Wang, 2004). Taking cultural factors into account in online education would seem to be an obviously important issue, but there is a surprising lack of research focused on this issue (Gunawardena et al., 2003). As noted by Macfadyen, Roche, Doff, Reeder, and Chase (2004), some research topics related to culture and online education that have been addressed include the:

1. Interactions in online courses which involve culturally diverse adult learners
2. Access to the Internet among different groups seeking access to online education

3. Assessment criteria applied in online education
4. Design of virtual learning environments to accommodate the needs of culturally diverse learners

One difficulty in conducting research on this issue is the lack of agreement on the concept of culture (Straub, Loch, Ev Aristo, Karahanna, & Srite, 2002). Multiple definitions of culture are reflected in the relatively few studies done concerning online communication and education. In order to establish a consensus about the concept of culture in this realm, some researchers have tried to develop theoretical frameworks to guide studies in this area. For example, Ess and Sudweeks (2005) discuss the application of frameworks in online communication and try to answer two questions:

1. To what extent are the now widely used—but also seriously criticized—frameworks for cultural analysis provided by Hall (1981) and Hofstede (1984, 2001) fruitful for cross-cultural and intercultural communication in CMC environments?
2. How have CMC scholars and researchers developed, modified, and/or created alternative frameworks for analyzing cultural dimensions of online communication?

The debate about the most appropriate frameworks for studying culture in online education is still ongoing among scholars (Ess & Sudweeks, 2005). Meanwhile, in addition to investigating the cultural backgrounds of online learners, we must also consider the potential influences of global cyberculture among participants in online education. As pointed out by Macfadyen et al. (2004), only a small number of studies actually document the impact of the Internet per se as the agent of cultural change. While current research interests in cyberculture are spread across different disciplines, the attempt to find a unifying theory to interpret the phenomena in cyberspace remains an elusive goal for many scholars (Silver, 2000). In terms of online education, we know almost nothing about how cyberculture interacts with the learners' local cultures and their online learning experience. In particular, if and how a globally shared cyberculture compensates for the cultural conflicts in online education remains unclear.

This chapter is an attempt to examine cultural issues in online education from a global perspective. The hypothesis underlying this chapter is that the cultural issues in online education do not only come from the diverse cultural backgrounds of the learners, but also from the cyberculture in which the learners participate during their daily lives. The questions of concern in this chapter are to what extent cyberculture influences online education, and how cyberculture interacts with the learners' cultural backgrounds.

WHAT IS CYBERCULTURE?

Cyberculture, simply put, is the culture that emerges in cyberspace. In alignment with the definition of online education described above, we are focusing on the cyberspace culture that has evolved as the Internet has evolved. However, like the thorny task of defining culture in the physical world, cyberculture generated via the Internet is also challenging to define. Le'vy (2001) actually argued that any attempt to assign a univocal meaning of digital technology is questionable because he maintains that digital technology lacks any stable essence. He posits that the culture of the future is associated with the concept of the "universal-without-totality." The universal refers to the virtual presence of humanity to itself. In terms of cyberculture, it means "to interconnect all speaking bipeds and encourages their participation in the collective intelligence of the species within a ubiquitous environment" (Le'vy, 2001, p. 233). On the other hand, totality refers to "the stabilized unity of meaning associated with diversity" (Le'vy, 2001, p. 234). Cyberculture does not have totality, because it is full of differences. Le'vy (2001) wrote, "Neither their laws (no 'rights of man'), their gods (no universal religions), their knowledge (no experimental procedures or reasoning that can be reproduced anywhere), nor their technology (no networks or global standards) is universal" (p. 234). The concept of universal-without-totality provides a way to express human diversity in cyberspace and at the same time acknowledges the fundamental reciprocity of human relations.

Given the complexity of cyberculture, Abdelnour-Nocera (2002) suggested the use of ethnographic strategies with a hermeneutic analysis to discover the structures and meanings of cyberculture. The objective of this strategy is to obtain "thick" descriptions (Geertz, 1973) to interpret the phenomena occurring in cyberspace. At the same time, a simple description of cyberspace is helpful. According to Mihalache (2002), David Silver provided a satisfactory provisional definition of cyberculture:

Cyberculture is a collection of cultures and cultural products that exist on and/or are made possible by the Internet, along with the stories told about these cultures and cultural products. Cyberculture, as well as the cybersociety that supports and develops it, has its own location, the mysterious realm of cyberspace. (p. 294)

To understand the mysterious realm of cyberspace, one early and still popular metaphor of cyberspace views it as an electronic frontier region where people break the limitations of their physical capacities, and it expands their lives to a broader level. As Cavallaro (2000) pointed out, the objects of cyberspace such as computers, mobile phones, and video recorders are comparable to *prostheses,* the medical technology to provide artificial supports for people lacking physical limbs. The prosthetic affordances of the Internet range from relatively simple ones (e.g., when school teachers use e-mail to communicate with students and their parents to document their assessments of student progress rather than making phone calls or having face-to-face meetings), to complex, even troubling ones (e.g., when international terrorists communicate via encrypted Web communication tools to plot an attack in either the real world or the cyber one).

While the metaphor of electronic frontier region sounds reasonable in cyberspace, Mihalache (2002) argued that the frontier metaphor associated with cyberspace has two major deficiencies. First, cyberspace should not be understood as a metric space or simply a spatial entity. Instead, it is an intellectual space formed by the interconnections of various information and knowledge bases, and it also involves time as in the real life. Second, cyberspace is not a preexistent territory, but an entity that emerges in the process of its development. In other words, cyberspace involves a process of space creation and knowledge building. By downplaying the perspective of viewing cyberspace as merely an electronic frontier, Mihalache (2002) shifted the focus toward cultural expression in cyberspace. Cyberspace, as defined by Mihalache (2002), is "the area wherein a distinctive population works, socializes, builds up friendships and love bonds, develops new skills and specific rules of behavior, ponders its identity, and experiments with new forms of organization" (p. 300).

In addition, behaviors in cyberspace, like other social phenomena, contain a range of psychological effects and open up the opportunities for identity play (Turkle, 1999). In her book, *Life on the Screen: Identity in the Age of the Internet* (1995), Turkle draws on resources in psychology, sociology, psychoanalysis, philosophy, and other fields, to examine how technology influences our identity and the "self" in cyberspace. Through several case studies, Turkle (1995) demonstrated how people create online identities to help find the way of balance in their off-line (real-world) lives. In Turkle's opinion, the computer is not only a tool, but also a mirror and gateway for people to reflect and transform themselves on the screen. In other words, virtuality is not just a space for alternative life but also a space for human growth.

In addition to the affordances of cyberspace, the term cyberspace as coined by the science fiction writer William Gibson (1984) implies the need for consensus while communicating in cyberspace. As Cavallaro (2000) described, this kind of consensus requires continuously sharing knowledge, opinions, and other forms of communication with large groups of people. This sharing process gradually transforms individual experiences into collective representations such as the Wikipedia phenomenon. Wikipedia is an online interactive encyclopedia to which anyone with Internet access can contribute. As of March 2006, Wikipedia contained over a million articles after less than five years of existence, while the *Encyclopedia Britannica* has only 120,000 entries after 200 years of existence (Schiff, 2006). This sharing process between individuals and the larger group is actually like the continual dialogue aiming to reach some degree of consensus in the cyberspace. With regard to the relationship between cyberculture and local culture, the questions are what kind of collective representation can be found on the Internet, and how this collective representation impacts local culture and further influences the online learning experiences of international students.

Regardless of how cyberculture is defined, we understand that cyberspace is an extension of physical life, and through the Internet people can get cognitive and emotional support. While cyberspace provides the prosthetic affordances to expand our physical life, it is so complex that a certain consensus is needed in order to communicate within it. With regard to understanding how cyberculture influences online education, further thinking about the influences of cyberculture is needed.

THE INFLUENCES OF CYBERCULTURE

Given the popularity of the Internet today, the effects of cyberculture formed through the Internet are becoming increasingly evident in our daily lives. The influences of cyberculture are related to the metaphors used to describe cyberspace. One early and still popular metaphor of cyberspace views it as an electronic frontier region where sovereignty is an issue requiring resolution (Kapor & Barlow, 1990). Who owns the information in cyberspace? How can human rights be protected in cyberspace? The importance of these issues can be seen in the contemporary struggle between corporations such as Google and the U.S. government concerning the release of search engine data. The Electronic Frontier Foundation (http://www.eff.org) was established specifically to deal with these issues in cyberspace. As in previous real-world frontiers, there is a certain degree of lawlessness in cyberspace. For example, many people share resources through the Internet in violation of copyright protection. Along with the increased risk of lawlessness, frontiers also feature a greater degree of freedom than traditional environments. The Massachusetts Institute of Technology (MIT) made a bold move several years ago to construct OpenCourseWare, an online environment that allows free and open access to educational resource by educators, students, and self-learners. MIT shares its curriculum materials with the public through the Internet and welcomes foreign countries to translate the content from English into their own native language. Although whether the resources are fully open is debated by some, it is clear that the liberty and freedom promoted in cyberculture is a double-edged sword. The value of freely sharing resources confronts the risk of plagiarism. Online learning environments can make plagiarism even more difficult to control, especially for international students who come from other educational systems where the problem of plagiarism is not as emphasized as it is in North American academic cultures. Hence, the instructor in online courses may need to spend more time teaching about issues related to academic honesty and copyright.

Another metaphor of cyberspace is neo-colonialism, as raised by Kolko, Nakamura, and Rodman (2000). Do Western values spread to other races in the world via cyberspace? Who is dominating cyberspace? Clearly, English is the dominant language used in cyberspace. The hegemony of English online, according to Warschauer (2000), is due to several factors. First, a high percentage of early users of the Internet were North Americans. Second, the Internet was designed on the basis of the American Standard Code for Information Interchange (ASCII), which made it inconvenient or impossible for other alphabets or character sets to compute. Third, English matches the need to have an international *lingua franca* to use on the Internet. Indeed, we need a language to communicate in cyberspace, and English happened to be the one chosen, partially because it is technically easier to use than other languages such as Chinese. To challenge the dominance of English, Le'vy (2001) recommends treating all languages as minority languages, and including the original text in which something was written in order to ensure the fairest possible distribution in cyberspace.

Virtual community is another metaphor used to describe the community formed in cyberspace. Howard Rheingold is the author of *The Virtual Community: Homesteading on the Electronic Frontier* (1993b), one of the major works cited in research on virtual communities. A virtual community, as described by Rheingold (1993a), is "a group of people who may or may not meet one another face-to-face, and who exchange words and ideas through the mediation of computer bulletin boards and networks" (p. 75). Interest in research related to virtual communities has increased as networked technologies have become more sophisticated and prevalent. Especially in the field of online education, how to build an efficient online learning community has become an area of significant interest (Hill, 2002). For example, Pan and Sullivan (2005) examine the impact of using Skype, a freeware program that provides high-quality audio communication via the Internet (skype.com), to supplement group chat discussion and contact with the instructor. The growth of virtual communities is changing the ways of human communication around the world. Those learners who have been involved in the experience of some kind of virtual communities before taking online courses are more likely to be successful in online courses. Studying the application of networked technologies in other countries may help educators understand how to better design online courses for international students because their communication styles will inevitably be influenced by their previous online experience in their own countries.

Another metaphor and important discourse regarding cyberculture is the debate between utopia and dystopia. The advocates who hold a utopian perspective believe that the Internet can bring new forms of democracy and create a global village (Ess, 2001) in cyberspace. Other protesters argue that utopian writers ignore the fact that the medium is not value free. Instead, the scholars holding dystopian visions think that cyberspace can extend existing social and economic inequality in current society, and promote and privilege Western cultural values via cyberspace (Macfadyen et al., 2004).

Clearly, cyberspace includes both utopian and dystopian elements (Friedman, 2005). On the one hand, it can bring together people from different cultural backgrounds around the world, and it provides the opportunity for them to learn from each other. In the real world, it is impossible to bring together the number of diverse people who can "gather" in cyberspace. Cyberspace is, in this sense, the dream of Utopia coming true. However, on the other hand, cyberspace is also full of conflicts. Just look at the languages used in cyberspace. How can people with different languages communicate efficiently in the global village? Do we need to identify and avoid certain culturally inappropriate elements on Web sites in order not to offend people with different cultural backgrounds? The dark side of cyberspace ranges from financial deception to sexual exploitation.

In addition, although the Internet has the potential to connect the world, the networked technology itself does not create increased globalization and cultural homogenization; instead, it becomes a communication tool which can be and often is manipulated by governments, corporations, and other entities. As argued by Goldsmith and Wu (2006), the technological barriers usually force people in any given nation to conform to the technological barriers rather than transcending the technological barriers in order to connect with other people. Take China for example where many Web sites are blocked by the government due to political considerations, and the only thing the Chinese people can do is to accept what access of the Internet they are allowed. Nations and large corporations have the power to shape and/or control the versions of Internet architecture provided to their citizens and customers, and as a result, the future of the Internet, as described by Goldsmith and Wu (2006), will be like a cold war where different nations and international corporations push their own version of the Internet's future.

It is no doubt that along with the development and popularity of networked technologies, the influence of cyberculture will increase globally. However, to what extent cyberculture will influence local culture is a difficult question to answer. Factors such as the access to the Internet, governmental and corporate controls, and economic issues are all playing a part in answering this question. To consider how cyberculture influences international students, it is necessary to further discuss the relation between cyberculture and local culture.

CYBERCULTURE AND LOCAL CULTURE

In this chapter, we have used the term *local culture* to signify the national or indigenous culture to which someone belongs as described by Hofstede (2001). When studying the relationship of local culture to cyberculture, we suggest that nation is the unit to measure a culture, and hence we conclude that the nationality of international students represents their local culture. No doubt each nation can include several different cultures, but in our experience, nation is still the best unit to study cultural issues in terms of international students in online education. On the other hand, cyberculture represents the global culture that is distributed through the Internet. Given the influence of cyberculture as discussed previously, how cyberculture interacts with international students' local culture becomes an interesting question. By understanding the interaction, or dialogue as some scholars put it, we may identify better strategies to accommodate the educational needs of international students in online learning environments.

As explained by Griswold and Nathan (2004), scholars have identified three processes to capture what may transpire when global culture encounters local culture. These three processes are world culture, localization, and hybridization. The view of world culture emphasizes the emergence of cultural forms that include a set of patterns on a global scale; the view of localization emphasizes how local culture interacts with global culture and ultimately transforms the latter; and the view of hybridization suggests that the interaction process will produce cultural hybrids and new perspectives. In terms of international students in online education, we believe that all of these processes could happen when international students enter cyberspace. In short, they may yield to cyberculture, resist cyberculture, or create a new cyberculture. One thing we have noticed in our research is that the cultural representation shown by the international students in online courses is a constantly changing one stemming from the interaction between local culture and global cyberculture. As a result, any attempt to investigate how cyberculture impacts international students in online education must deal with complex and dynamic interactions. Although the concept of interaction sounds logical, it is difficult to predict and analyze to what extent cyberculture can influence local culture. Nevertheless, we argue that the interaction between cyberculture and local culture will help to decrease potential cultural conflicts in online education. This seems reasonable because before learners get involved in online education, the process of assimilation and accommodation between their own cultural background and cyberculture in daily life will have prepared them to anticipate and deal with cultural conflicts in online education.

Consider, for example, that using blogs to share personal thinking with the public has become popular not only in North America, but also in East Asian countries like Japan and Taiwan. Through their blogging experiences, the new generation of potential online learners may already be accustomed to an environment where they frequently communicate online, and interact with people from different social and cultural backgrounds. Nevertheless, by arguing this, we are not saying that the pedagogical differences embedded in culture or language barriers are not still important in online education. On the contrary, understanding to what extent cyberculture can decrease the conflicts inherent in online education requires further study. How might cyberculture values such as equality and liberty conflict with the values embedded in a specific pedagogy? How are language differences in cyberspace overcome? With the growth of networked technologies, one trend is clear: the younger generations will have more cyber-experiences than today's students, and this may have important implications for online teaching and learning (Oblinger & Oblinger, 2005).

Although we think the salience of cultural issues in online education will be decreased with the increased global Internet access experienced by future generations, some researchers suggest that Internet access is the channel for people to remain connected to their own culture when they are overseas. For example, Zhang and Xiaoming (1999) argued that online technologies can help ethnic groups be assimilated into the mainstream culture without losing their own cultural roots and ethnic identity. Especially with the increased global Internet access, different cultural media play an even more important role than ever before in maintaining contacts between minority ethnic groups overseas and their cultural roots. In terms of international students, this may mean that Internet access will help international students to surf Web sites of their own culture, and thus the cultural conflicts in online education may actually not decrease as global Internet access increases.

Will the interactions among local culture, cyberculture, and the culture of online learning be positive or negative? As previously suggested, more research is needed to examine these interactions and delineate the results of these interactions. Leaving the discussion of the interactions between cyberculture, local culture, and online education, in the next section of this chapter we redirect our discussion to more concrete ideas about how to improve international students' online learning experiences from the understanding of cyberculture.

MAKING THE CONNECTION

As discussed previously, the virtual community in cyberspace is actually an extension of our real society where there are diverse perspectives and subcultures. People in the cybersociety are highly unlikely to reach a universal agreement about anything, and they need to forge a consensus while communicating in cyberspace. The cybersociety, as termed by Jones (1995), is consistently dissolved and reconstructed in tandem with the real society, and the distinctions between cybersociety and real society are becoming vague with the increased access to the Internet. Consequently, future emphasis will be on the interfaces between real society and cybersociety (Robins, 2000). Compared to previous, more visionary discussion about cyberculture, redirecting thinking to visible things such as computer interfaces or Internet products (Web sites, e-mail, etc.) certainly makes our attempt to discuss the relationship between cyberculture and online education more understandable. Given that it is through the interfaces that cyberculture is created and introduced to people, the future of cyberspace will evolve as technology evolves. Cyberculture is primarily limited to the senses of sight and hearing at this time, but the time when people will be literally "jacked in" to the Internet is not far off, and our sensual interactions are likely to engage all the senses including touch, taste, and smell.

Computer interface designs and Internet products can be viewed as the characteristics of cyberculture, because people learn to live within cyberspace by the rules or functions of these interfaces and products. For example, the rapid growth of blogs has impacted politics, business, and cultures (Kline, Burstein, De Keijzer, & Berger, 2005). Through their blogging experiences, people not only get the chance to express ideas to a broader audience, but also learn to communicate in cyberspace. However, as the number of blogs proliferates, the audience for any given blog may actually decrease.

Whether blogging experiences can actually change cognition and learning to an extent are as yet unknown. In terms of online education, how blogging activities can be designed to improve or prepare students in online education becomes a reasonable question. To date, there has been little analysis of intercultural communication via blogs in cyberspace (Macfadyen, 2005).

When making the connections between cyberculture, the interfaces of Internet products, and online education, the question about how cyberculture affects the online learning experiences of international students can be transformed into a question asking how these Internet products can improve the experiences of international students in online education or prepare them for success. As noted above, it seems plausible that international students will be increasingly exposed to Internet products and thus cyberculture before actually taking online courses that include people from diverse cultures. In light of this development, other questions become important to address: Is cyberculture a more important factor than students' local culture to explain success or failure in online education? How can cyberculture compensate for the cultural conflicts in online education? These questions are complex and interdependent. As previously discussed, the key point is to investigate how cyberculture, along with the force of globalization, interacts with the influence of local cultures.

In addition, most new Internet technologies spread globally quite rapidly. From the perspective of international corporations, when an Internet phenomenon such as blogging develops, the important thing is to market some aspect of it globally. Although globalizing their products is an important step for international corporations focused on maximizing their profits, they may also need to localize the products in order to adjust to robust cultural and social differences. For example, developing different language versions of a new product is almost always the first thing corporations do. As access to the Internet increases globally, we can predict that international students in their own domestic countries will have more opportunities to get familiar with the technology and the cyberculture generated using this technology. In addition, most of these online experiences may utilize their native language, making them even more comfortable interacting online.

Furthermore, the increased global access to the Internet along with the advance of technology are enabling stronger channels for ethnic groups to share information and construct knowledge in cyberspace. Linguistic and cultural issues in cyberspace arise during the process of localization or when there is a need for intercultural communication such as international students in online courses. Other chapters in this book as well as some other studies are dealing with these issues directly. Our way of viewing the influences of cyberculture in online education requires zooming in from the technological issues in the global level to the linguistic and cultural issues in the local level. We are engaged in research to examine the cultural issues in online education from a global perspective and investigate how increased global Internet access for online education affects international students. Can educators take advantage of the younger generation's growing familiarity with cyberculture on the technological level and apply this knowledge to designing better online learning environments for international students? The last section in this chapter points out additional studies needed to answer this question.

FUTURE STUDIES

Our discussion of cyberculture reveals the importance of Internet products in preparing and improving online education for international students. Although a very important assumption underlying our discussion is that the potential learners in online education are from places where there is access to the Internet, we do not want to exclude the possibilities that even learners without Internet access can have cyber-experiences from other media like TV or radio. We think that Internet products—or broadly speaking, Internet technologies—enable powerful connections among people cross-culturally. We believe that, even without noticing, the younger generation is actually qualifying themselves to survive in a world where international online education may increasingly replace traditional on-campus instruction. The younger generation can be viewed as citizens of a new cyber-nation which is cross-cultural, one that promotes its own global youth culture. As a result, to better prepare international students for online education, it is helpful to investigate how young people are employing these new technologies to extend and improve their online experiences, and the implications of these findings for online education. Below we list some category of technologies which should be studied in this context:

- **Instant messaging:** The instant messages services such as MSN have become very popular in many areas around the world. How can this online communication experience be situated in online education?
- **Internet (video) phone service:** Free Internet phone service is expanding. For example, Skype already had roughly 79 million registered users in over 240 countries in 2006 (data from the Skype Support). How can this technology be applied into online courses?
- **Networked video games:** An emerging research area in education is to investigate how video games can be applied into instructional design (Dickey, 2005). Networked video games, especially Multi-User Virtual Environment (MUVE) games, are definitely a popular global Internet product among the younger generation. How can we take advantage of this social phenomenon to improve the online learning experience?

- **Weblog:** As mentioned earlier, there is little or no analysis of intercultural communication via blogs in cyberspace (Macfadyen, 2005). Blogs are being applied into online courses, but we know little about their effects (Oravec, 2003; Stahmer, 2006).
- **Course management systems:** Not every country has course management systems such as WebCT or Blackboard, but knowing how the course management systems are being used can help to design a better online learning environment.
- **E-mail:** Knowing the cultural differences in the expectations and expressions of e-mails can be helpful for the participants in online courses to communicate better.
- **Web sites:** How are Web sites designed in different cultures? This information can help us to get a sense of how to accommodate cultural differences in Web design. For example, Chu (1999) investigated the differences of Web design between American and Taiwanese college students.

We would also like to promote cross-cultural educational programs via network technologies not only in higher education, but also in K-12 levels. Through these programs, the younger generations can increase their cultural sensitiveness and get used to communicating with people from other cultures through the Internet. Meanwhile, instructors and designers can learn lessons from these experiences and eventually help to improve the design of online courses for international students. We predict that online education has a bright future. However, knowing that younger generations will feel more comfortable learning online is not enough; we also should devote more effort to investigate what pedagogies and strategies can be applied to online education, especially when it involves international students from diverse cultures.

CONCLUSION

This chapter speculates about the impact of cyberculture in online education and examines the influence of increased global Internet access for online education for international students. In light of our discussion of cyberspace and cyberculture, we recognize the importance of computer interfaces and Internet products as a way to understand the impact of cyberculture. We argue that the interactions between cyberculture and local culture may have the potential to decrease the cultural conflicts for international students as a result of increased familiarity with technology. But more research is needed to examine how these technologies are used and can be applied in designing better online learning environments. With the growth in global Internet access, online learning is becoming more and more popular and feasible. Friedman (2005) argues that the flattening of the world through new technologies will have more impact on mankind than even the printing press. It is already obvious that the interaction between cyberculture and local culture of the younger generation is fostering a new global youth culture. We need to take advantage of the younger generation's familiarity with new technologies and study the application of these new technologies in order to better design online education. In other words, we must adjust our instructional designs to take advantage of the ways that young people are interacting online to better facilitate online learning. Although in this chapter we emphasize the need to think of the application of cyberculture from the aspect of technology, we also recommend studying local linguistic and cultural issues in considering the relationship between cyberculture and online education. A popular environmental expression says "Think Globally, Act Locally." Perhaps with respect to online learning, we should teach and learn locally, globally, and "cyberly."

REFERENCES

Abdelnour-Nocera, J. L. (2002). Ethnography and hermeneutics in cybercultural research accessing IRC virtual communities. *Journal of Computer-Mediated Communication, 7*(2). Retrieved from http://jcmc.indiana.edu/vol7/issue2/nocera.html

Bell, D. (2001). *An introduction to cybercultures.* London/New York: Routledge.

Cavallaro, D. (2000). *Cyberpunk and cyberculture: Science fiction and the work of William Gibson.* London: Athlone Press.

Chu, G.-L. (1999). *The relationships between cultural differences among American and Chinese university students and the design of personal pages on the World Wide Web.* Unpublished PhD thesis, University of Georgia, USA.

Collis, B. (1999). Designing for differences: Cultural issues in the design of WWW-based course-support sites. *British Journal of Educational Technology, 30*(3), 201-215.

Dickey, M. D. (2005). Engaging by design: How engagement strategies in popular computer and video games can inform instructional design. *Educational Technology Research and Development, 53*(2), 67-83.

Ess, C. (2001). Introduction: What's culture got to do with it? Cultural collisions in the electronic global village, creative interferences, and the rise of culturally-mediated computing. In C. Ess & F. Sudweeks (Eds.), *Culture, technology, communication: Towards an intercultural global village* (pp. 1-50). Albany: State University of New York Press.

Ess, C., & Sudweeks, F. (2005). Culture and computer-mediated communication: Toward new understandings. *Journal of Computer-Mediated Communication, 11*(1). Retrieved from http://jcmc.indiana.edu/vol11/issue11/ess.html

Friedman, T. L. (2005). *The world is flat: A brief history of the twenty-first century* (1st ed.). New York: Farrar Straus and Giroux.

Geertz, C. (1973). Thick description: Toward an interpretive theory of culture. In *The interpretation of cultures: Selected essays* (pp. 3-30). New York: Basic Books.

Gibson, W. (1984). *Neuromancer.* New York: Berkley Publishing Group.

Goldsmith, J. L., & Wu, T. (2006). *Who controls the Internet? Illusions of a borderless world.* New York: Oxford University Press.

Griswold, W., & Nathan, W. (2004, May). Cowbird, locals, and the dynamic endurance of regionalism. *The American Journal of Sociology, 109,* 1411-1451.

Gunawardena, C. N., Wilson, P. L., & Nolla, A. C. (2003). Culture and online education. In M. G. Moore & W. G. Anderson (Eds.), *Handbook of distance education* (pp. 753-775). Mahwah, NJ: Lawrence Erlbaum.

Hall, E. T. (1981). *Beyond culture.* Garden City, NY: Anchor Books.

Hill, J. R. (2002). Strategies and techniques for community building in Web-based learning environments. *Journal of Computing in Higher Education, 14*(1), 67-86.

Hofstede, G. (1984). Hofstede culture dimensions—an independent validation using Rokeach Value Survey. *Journal of Cross-Cultural Psychology, 15*(4), 417-433.

Hofstede, G. (2001). *Culture's consequences: Comparing values, behaviors, institutions and organizations across nations* (2nd ed.). Thousand Oaks, CA: Sage.

Jones, S. (1995). *Cybersociety: Computer-mediated communication and community.* Thousand Oaks, CA: Sage.

Kapor, M., & Barlow, J. P. (1990). *Across the electronic frontier.* Retrieved November 26, 2005, from http://www.eff.org/legal/cases/SJG/?f=eff_creation.html

Kline, D., Burstein, D., De Keijzer, A. J., & Berger, P. (2005). *Blog!: How the newest media revolution is changing politics, business, and culture*. New York: CDS Books.

Kolko, B. E., Nakamura, L., & Rodman, G. B. (2000). *Race in cyberspace*. New York: Routledge.

Le'vy, P. (2001). *Cyberculture*. Minneapolis/London: University of Minnesota Press.

Macfadyen, L. P. (2005). Internet-mediated communication at the cultural interface. In C. Ghaoui (Ed.), *Encyclopedia of human-computer interaction.* Hershey, PA: Idea Group, Inc.

Macfadyen, L. P., Roche, J., Doff, S., Reeder, K., & Chase, M. (2004). *Communicating across cultures in cyberspace: A bibliographical review of intercultural communication online*. Piscataway, NJ: Transaction Publishers Distributor.

McLoughlin, C., & Oliver, R. (2000). Designing learning environments for cultural inclusivity: A case study of indigenous online learning at tertiary level. *Australian Journal of Educational Technology, 16*(1), 58-72.

Mihalache, A. (2002). The cyber space-time continuum: Meaning and metaphor. *Information Society, 18*(4), 293-301.

Moore, M. G., & Anderson, W. G. (2003). *Handbook of distance education.* Mahwah, NJ: Lawrence Erlbaum.

Oblinger, D., & Oblinger, J. (Eds.). (2005). *Educating the Net gen.* Washington, DC: EDUCAUSE.

Oravec, J. A. (2003). Blending by blogging: Weblogs in blended learning initiatives. *Journal of Educational Media, 28*(2-3), 225-233.

Pan, C.-C., & Sullivan, M. (2005). Promoting synchronous interaction in an e-learning environment. *THE Journal, 33*(2), 27-30.

Pittinsky, M. S. (2003). *The wired tower: Perspectives on the impact of the Internet on higher education.* Upper Saddle River, NJ: Financial Times/Prentice-Hall.

Rheingold, H. (1993a). A slice of life in my virtual community. In L. Harasim (Ed.), *Global networks: Computers and international communication* (pp. 57-80). Cambridge, MA: MIT Press.

Rheingold, H. (1993b). *The virtual community: Homesteading on the electronic frontier.* Reading, MA: Addison-Wesley.

Robins, K. (2000). Cyberspace and the world we live in. In D. Bell & B.M. Kennedy (Eds.), *The cybercultures reader* (pp. 77-95). London/New York: Routledge.

Schiff, S. (2006). Know it all: Can Wikipedia conquer expertise? *New Yorker,* (July 31). Retrieved August 9, 2006, from http://www.newyorker.com/printables/fact/060731fa_fact

Silver, D. (2000). Looking backwards, looking forward: Cyberculture studies 1990-2000. In D. Gauntlett (Ed.), *Web studies: Rewiring media studies for the digital age* (pp. 19-30). New York: Oxford University Press.

Sloan-C. (2004). *Entering the mainstream: The quality and extent of online education in the United States, 2003 and 2004.* Retrieved May 23, 2005, from http://www.sloan-c.org/resources/entering_mainstream.pdf

Stahmer, T. (2006). Think outside the BLOG. *Technology and Learning, 26*(6), 28-30.

Straub, D., Loch, K., Ev Aristo, R., Karahanna, E., & Srite, M. (2002). Toward a theory-based measurement of culture. *Journal of Global Information Management, 10*(1), 13-23.

Tapscott, D. (1998). *Growing up digital: The rise of the Net generation*. New York: McGraw-Hill.

Turkle, S. (1995). *Life on the screen: Identity in the age of the Internet*. New York: Simon & Schuster.

Turkle, S. (1999). Looking toward cyberspace: Beyond grounded sociology. *Contemporary Sociology, 28*(6), 643-648.

Wang, C. M. (2004). Taking online courses in the United States: The perspectives of Asian students from China, Korea, Singapore, and Taiwan. In J. Nall & R. Robson (Eds.), *Proceedings of the E-Learn 2004 Conference: World Conference on E-Learning in Corporate, Government, Healthcare, & Higher Education* (pp. 2466-2468). Norfolk, VA: Association for the Advancement of Computing in Education (AACE).

Warschauer, M. (2000). Language, identity, and the Internet. In B. E. Kolko, L. Nakamura, & G.B. Rodman (Eds.), *Race in cyberspace* (pp. 151-170). New York: Routledge.

Zhang, K., & Xiaoming, H. (1999). The Internet and the ethnic press: A study of electronic Chinese publications. *Information Society, 15*(1), 21-30.

Chapter XVI
Teaching, Learning, Negotiating:
The World Wide Web as a Model for Successful Cross-Cultural Communication

Tatjana Chorney
St. Mary's University, Canada

ABSTRACT

New technologies and computer-mediated communication (CMC) in general seem inherently suited to result in constructive cross-cultural communication. Yet researchers note that students and teachers, both of whom are instructional planners, lack the skills necessary to function in environments where they are "collaborative designers, rather than transmitters of knowledge" (Campbell, 2004b). As a result, the new possibilities for cross-cultural teaching and learning through dialogue and negotiation in the online environment compel us to reconceptualize the traditional role of the instructor and to ask, what does it mean to teach collaboratively, interactively, open-endedly? This chapter examines several central questions related to this situation as well as provides an overview of the dialogue-enabling properties of the Internet environment and its potential to support multiple learning styles.

INTRODUCTION

New technologies and computer-mediated communication (CMC) in general seem inherently suited to result in constructive cross-cultural communication due to their potential to encourage collaboration in an environment of diverse users with multiple perspectives. However, researchers note that students, but more importantly teachers, who are instructional planners, lack the skills necessary to function in environments where they are "collaborative designers, rather than transmitters of knowledge" (Campbell, 2004b), which is how their role has been defined traditionally. The increasing number of virtual universities, online training with a global reach, and international degree programs offered online by many traditional universities around the world indicates that the opportunities and demands for successful cross-cultural communication expand exponentially. The new possibilities for cross-cultural teaching and learning through dialogue and negotiation in the online environment compel us to reconceptualize the traditional role of the instructor and to ask, what does it mean to teach collaboratively, interactively, open-endedly?

In this chapter I will attempt several preliminary answers to this question. First I will provide a brief overview of the dialogue-enabling properties of the Internet environment and its potential to support multiple learning styles, some of which may be culturally determined. I will outline the World Wide Web's inherent affinity with recent social theories of learning. I will also compare the old instructional model according to which knowledge is seen in static terms, as a transfer of facts from specialist to learner, to the new model emphasizing knowledge as negotiation. Then I will interpret the differences between the two models in practical terms and in their significance for successful cross-cultural communication in the context of distance education. My purpose will be to demonstrate that the role of the instructor in the new model encouraged by the online environment needs to be defined in terms of mentorship and facilitation rather than authority in the traditional sense. In many cases in cross-cultural teaching and learning, students may possess knowledge that the instructor does not. I will suggest a number of practical approaches that can help with designing an interactive online learning unit. In the chapter I will also show that the open-ended and interactive nature of the World Wide Web, where multiple perspectives coexist, can serve as a conceptual model to help both teachers and students overcome common challenges in cross-cultural communication.

BACKGROUND

We live in an age when advances in information and communication technologies are significantly changing attitudes toward teaching and learning. Large investments have been made across educational sectors to provide more flexible curricula and extend the use of e-learning beyond pockets of innovation. The diversity of the student body, only intensified by the number of distance education programs offered worldwide, places new demands on educators' knowledge and skills. However, researchers have pointed out that educational sectors aiming toward increased flexibility have yet to develop a work ethos that will promote the development of new pedagogies (Wiles & Littlejohn, 2003; Campbell 2004). Scholars note that there is still a major gap between theory and practice that continually burdens the progress of education for an increasingly diverse student population (Bruch, Jehangir, Jacobs, & Ghere, 2004).

Due to the global reach of distance education and the cross-cultural interaction that is a component of today's education, instructional paradigms are shifting. Educational practices invented when higher education served only the few, and in a monocultural context, are increasingly seen as disconnected from contemporary realities and the needs of contemporary students (cf. Association of American Colleges and Universities, 2002, p. viii). Especially, there has been a growing demand for change in learning technologies and the way education is delivered online: from environments in which highly structured information is presented electronically and in a linear way, to environments in which the learner is supported in the making of meaning and knowledge construction. No doubt, there is a big difference between the earlier and current forms of online learning. The earlier type, computer-based instruction, focused on the interaction between the student and computer drills, tutorials, or simulations. Today, the prevailing paradigm is computer-mediated communication, "where the primary form of interaction is between students and instructors, mediated by the computer." Studies suggest that "without any doubt, the biggest single change that online education brings about is an increased tendency towards collaboration amongst students and teachers" (Enhance Learning, n.d.).

At the same time, however, it has been recently reported that many cultural groups feel excluded from e-learning opportunities because content is "designed by the dominant culture, and reflects mainstream values, and thus is not culturally inclusive" (Campbell, pp. 49, xvi). As e-learning is gaining prominence, and as distance education turns our world into a "global village," compelling us to interact meaningfully with individuals from different cultures daily, there is evidence that teachers are having trouble reconceptualizing their traditional role of "transmitter of knowledge" to "instructional planner" and facilitator (Campbell, 2004).

In this chapter I take the view that all education is based on interaction and communication, and that if cultural differences are often at the root of communication challenges, our ability as educators to deal with those differences in a global world will determine largely how successful we are in practice. One of the objectives of this chapter is to help educators who teach online courses in a cross-cultural context review their pedagogical assumptions and plan for diversity. The first step toward this objective is to acknowledge that both learning and teaching are culturally based processes, and that instructional design is not culturally neutral (Campbell, 2004, p. 49). Instructional content, and the way that content is experienced, reflects the values and practices of a particular cultural group—most commonly, English-speaking western cultures. The term "culture" is used here to define sets of social relationships and patterns of behavior reflecting traditional ideas and actions. It is the "sum total of ways of living, including values, beliefs, aesthetic standards, linguistic expression, patterns of thinking, behavioral norms, and styles of communication" (Campbell, 2004).

Understanding that both teaching and learning are not value neutral means that e-educators will become more self-reflective about the cultural assumptions guiding their pedagogical practice, and consequently more likely to modify their approach to online communication and interaction accordingly. Traditional pedagogy rests on the idea of "authority," supported and regulated by dominant bodies of knowledge and institutionalized academic structures (Chan, Fortunato, Fortunato, Oakes, & RyanMann, 2001). Teachers are traditionally viewed as figures with access to specialized knowledge, and as possessing the prerogative to identify not only what is important to learn and know, but also how it will be learned. The interaction and communication that ensue in this learning environment are implicitly controlled by the same set of boundaries. Assignments are designed as self-enclosed units, and problem solving rests on a linear logic, operating according to pre-determined, teacher-set rules. In this situation, what is being communicated by the teacher, often even without words, is that there is a preferred version of knowledge—the knowledge that the teacher is in possession of—and that learning is circumscribed and measured by the students' passive ability to assimilate that knowledge according to the given parameters.

A number of global changes are challenging this model of instruction. In response to the changing demographics of the student body and increasing cross-cultural interconnectedness effected through distance learning, institutions are beginning to change. The role of the teacher needs to and is likely to change as well, from one who is "dishing out knowledge on a silver plate," to a negotiator, a facilitator, "giving hints and directions on how to find knowledge" (Tripathi, 2006. p. 6) and encouraging forms of active learning compatible with various contexts. Teachers also need to acquire greater familiarity with different learning styles, as well as an understanding that many of the components determining the nature of learning styles and attitudes toward learning are culture based. In their new role, teachers know and model the idea that skills and knowledge are acquired in context and do not exist disconnected from context. Of particular relevance to communicating and teaching in cross-cultural contexts, teachers in their new role acknowledge that learners bring prior knowledge and experience to the learning environment. Teachers can no longer see themselves as exclusive sources of knowledge, because in cross-cultural contexts in particular, students will posses skills and knowledge that teachers will not, and also because of the variety of information readily available to all on the Internet. Rather, they would see themselves as guides who facilitate the learner's navigating through networks of existing meaning to create new ones. They would encourage learners to make connections between previous and new knowledge, to integrate previous knowledge with new knowledge and transfer it from one context to another. Teachers in this new paradigm thus teach "for transfer." In other words, in their new role teachers embrace collaborative teaching. Collaborative teaching rests on the assumption that learning is "more of a process than a product, in which internal meaning is made through the building and reshaping of personal knowledge through interaction with the world" (Campbell, 2004, p. 152).

The nature of the online environment can support the features of active and collaborative learning and the reconceptualization of the teacher's role. I am using the World Wide Web as a conceptual model for cross-cultural "teaching, learning, and negotiating" in the online environment because it is a good example of a "space" allowing for various "narratives" and "knowledges" to circulate, and to be added to the existing collections of information and systems of meaning (cf. Seely Brown, Collins, & Duguid, 1989, p. 39). The Web, as a platform in which e-learning happens, is open, flexible, multimodal, and networked, and as such it provides the "richest opportunity to date to bring the elements of active learning together" (Campbell, 2004, p. 154). The diversity of materials that can be found on the Web, and their coalescence, is similar to the diversity of perspectives and knowledges that coalesce within a cross-cultural community of learners. One of the reasons why the Web is a good conceptual model for collaborative cross-cultural communication is that it resists systematization in a traditional sense, it allows for multimodal teaching and learning, and it accommodates multiple "literacies" and learning styles. The World Wide Web so exemplifies the central idea inherent in collaborative teaching and learning, which is that knowledge and heuristics are not absolute, but situated within various communities of knowing, each operating according to particular cultural dynamics, yet sharing in universal issues affecting us all.

In global, Web-based learning environments, students and the instructor together enter a process of learning and negotiating. Through this process they will learn how to negotiate their way through a multiplicity of perspectives and to engage with difference productively (not merely tolerate it). They will also learn how to acknowledge difference openly without devaluing its implications, how to explore it in ways that result in new understanding, and how to come to flexible solutions appropriate for given contexts.

ISSUES

Learning Styles

Students process information and approach learning in different ways. This fact acquires even greater significance in the context of cross-cultural instruction, where differences of this kind have to be assumed as a starting point. In a recent study on learning in electronic environments, author Katy Campbell (2004) argues that the Web supports and encourages "diverse learning and cognitive styles" due to its interactive nature: it offers multiple presentations of information, encourages creativity, and enhances learning outcomes that develop from collaborating and working in groups. However, studies reveal that the potential of the hypermedia to engage in and adequately support dynamic, interactive, and diverse models of thinking and learning in diverse learners has not been realized. This is due to the fact that the technology enabling online instruction is still new to most instructors. Many of those who teach online courses tend to transfer unknowingly traditional pedagogical assumptions and practices to the new environment. However, traditional perspectives on structuring learning tasks and teaching practice do not correspond easily to the online environment (Ewing, Dowling, & Coutts, 1998).

While in a traditional learning environment information tends to be presented to students primarily via text and in a predominantly linear fashion, research shows that people learn and process information in different ways, and indicates the existence of so-called multiple intelligences and even "mind styles." Keefe (1979) defines learning styles as "the composite of characteristic cognitive, affective, and physiological factors that serve as relatively stable indicators of how a learner perceives, interacts with and responds to the learning environment" (Keefe, 1979, p. 4). Individuals "differ in their general skills, aptitudes, and preferences for processing information, constructing meaning from it, and applying it to new situations" (Jonassen & Grabowski, 1993, p. 3). Their responses to the learning situation will be influenced by the way they think and their past experience. The ways they think and past experiences, in turn, will depend in some definite measure upon each person's background, including their culture.

A number of studies indicate that different learners will process and appropriate information in different ways. According to the work of Howard Gardner, for example, there are at least seven modalities or intelligences that can be used to describe and assess each student's individual learning style (cf. Gardner, 1983, 1993, 1999):

1. **Verbal linguistic:** Sensitive to meaning and order of words, and likely to be text oriented
2. **Musical:** Sensitive to pitch, melody, rhythm, and tone, and likely to assimilate information through a mixture of cognition and effect
3. **Logical-mathematical:** Able to handle chains of reasoning and recognize patterns and order, and likely to be partial to sequential ordering of information
4. **Spatial:** Perceives world accurately and tries to recreate or transform aspects of world
5. **Bodily-kinesthetic:** Uses body skillfully and handles objects adroitly, likely to be partial to any model of acquiring information that enables physically based interactivity
6. **Interpersonal:** Understands people and relationships, and able to learn through reflection and empathy
7. **Intrapersonal:** Possesses access to one's emotional life as means of understanding self and others, and partial to learning from experience and effect

Another classification was suggested by Kolb (1984, 1985). In this classification, learners and their cognitive differences are classified in the following ways:

- According to the way in which they take in information, through concrete experience or abstract conceptualizations
- According to how they internalize information, through active experimentation based on the method of scientific, deductive reasoning, or reflective observation[1]

Some researchers suggest that there are "sequential" learners—those who tend to gain understanding through step-by-step, logic-guided approaches—and "global" learners." Global learners are those who will learn better with an approach emphasizing overviews of topics (Hao, 2004, p. 32) and holistic forms of thinking emphasizing the interconnectedness between cognition and affect. Another way of thinking about differences in how learners process and adopt information that is of relevance to the online environment is according to the learners' tendency to prefer and appropriate knowledge and information offered through text, or through images and graphic representations.

The World Wide Web therefore offers unprecedented opportunities for both instructor and student to accommodate either of these preferences, and to integrate text and image in complementary ways that could support virtually any learning outcome. Teachers need to acknowledge that while most students have become proficient in interpreting text or print, in reality, only a portion of these students is actually composed of so-called "verbal learners," those who prefer to learn from texts and lectures. Many prefer to have access to knowledge presented visually, orally/aurally, or to learn by manipulating objects or performing a physical activity (Campbell, 2004, p.178).

Although the models trying to document and classify different learning styles may vary, they all point to the fact that learners think and learn in different ways. Their approach to learning is affected by their past experience and background, including their culture. This reality suggests that teaching styles need to be flexible and varied, responsive to circumstances in all cases, but especially so in explicitly cross-cultural contexts.

Learning Styles and Cultural Backgrounds

Researchers have suggested that individual differences, prior knowledge, and learning styles have significant effects on learning in the online environment (Hao, 2004; Kim, 2001). The relationships between learning styles and cultural backgrounds are therefore strong and complex. Models of cognition are not entirely predetermined, but are also shaped through social interaction (Nations Johnson, 1993, p. 141; Smetana, 2002; Helwig, 2005; Oishi et al., 2005). Since individual development is mediated by social interaction in a culture-specific, historical setting, and since culture influences one's cognitive processes including the attitudes governing the assimilation information, there is common ground on which culture and cognition can be studied (Abi-Nader, 1993, pp. 214-215; Neff & Helwig, 2002). The nature of these relationships can be understood in the context of often-cited differences among cultures and the attitudes they shape. These differences and their implications for cross-cultural education have been articulated mainly in terms of a "non-western" vs. "western" worldview. They have also been interpreted in terms of a "global learning style," associated with Japanese learners, vs. an "analytical learning style" associated with learners from Europe, North America, Australia, and New Zealand.

According to the first classification, the "non-western" worldview is seen as being based on the following traits:

- Emphasis on group cooperation
- Achievement as it reflects group
- Holistic thinking
- Social orientation in general
- Acceptance of affective expression
- Understanding that religion permeates culture
- Understanding that time is relative
- Placing value on the individual's harmony with nature

In contrast, the "western" worldview is seen as being based on the following traits:

- Emphasis on individual competition and individual achievement
- Dualistic thinking
- Limited affective expression
- Task orientation
- The assumption of a rigid time schedule partially captured in the expression "time is money"
- The understanding that religion should be distinct from other parts of culture
- The assumption that human beings have, and need to have mastery and control over nature (McCutcheon, 1993, p. 248)

Similarly, according to the classification "global learning style" vs. "analytical learning style," Japanese students, who are seen as exemplifying the "global learning style," are generally image oriented, cooperative, learn by experience, depend on insight and intuition, prefer indirect expressions, value the subjective, and avoid standing out. On the other hand, students from Euro-America, Australia, and New Zealand, who are seen as exemplifying the "analytical learning style," generally learn by reasoning; they compete, assert themselves, value the objective and the rational, are text oriented, and prefer direct expressions (Ito, 2005).

What is clear is that attitudes toward learning are built into the education system, and that as difficult as they are to systematize, they will influence the learners' own approaches to learning and knowing. In turn, this will likely result in different communication styles, different attitudes toward task completion, disclosure, assertiveness, and so forth. However, it is very important that these classifications, like any other attempts at systematizing human thought and behavior, are understood in a very broad, nonrestrictive sense. Individual characteristics will always vary and overlap, and no single model can completely define any one person.

Although we can and should continuously strive to learn as much as we can about the values and assumptions of different cultures, we can never hope to learn everything there is to learn about all cultures. The issue here, therefore, is not so much about learning cultural content, as much as it is about learning a method of approaching and dealing with cultural differences in general. Since we tend to design "from our own experience and based on our own needs and values" (Ito, 2005), educators who teach online and in a cross-cultural context need to reflect on the diversity among the students they teach, and the implications diversity has on course and program design. But, primarily, instructors need to reflect on their own cultural and pedagogical assumptions. Seeing that our own culture provides the lens through which we see the world, others, and ourselves in the world, it is of paramount importance to reflect purposefully and specifically on the values and experiences that have shaped our own pedagogical approach. This kind of reflection with the purpose of attaining greater understanding of ourselves as members of a particular culture will aid in instructional planning and design in a cross-cultural context. We cannot fully understand the rules and assumptions of a different culture and approach those differences in a constructive way until we understand the rules of assumptions of our own culture.

Given the global reach of the World Wide Web and the simultaneous increase in distance education offered through various institutions, the instructor's values and experiences are likely to be very different from the values and experiences of many, if not most distance learners. This reality makes Bernstein's (1996) claims regarding the inherent dangers present in a dominant but unexamined pedagogical perspective have increasing relevance. He points out that all too often institutional education continues to carry a promise of universal value and relevance despite social and demographic realities of cultural and global diversity. Bernstein cautions educators not to perpetuate unwittingly the "imaginary certainties of a monoculture," but rather to work on a new pedagogical discourse focused on communities of difference.

Social Theories of Learning and Collaborative Technologies

Whatever the style of interaction, all education, teaching, and learning *are* based on interaction and some form of communication. With the advent of distance education, the communicative and interactive aspect of education is an increasingly relevant area of study. Many researches now accept the view that interaction is the origin of all mental activity. According to a recent initiative by physics education researchers intended to measure student learning gains, the so-called "interactive engagement" method (microcomputer-based labs, concept tests, modeling, active learning problem sets, overview case studies, and Socratic dialogue-inducing labs) is more effective in promoting higher-level learning. Higher-level learning includes more critical thinking, conceptual understanding, reasoning, and creative problem solving than traditional models of teaching and evaluation (Hake, 2005).

Pedagogical theory increasingly sees learning as a process that takes place in a social context, and as heavily influenced by social interactions and factors such as culture, technology, and instructional practice (Vygotsky, 1978; Bruner, 1986; Jonassen, 1999; Campbell, 2004). These new theories of learning, which are gaining increasing prominence as the new media is increasingly beginning to play an important role in education, reveal that cognitive development is profoundly influenced by social interaction, and that learning itself is a product of the connection between people and the cultural context in which they interact in shared experiences (Crawford, 1996). The activity in which knowledge is developed and used is neither separable nor ancillary to what is learned, but an integral part of it, as both learning and cognition are fundamentally situated and embedded in the context in which they take place (Seely Brown et al., 1989, pp. 32-33).

Proponents of so-called "situated learning" argue that conceptual tools, the basis of different kinds of knowledge, reflect the "cumulative wisdom of the culture in which they are used and the insights and experience of individuals" (Seely Brown et al., 1989, p. 35), including the instructor's. The experiential learning theory defines learning as the process in which knowledge is created through the transformation of the learners' experience and the holistic engagement of all cognitive and emotional processes (Kolb, 1984). Teaching and learning, in both content and methodology, are interdependent processes, just as "activity, concept and culture," and effect and cognition are interdependent and inextricably linked (cf. Seely Brown et al., 1989, p. 35). This research clearly shows that teaching, learning, and knowledge itself are not value-neutral. From the socioconstructivist perspective, "meaning is seen as a negotiation" and knowledge building as an open-ended, interactive process within one learning community (Bonk & Reynolds, 1998). The process of making meaning, according to the socioconstructivist view of learning, requires "articulation and reflection" related to one's existent knowledge, which involves simultaneous "internal negotiation and social negotiation" (Hao, 2004, p. 24).

Learning thus conceived becomes a process in many ways exemplified by the open-ended structure and non-linearity of the Web, which has great potential to support communication across and among diverse communities of knowers (Campbell, 2004). Like collaborative teaching and learning, the loose organization of the World Wide Web demonstrates the co-existence of multiple and apparently contradictory perspectives on a single issue. One Google keyword search will retrieve hundreds of documents linked by one single term but applied in a variety of contexts. This characteristic of the Web environment, just like collaborative and cooperative teaching, emphasizes the importance of individual contexts to which a single concept can be applied productively. In the fluid online environment, supported by teaching that is social and interactive, learners no longer rely on a situation where the instructor provides a single learning context and suggests the desired connections among the offered concepts. They are encouraged to generate connections between problems and solutions, and apply the findings in a transformative way to a variety of contexts meaningful to them.

In addition, the non-linearity and interactive multimedia capacity inherent in the Web and forms of CMC can present knowledge and information in ways that combine orality, literacy, and "videocy" (Ulmer, 1989, p. vii). In this way, CMC and the properties of the World Wide Web can support the complex range of learning needs, characteristics, and preferences. Internet-based communication tools such as e-mail, Internet relay-chat, forums, and synchronous conferencing are seen as potentially enabling dialogue. In turn, dialogue encourages critical thinking and cooperative learning, and enhances opportunities for "generative learning, wider diversity of ideas, most reflective thinking, and increased creative responses" (Oliver, Omari, & Herrington, 1998, in Oliver, 1999, p. 10). Many theorists believe that hypermedia most effectively supports tasks and forms of interaction requiring the acquisition of high-level reasoning, problem-solving skills, and critical thinking (Nunes & Fowell, 1996; Ryser, Beeler, & McKenzie, 1995).

Forms of online teaching and learning and CMC, in general, have been termed "collaborative technologies" since they have an inherent affinity with definitions of learning emphasizing social, interpersonal, and collaborative interaction. Some researchers suggest that even simple text-based CMC "equalizes the participants to the extent that everyone, regardless of gender, race, authority, age, etc. is limited to exchanging texts" (Markham 1998, p. 155). At the same time, others have pointed out that other factors, such as economic status or individual facility with manipulating texts in CMC, need to be taken into consideration as they may affect that potential equality (Mason, 2004). It is clear, however, that new models of communication enabled by the "universal network of the Internet" lead to breaking down, or at least diminishing the distinction between "private and public writing" (Bolter 1991, p. 102), and thus potentially create a space for constructive communicative intimacy. This trait can be very valuable in cross-cultural CMC, where the onus is on the instructor to create a sense of trust and open-mindedness in an inclusive and culturally diverse learning community. The potential of the Web and online learning to support collaborative teaching and multiple learning styles, and to foster critical thinking is especially important for cross-cultural teaching, where the diversity in learning styles is matched and often conditioned by diversity in cultural perspectives.

Traditionalist Assumptions in Teaching Practice

While collaborative technologies can support collaborative learning by supporting various forms of interaction, the mere use of these technologies in distance education does not guarantee either that the technology will facilitate learning or that it will automatically improve cross-cultural communication. Part of the difficulty lies in the assumptions of traditional educational practice, which are often unknowingly transferred to the online environment. It is true that most educators today, if asked, will not identify their practice with something called "traditional pedagogy." However, the relative newness of technologies enabling online education, the "trial-and-error" process that accompanies the designing and delivering of online courses, and the increasing cultural diversity of the student body, all suggest that we are still engaged in a process of learning. This learning, if it is to be successful, has to include a critical and thorough assessment of the assumptions underlying our current practice: those aspects of teaching that we take for granted. We have to examine the pedagogical assumptions of a practice formed in a time preceding online education, which for most of us is the case, and which is what I refer to as "traditionalist."

According to the assumptions that underlie the traditionalist model, the concept of knowledge remains implicitly understood as a more or less "closed" set of ideas and facts. Meaning is seen as a set of predetermined "nuggets" of knowledge transferred from the expert teacher in possession of this entity to the student who lacks it. Studies of professional teacher behavior and cognition have revealed a tendency toward routinization of thought and instruction, as well as avoidance of critical analyses of teaching routines. Many secondary teachers, for example, appear to equate much of teaching with classroom control, an observation that researchers found to be consistent with empirical literature concerned with preservice and beginning teachers (Kagan & Tippins, 1993, p. 109; Davis, Sumara, & Luce-Kapler, 2000). In traditional teaching practices it is still tacitly assumed that conceptual knowledge can be abstracted from the situations in which it is learned and used, and that it is an integral, self-sufficient, decontextualized substance (Seely Brown et al., 1989, p. 32), theoretically independent of the learning context or the learner's own cognitive and affective domains of implicit knowledge.

A corollary of this assumption especially relevant to cross-cultural communication in an educational context is what Bernstein (1996, p. 9) calls the "symbolic violence" of dominant pedagogical discourses that tend to present themselves as value-neutral and objective. The pedagogical assumptions of this model—according to which instructional discourse is implicitly passed on as regulative discourse (Bernstein, 1996)—inherently leaves little room for questioning and creative, innovative, and transformative collaboration. This is especially the case when that collaboration is meant to happen between teacher and students, but also between student and other students. Even when a degree of dialogue or interaction is built into the traditionalist model, students often have the impression that there exists a sort of implied hierarchy of answers and acceptable dialogue cues. Thus, the local organization of discourse, in this case pedagogical discourse in an instructional situation, may systematically but covertly exert a form of control and power over the learners. If there is such a thing as a clearly delineated body of knowledge, legitimized by the teacher's attitude and practice, then the number of "legitimate" questions and their answers will be delimited to its circumference. Dialogue and communication in this context often implicitly appear as demonstration of "knowledge already in place" (cf. Chorney, 2005, p. 52). This knowledge comes from a place of prior possession rather than creative and collaborative processes through which multiple solutions may be found and new thinking and new ideas may be developed. If knowledge and learning are by default defined as fixed entities and in terms of one-directional transfer of information, and if the teacher is implicitly defined as the sole possessor of this commodity, then the teacher's role and responsibility can be described in terms of control.

The subtle and often implicit ways in which control becomes associated with one of the primary roles of the teacher is evident in the phrase "managing the classroom." The notion that the teacher needs to "manage the classroom" implies either that there is an inherent disorder that typifies the classroom if left unmanaged, or that the implicit assumption of different perspectives that exist in the classroom pose a potential threat to "order," which is thought of as the desired goal and state. Although the goal of traditional instruction is to encourage problem solving and critical thinking, it "disregards most of the inventive heuristic that students bring to the classroom," implicitly devaluing "not just individual heuristics, which may be fragile, but the whole process of inventive problem solving (Seely Brown et al., 1989, p. 36). It has been documented that in this kind of environment, where hierarchies of methods and solutions exist, students feel it is necessary to disguise potentially effective strategies so that teachers believe the problem has been solved in the approved way.

Although I have sketched out this dynamic in very broad and thus crude terms, this is the underlying, and often subtle operative dynamic in the so-called "banking concept" of education (Freire & Shor, 1987) in which learned teachers in possession of the substance of legitimate knowledge deposit it into passive students, or expect to hear parts of that knowledge echoed back to them in a fairly formalized and carefully controlled dialogue, which is part of the so-called "recitation teaching." In the context of cross-cultural communication, this dynamic can be especially detrimental as it will tend to alienate large groups of students; it may be perceived that the dominant instructional framework, which has its own culturally embedded values and assumptions, presents those values either as normative or as more legitimate than those existing in any other framework.

SOLUTIONS AND RECOMMENDATIONS

Collaborative and Open-Ended Teaching: The Importance of Reflection

In contrast to practice informed by traditionalist assumptions, collaborative and open-ended teaching means that the implicitly hierarchical, discipline- and control-based interaction between teacher and students is replaced by a flexible model where the teacher is not only a practitioner of knowledge but also a learner, and students are not only learners but practitioners as well. This kind of interaction is especially needed and meaningful in a cross-cultural context, in cases where the students will often likely know more about a particular approach or aspect of knowledge than the instructor. The instructor's role here would be not to maintain his or her traditional status as "provider of definite knowledge," but to act as "catalyst," facilitator, and "inspirer of learning." He or she would be the person who encourages and acknowledges difference in perspective, communication style, and so forth, and who models how enjoyable and productive it can be to discover, analyze, and learn from that difference: "so, let them teach you, and you teach them" is the way one researcher describes this interaction (Coghlan, 2002). The redefinition of the traditional role of the educator also models the multiple roles required of each one of us in the successful execution of cognitive tasks, including reflecting productively upon one's own activity (cf. Seely Brown et al., 1989, p. 39).

Collaborative and open-ended teaching and learning emphasize reflection, since at their basis stands collaborative social interaction, social construction of knowledge in a context of cultural diversity, and a new role of the instructor as mentor and facilitator who models appropriate and productive ways of engaging with diversity and difference. Through ongoing dialogue and reflection, students will be allowed and encouraged to assimilate knowledge in accordance to their own needs, values, and cultural contexts. Reflection and critical thinking are crucial to this reconceptualization of the traditional role of the teacher as well, because the basis of critical thinking is the willingness to examine assumptions. Critical thinking requires radical reflection since it "involves calling into question the assumptions underlying our customary, habitual ways of thinking and acting, and then being ready to think and act differently on the basis of this critical questioning" (Brookfield, 1987, p. 1). The importance of ongoing reflection for the instructor as the instructional planner and facilitator cannot be overstated when it comes to cross-cultural communication in a global world increasingly connected through technology.

Collaborative Teaching: Theoretical Basis

Understanding the theoretical basis of this flexible approach to learning will help clarify what collaborative teaching means. Collaborative e-learning has been described as being based on five assumptions:

1. Learning is an active, constructive process.
2. Learning depends on rich contexts.
3. Learners are diverse.
4. Learning is inherently social.
5. Learning has affective and subjective dimensions. (cf. Campbell, 2004, p. 10)

These five assumptions emphasize the collaborative aspect of information sharing and knowledge building. From the instructor's point of view, taking into consideration these five assumptions when beginning to plan a course or program is likely to change crucial aspects of his or her teaching practice and communicative style that inherently stand in opposition to these assumptions.

Teaching collaboratively and open-endedly means being willing to focus actively on innovation and flexibility. Both of these terms imply a spontaneity arising from responsiveness to each communicative situation. Teaching collaboratively means being willing to recognize and practice explicitly the reality that there is always more than one way to solve a problem and more than one point of view in interpretation: one successful interpretation does not negate the validity of others. Collaborative teaching means engaging learners in the learning process and encouraging them through various activities to construct knowledge in a way that is meaningful to them. This teaching model is inherently non-linear, as it encourages a multiplicity of perspectives on the same issue and models this way of flexible thinking. It also means that instructors as facilitators will not exhibit, explicitly or implicitly, defensive behavior when student solutions differ from their own, but when the same learning goal was achieved.

Teaching collaboratively begins by a crucial shift in perspective and the recognition that there is no need to be "in control" of either the class as a whole or each class discussion. This model of teaching is reflective, as it implies that the instructor will be willing to reflect on his or her teaching, words, claims, and so forth on an ongoing basis, and be prepared to change his or her perspective at any given point if change is needed. The instructor who is committed to teaching collaboratively will be teaching students the nature and value of communication and collaboration by example. This process also implies that both sides are continuously engaged in learning and that learning is not something reserved for students. This is the biggest change from the traditional model of teaching: the instructor who teaches collaboratively in a cross-cultural context in particular is explicit about his or her own ongoing learning and the natural evolution of perspectives and ideas. The collaborative model is thus inherently opposed to the hierarchy implied in traditional teaching based on the idea of control and the student's submission to authority.

The collaborative model is best achieved through the practice of so-called "transformative communication." Researchers list a number of indicators that transformative communication is happening. Among them are the following:

1. The student teaches the instructor something he or she did not know before, either about the technology or about the content.
2. More emphasis is placed upon finding support or backing for a position than on conforming to an authority.

3. Students participate in setting the agenda for the class by helping to choose content or learning methods or both.
4. Students are calling the instructor's attention to valuable learning resources.
5. While the instructor helps establish expectations and articulates a clear assessment standard, the students collaboratively guide much of their own learning.
6. The instructor finds himself or herself saving student work—not merely as examples of student work, but as content resources for later reference. (cf. Sherry & Wilson, 1997, p. 69)

The instructor who teaches collaboratively engages in ongoing critical thinking and practice. He or she is learning and modeling simultaneously how to recognize and productively link apparently different solutions or perspectives, and how to point out both the differences and the connections among them without losing sight of learning goals. In this sense, while explicitly acknowledging in discussions students' individual approaches and past experience, the instructor implicitly acknowledges the inherent interdependence and interconnectedness among all forms of human interaction, experience, and discourse. It is this willingness and ability to allow for the peaceful and productive coexistence of multiple points of view that results in productive communication and learning. This approach is of crucial relevance in cross-cultural communication where the existence of multiple perspectives will be the beginning point of interaction.

The collaborative instructor who teaches in a cross-cultural environment will invite and actively encourage different approaches and strategies for problem solving, acknowledging that there may be, indeed that there are likely to be, such strategies that he or she may not have mentioned. Although it may seem at first that the instructor here teaches only method and not content, this is not the case. This practice, on the one hand, demonstrates the teacher's engagement with the practice of his or her discipline or subject (an active process, which is the basis of scholarship), and on the other, enables students to develop a composite understanding of the subject matter and understand the connections that exist among different approaches. It shows the students the "legitimacy of their implicit knowledge"; it stresses that "heuristics are not absolute but assessed with respect to particular tasks" and thus always situated. It also helps students to "generate their own solution paths...making them conscious, creative members" in the process of knowledge creation (Seely Brown et al., 1989, p. 40).

Collaborative Teaching: Practical Issues

There are a number of practical issues that instructors who teach collaboratively in distance education could consider when preparing to engage in cross-cultural online education. What follows are guidelines arising from the ideas presented in this chapter that readers can use to prepare their own online programs:

1. **Success in communication depends on the context of reception:** Although it may seem self-evident, it is important to reflect on the notion that the outcome and success in communication in general is largely dependent on the contexts of reception. In a face-to-face interaction, body language allows interlocutors to interpret meanings that may not be present in the actual verbal content. This context is mostly absent in online communication. However, "emoticons," visual representations of emotional responses communicated through body language and facial expression, have been created to supplement for this lack. Emoticons may be used effectively to indicate the

"spirit" or intention in which a statement is made and the hoped-for response. Nonetheless, a certain amount of linguistic and conceptual ambiguity remains in the context of online communication, especially when the participants belong to different cultures. This reality necessitates the acknowledgment that even those statements that seem perfectly clear from one point of view can in fact be ambiguous to a member of a different culture. Cross-cultural educators also warn that "we should become conscious of the fact that cultures vary in what they consider humor and taboo, which might give rise to misinterpretations and resentment, and that speed of delivery...as well as turn taking should also be respected, as much the rules for entering conversations in progress" (Zeinstejer, 2002). They note that only "by becoming aware of the difficulties leading to possible misunderstandings in online communication practice will teachers be able to create a similar atmosphere of awareness and respect both in their face-to-face and in their cyber classrooms, and teach students how to communicate more effectively across cultural lines and how to address and negotiate differences" (Zeinstejer, 2002).

This caution does not mean, however, that instructors need to be elaborate in their descriptions in an attempt to eliminate potential ambiguity. In some cases, the more elaborate the description is, the more "opaque it is in danger of becoming' (Seely Brown et al., 1989, p. 35). A good way of dealing with potential ambiguity resulting from cross-cultural communication online would be to acknowledge openly at the beginning of the learning unit that language as a communicative tool is embedded in a number of culture-specific uses, and as such it can be ambiguous.[2] The instructor could emphasize that in many instances what one person understands as self-evident meaning in fact depends on context, situational and cultural alike. Stating this reality openly, and encouraging ongoing clarifications, models that learning and meaning are situated, and ensures that students do not feel awkward, self-conscious, and reticent when something said does not appear perfectly clear to them for whatever reason.

2. **Design for interaction and collaboration by gaining a clear sense of "audience":** The disembodied nature of online communication, when added to the inherent challenges already present in cross-cultural face-to-face communication, such as different communication styles, different approaches to knowing or to completing tasks, different decision-making styles, or different attitudes toward participation and disclosure, may appear as detrimental to successful cross-cultural communication online, but that does not have to be the case. In fact, there are a number of practical suggestions that educators could make use of when beginning to teach online using the collaborative and open-ended model. Given that collaborative technology itself does not guarantee that successful communication and learning will take place, instructors will need to devote some time to designing for interaction and collaboration.

One of the first things to consider in interaction design would be how to gain a clear sense of audience. One way to address this would be to ask participants to fill out a "pre-course" questionnaire. In designing the questionnaire the instructor may consider how to obtain answers to questions such as: Who are the learners? What is their attitude toward interaction? What do they need or want to learn? In what contexts or environments will the learning be applied? What do they already know? Obtaining answers to these and similar questions will help the instructor map out the skills, attitudes, or aspects of knowledge and meaning that he or she is trying to help learners develop (Faradouly, 1998).

3. **Organize material in a non-linear, open-ended format and provide opportunities for learning in context:** Although there has to be a certain amount of interaction design to enable its general purposeful role, this does not mean that the design should be too rigid so as to constrain a necessary degree of flexibility. Part of the interaction and structuring of materials or methods should be done in the form of scaffolding, which is a non-linear, open-ended way of organizing data allowing learners to adopt the suggested material but in a way that also allows them to apply it to their own needs, values, and contexts. The instructor would enable at all levels of instruction an open-endedness of approach in that he or she will involve students as much as possible in "the choice of content, method, medium, reward, assistance, feedback, quantity, pacing, sequencing, or difficulty of instruction" (Sutton, 2000, p. 34).

   As a collaborative teacher, the instructor will provide opportunities for learning in context. This means creating group learning opportunities so that learning can occur through dialogue with peers and experts, and generally directing the learners' "cognitive resources toward activities that are relevant to learning rather than to processes that are adjunct to learning" (Hao, 2004, p. 22). The instructor also encourages the application and transfer of method or aspects of knowledge to the learner's unique context. Scaffolding enables this process. While traditional school research placed too much emphasis upon collection, scaffolding encourages "continuous sorting and sifting as part of a "puzzling" process—the combining of new information with previous understandings to construct new ones." Students are adding on, extending, and refining their existing knowledge, while acquiring a "more astute view of whatever truth matters for the question or issue at hand" (McKenzie, 1999).

4. **Set clear and explicit expectations for online behavior and communication:** Because of different cultural attitudes toward communication and participation in a group, courses relying on collaboration and interaction should indicate the requirement for collaboration prior to the beginning of class, or alternative accommodations need to be made for students unable to participate fully (Hao, 2004, p. 26). In addition, it will be the instructor's role at the beginning of the session, course, or discussion to express general expectations for online behavior and communication. These expectations should be explicit about removing language that appears to stereotype learners and reducing the violations of cultural rules during discussions (Zeinstejer, 2002). The instructor should model a way of conversing that is neutral and non-judgmental, understanding of difference. One way of teaching that cross-cultural communication online can be open but respectful would be to encourage students to write about how their ideas are different from those of others, and why. The instructor may also model continuously the idea that many of the beliefs we take for granted are in fact culturally determined, and that successful communication is based on an honest, benevolent, and non-judgmental approach to cross-cultural differences.

5. **Provide opportunities for different kinds of online interaction:** Studies have shown that learners experience greater levels of satisfaction in distance learning environments when they engage in meaningful and frequent interaction with the instructor (Hao, 2004, p. 37). There is no reason to believe that this would not almost equally pertain

in cross-cultural educational contexts; most participants in distance education tend to judge the course based on how interactive the course is (Hao, 2004, p. 38). Collaborative interaction occurs when learners have the opportunity to discuss issues on a bulletin board, to solve problems by working together, to take part in discussion activities, such as sharing ideas and information. And so it is clearly very important to build into the course forms of activity enabling individual learning, those done in pairs or through e-mail (one-to-one), those relying on the use of bulletin boards (one-to-many) and the use of computer conferencing techniques (many-to-many) (Hao, 2004, p. 25). This variety of interaction will successfully address the variety of different learning styles of students, as well as their culturally determined approaches to communication and to knowing. Team-learning activities also promote collaborative learning, because when working in teams students are likely to form bonds with one another, develop positive interdependence, group and individual accountability, and interpersonal skills, and thus would be more likely to participate in other interactive activities with less inhibition.

In fact, an overwhelming number of international students enrolled in a first-year literature course delivered online at a Canadian university said that they definitely participated more in the online format in a way that they never would in the classroom. Among the factors contributing to this attitude they cited the flexibility and time that the online format enabled them, since no one knows how much time they spent writing their discussions points before they felt comfortable posting them.[3]

6. **Asynchronous communication is preferable for online learning:** Many researchers believe that asynchronous communication is preferable for online learning, as it provides opportunities for active input from all members of the learning community with flexibility in time and place, so learners have greater control over the learning environment (Hao, 2005; Carr, 1998; Graham, Scarborough, &Goodwin, 1999). They advocate using synchronous interaction as a good supplement to asynchronous delivery medium (Hao, 2004, p. 15). Studies indicate that interpersonal interaction and social integration are prominent in asynchronous online communication and are among the most influential factors in online learning success (Hao, 2004; Jung & Rha, 2000; McDonald & Gibson, 1998). What emerges from research on collaborative learning in general is a need for the instructor to step in when needed to act as a facilitator of discussions, to increase feedback, and to build reciprocity, encouragement, and support in interpersonal interaction in a way that validates the learners' individual values and experience. This kind of collaborative interaction allows learners to discuss, debate, negotiate, and reflect upon their existing beliefs and knowledge (Hao, 2004, p. 21). Collaborative interaction also allows for the kind of learning and interaction researchers call "vicarious interaction." Vicarious interaction takes place when "a student actively observes and processes both sides of interaction between two other students or between another student and the instructor (Hao, 2004; Sutton, 2001, p. 227). This type of interaction is of special value in a cross-cultural context as it promotes indirectly, but through concrete conversational situations, awareness and understanding of the issues involved in cross-cultural communication. It also provides an additional tool for acquiring knowledge and skills that enhance communication among members of diverse cultures, but from a safe place of

non-participation, thus allowing for productive reflection and assimilation of effective communication techniques and models, and encouraging them to participate themselves. In addition, studies indicate that vicarious interaction is especially suitable for learners who may be apprehensive about interacting directly or "standing out," so that if this particular model of interaction is recognized and actively pursued, cross-cultural online communication in an education setting will provide more choices for learners of various learning styles and from various cultural backgrounds. In general, in this "team-based" flexible structure, and through the process of discussing and interacting with other learners and the instructor, learners are encouraged to construct new knowledge.[4]

7. **The evaluation scheme should be varied and flexible:** Once knowledge is redefined from being understood as the possession of facts and correct methodology that is passed one-directionally from teacher to student, to an active and flexible process of "meaning negotiation" in which multiple ways of arriving at different "knowledges" exist, are validated, and can lead to the same desired learning goals, the nature and design of evaluation of student learning will shift.

   Content itself acquires a different meaning in online education as well as in a cross-cultural context: it is not only "what," but it is also "how." Content can be information, as well as interpretation of information by experts, novices, and students. It can be in the form of research reports that are generated individually or with a partner/group, arguments, journalistic accounts and essays, represented through text, graphics, or any other multimedia format. Similarly, feedback should not be restricted to that of the instructor, but should include types of peer evaluation. One way of doing this would be to provide evaluative opportunities in the form of "collegial, and often interdisciplinary responses from learners in various geographic locations" (Relan & Gillani, 1997, p. 45). Researchers remind us that "just as the application of hypertext and hypermedia created novel modes of learning and contributed to the restructuring of the instructional environment, the World Wide Web has the power to generate novel learning strategies which will eventually be embedded in cognitive, social, and cultural contexts" (Relan & Gillani, 1997).

   In addition to articulating clearly at the beginning of each distance-learning opportunity what critical concepts the students should be able to understand by the end of the unit, instructors may evaluate student work on the basis of various kinds of tasks. As in traditional evaluation, the evaluation criteria should be outlined clearly at the beginning of each learning unit and in relation to all assignments. Evaluation may continue to be based on the student's demonstrable ability for reasoning, critical thinking and writing, and conceptual understanding supported with convincing detail. However, in the evaluation criteria the instructor will acknowledge that convincing detail and relevant information may be drawn from either cognitive or affective fields of knowledge and experience, and various content sources the access to which is facilitated and encouraged by the World Wide Web (cf. Hake, 2005; Campbell, 2004; Relan & Gillani, 1997).

FUTURE TRENDS

Just as in one culture ideas are exchanged and modified, and belief systems developed and appropriated through conversation and social interaction, so collaborative cross-cultural teaching promotes meaningful and transformative conversations among cultures. The non-linearity in the presentation of material and ideas on the World Wide Web is a useful model for collaborative cross-cultural teaching and learning. The non-linearity of the Web models scaffolding as a way of grouping materials, and it promotes the idea that learning works best when it is situated in the individual learner's implicit knowledge. The long-term value of this approach can be seen in the correlation between the dominant media and cultural practices, including education. Whereas traditional "mass education tended to see life in a linear fashion based on print models and developed pedagogies which broke experience into discrete moments and behavioral bits," new critical pedagogies enabled through the online medium could produce "skills that enable individuals to better navigate the multiple realms and challenges of contemporary life" (Kellner, n.d., p. 9).

Online communication, due to its multimedia capacity, facilitates the use of tools promoting connectivity and audio visualization (Campbell, 2004, p. 10). It enables the creation of variable contexts in instruction, relevant given the cultural differences in learning and communication styles. A more systematic study of the ways diverse information can be linked and presented in online communication can take us one step further toward re-visioning the goals of education in the 21st century. Most importantly for the future, further study of collaborative online education in relation to cross-cultural contexts has the potential to provide concrete answers to the call for meaningful reform in higher education (cf. Association of American Colleges and Universities, 2002). Not only are technology and international, distance education already an undeniable presence and a growing trend in higher education, but the student body increasingly reflects an extraordinarily diverse array of cultural backgrounds. In market-dominated societies in North America, adopting information technologies for online instruction allows universities to respond to pressures from politicians and students-cum-customers to decrease the cost of delivering higher education (cf. Cole, 2000). These facts place an increasing need on institutions and instructors to develop new approaches to educational quality in a way that will serve meaningfully the needs of contemporary students who live in a complex, technologized, interconnected world.

And finally, practicing cross-cultural collaborative teaching in the online environment responds in practical ways to the challenges posed by diversity and cultural differences, and models a positive, constructive way of dealing with them. This practice clearly promotes the forms of learning needed for the 21st century as they have been identified by the "Greater Expectations National Panel Report" (Association of American Colleges and Universities, 2002). Taking into consideration that the intellectual and practical skills today's students need are extensive, sophisticated, and expanding with the explosion of new technologies and the changes they have brought to education, cross-cultural collaborative teaching in the online environment will help students become "intentional learners." Intentional learners are those who can "adapt to new environments," who can integrate different kinds of knowledge from a variety of sources, who can "demonstrate intellectual agility and the ability to mange change," and who have the skills to engage in meaningful dialogue and deal with the interrelations within and among global and cross-cultural communities (adapted from the recommendations of the National Panel Report, pp. xi-xii). By valuing and promoting "cooperative as well as individual performance, diversity as a resource for learning, real solutions to unscripted problems and creativity as well as critical thinking," teaching, learning, and negotiating in this way will help prepare students for participation in dynamic global economies and build their sense of civic responsibility in a diverse world (Association of American Colleges and Universities, 2002).

Also, the study of cross-cultural communication and collaborative strategies acquires urgent significance in a world increasingly interconnected through technology, yet struggling to deal with differences among cultures. Future research may investigate whether systematization of training in cross-cultural communication and collaboration in a variety of contexts and social sectors, including distance and online education, would help solve the crises in cross-cultural communication we are faced with today.

CONCLUSION

In this chapter I have tried to outline a model of cross-cultural teaching in the online environment using the World Wide Web as a concept that can illustrate many of the philosophical aspects that should accompany cross-cultural communication. Many of the practices associated with collaborative teaching are not new and are being used by many instructors in any classroom. However, the increase in online educational programs compels us to thoroughly examine dominant pedagogical practices and the goals of teaching in an increasingly interconnected and diverse global world. These new educational trends also encourage us to explore and apply systematically certain teaching practices that can improve the quality of cross-cultural education, especially when delivered by mono-cultural teachers.

One thing that we can cay with a degree of certainty about cross-cultural communication in distance learning is that by compelling us to interact in an environment characterized by cultural pluralism, it also compels us to acknowledge that "all knowledge is *ultimately* situated and contextualized, relative and limited" (Gross, 1999, p. 191). In method, then, collaborative and open-ended teaching and learning extends the relevance of the subject matter beyond the classroom, it encourages a process of rigorous reflection, and it can be effective in combating the unwitting presupposition in traditional pedagogical practice of a static, unchanging world or a classroom, where knowledge is framed as pre-existing, as something that one receives, holds, and then releases (cf. Chorney, 2005, p. 63). Although technology itself is not inherently interactive, the online environment, usually synonymous with the Internet and the World Wide Web, could be used as a tool and as a model to encourage cross-cultural interactivity and to support constructivist, situated, and open-ended learning. The structural flexibility that is the hallmark of collaborative and open-ended instruction is similar to the structural flexibility and conceptual "open-endedness" of the Web. The multiplicity of perspectives represented on the Web and their coexistence indicate not only that culture is often at the root of communication challenges, but also that by being exposed to and learning to see the world from another's point of view is a process of discovery and cognitive and social growth that can deepen and enrich our understanding of ourselves and others.

ACKNOWLEDGMENTS

This chapter benefited a great deal from the editors' and the anonymous readers' helpful comments.

REFERENCES

Abi-Nader, J. (1993). Meeting the needs of multicultural classrooms: Family values and the motivation of minority students. In M. J. O'Hair & S. Odell (Eds.), *Diversity and teaching: Teacher education yearbook I* (pp. 212-228). Fort Worth, TX: Harcourt Brace Jovanovich.

Association of American Colleges and Universities. (2002). *Greater expectations: A new vision for learning as a nation goes to college.* Retrieved

June 20, 2006, from http://www.greaterexpectations.org

Banks, J. A. (2004). Remembering brown: Silence, loss, rage and hope. *Multicultural Perspectives, 6*(4), 6-8.

Bernstein, B. (1996). *Pedagogy, symbolic control and identity: Theory, research, critique.* London: Traylor & Francis.

Bolter, J. D. (1991). *Writing space: The computer, hypertext and the history of writing.* Hillside, NJ: Lawrence Erlbaum.

Brookfield, S. D. (1987). *Developing critical thinkers: Challenging adults to explore alternative ways of thinking and acting.* San Francisco: Jossey-Bass.

Brookfield, S. D., & Preskill, S. (1999, 2005). *Keeping discussion going through questioning, listening, responding.* Retrieved June 27, 2006, from http://ctl/stanford.edu/Tomprof/736.html

Bruch, P. L., Jehangir, R. R., Jacobs, W. R., & Ghere, D. L. (2004). Enabling access: Toward multicultural developmental curricula. *Journal of Developmental Education, 27*(93), 12-14.

Campbell, K. (2004a). *E-ffective writing for e-learning environments.* Hershey, PA: Information Science Publishing.

Campbell, K. (2004b). E-ffective writing for e-learning environments (and other online spaces). In *Proceedings of Instructional Design Conference.*

Carr, K. (1998). *Reflective judgment and cognitive interaction in an electronically distributed learning environment.* Unpublished doctoral dissertation, University of Idaho, USA.

Chan, M. V., Fortunato, A. M., Oakes, S., & Ryan-Mann, D. (2001). Reconceptualizing the faculty role: Alternative models. In B. Leigh Smith & J. McCann (Eds.), *Reinventing ourselves: Interdisciplinary education, collaborative learning, and experimentation in higher education.* Bolton: Anker Publishing.

Chariot. (2002). *Cross-cultural communication online: Perspectives from around the globe.* Retrieved October 4, 2005, from http://users.chariot.net.au/~michaelc/ccc/press.htm

Chorney, T. (2005). Milton's *Paradise Lost* and global culture: A Christian epic in the multicultural classroom. *Professional Studies Review, 2*(1), 48-65.

Chorney, T. (2006, March 31-April 1). Difference as crisis in cross-cultural communication. In *Proceedings of the Communication in Crisis Conference,* Amherst, MA.

Coghlan, M. (2002). Introduction. *Cross- cultural communication online: Perspectives from around the globe.* Presented by the Webheads Community at Networking. Retrieved October 4, 2005, from http://users.chariot.net.au/~michael/ccc/pres.htm

Cole, R. A. (Ed.). (2000). *Issues in Web-based pedagogy: A critical primer* (The Greenwood Educators Reference Collection). Westport, CT.: Greenwood Press.

Crawford, K. (1996). Vygotskian approaches to human development in the information era. *Educational Studies in Mathematics,* (31), 43-62.

Crisp, G., Thielle, D., Scholten, I., Barker, S., & Baron, J. (Eds.). (2003, December 7-10). *Interact, Integrate, Impact: Proceedings of the 20th Annual Conference of the Australasian Society for Computers in Learning in Tertiary Education (ASCLITE),* Adelaide.

Davis, B., Sumara, D., & Luce-Kapler, R. (2000). *Engaging minds: Learning and teaching in a complex world.* Hillside, NJ: Lawrence Erlbaum.

Enhance Learning. (n.d.). *Enhancing learning with technology: Paradigm shift.* Retrieved June 30, 2006 from http://members.shaw.ca/priscillatheroux/paradigmshift.html

Ewing, J. M., Dowling, J.D., & Coutts, N. (1998). Learning using the World Wide Web: A collaborative learning event. *Journal of Educational Multimedia and Hypermedia, 8*(1), 3-22.

Fardouly, N. (1998). *Instruction design of learning materials.* University of New South Wales, Australia.

Freire, P., & Shor I. (1987). *A pedagogy for liberation.* London: Macmillan.

Gardner, H. (1983, 1993). *Frames of mind: The theory of multiple intelligences.* New York: Basic Books.

Gardner, H. (1999). *Intelligence reframed: Multiple intelligences for the 21st century.* New York: Basic Books.

Graham, M., Scarborough, H., & Goodwin, C. (1999). Implementing computer mediated communication in an undergraduate course—a practical experience. *Journal of Asynchronous Learning Networks, 3*(1), 32-45. Retrieved February 20, 2006, from http://www.sloan-c.org/publications/jaln/v3n1/v3n1_graham.asp

Gross, R. (1999). A rose by any other name … A response to Katherine K. Young. *Journal of the American Academy of Religion, 67,* 185-194.

Hake, R. (2005). *A possible model for higher education: The physics reform effort.* Retrieved February 14, 2006, from http://ctl.stanford.edu/Tomprof/postings

Hao, Y.-W. (2004). *Students' attitudes toward interaction in online learning: Exploring the relationship between attitudes, learning styles, and course satisfaction.* Unpublished doctoral dissertation, University of Texas at Austin, USA.

Helwig, C. C. (2005). Culture and the construction of concepts of personal autonomy and democratic decision making. In J. E. Jacobs & P. A. Klaczynski (Eds.), *The development of judgment and decision making in children and adolescents.* Hillside, NJ: Lawrence Erlbaum.

Ito, S. (2005). Cultural inclinations in learning styles. *Cross cultural communication online: Perspectives from around the globe.* Presented by the Webheads Community at Networking. Retrieved October 14, 2005, from http://users.chariot.net.au/~michael/ccc/pres.htm

Jonassen, D. (1999). Designing constructivist learning environments. In C. M. Reigeluth (Ed.), *Instructional-design theories and models: A new paradigm of instructional theory* (Vol. 2, pp. 215-237). Hillside, NJ: Lawrence Erlbaum.

Jonassen, D., & Grabowski, B. L. (1993). *Handbook of individual difference, learning, and instruction.* Hillsdale, NJ: Lawrence Erlbaum.

Jung, I., & Rha, I. (2000). Effectiveness and cost-effectiveness of online education: A review of the literature. *Educational Technology, 40*(4), 57-60.

Kagan, D. M., & Tippins, D. J. (1993). Classroom cases as gauges of professional growth. In M. J. O'Hair & S. Odell (Eds.), *Diversity and teaching: Teacher education yearbook I* (pp. 98-110). Fort Worth, TX: Harcourt Brace Jovanovich.

Keefe, J. W. (1979). Learning style: An overview. In *NASSP, Student learning styles: Diagnosing and prescribing programs* (pp. 1-17). Reston, VA: National Association of Secondary School Principals.

Kellner, D. (n.d.). *Technological transformation, multiple literacies, and the re-visioning of education.* Retrieved February 13, 2006, from http://www.gseis.ucla.edu/faculty/kellner

Kim, K.-S. (2001). Implications of user characteristics in information seeking on the World Wide Web. *International Journal of Human-Computer Interaction, 13*(3), 323-340.

Kolb, D. (1985). *Learning style inventory.* Boston: McBeer and Company.

Kolb, D. (1984). *Experiential learning: Experience as a source of learning and development.* Englewood Cliffs, NJ: Prentice-Hall.

Markham, A. (1998). *Life online.* Walnut Creek, CA: Altamira Press.

McCutcheon, G. (1993). Curriculum: Overview and framework. In M. J. O'Hair & S. Odell (Eds.), *Diversity and teaching: Teacher education yearbook I* (pp. 237-268). Fort Worth, TX: Harcourt Brace Jovanovich.

McDonald, J., & Gibson, C. (1998). Interpersonal dynamics and group development in computer conferencing. *The American Journal of Distance Education, 12*(1), 7-25.

McKenzie, J. (1999). Scaffolding for success. *From Now On: The Educational Technology Journal,* (9), 4. Retrieved June 28, 2006, from http://www.fno.org/dec99/scaffold.html

Nations Johnson, L. (1993). In M. J. O'Hair & S. Odell (Eds.), *Diversity and teaching: Teacher education yearbook I* (pp. 138-151). Fort Worth, TX: Harcourt Brace Jovanovich.

Neff, K. D., & Helwig, C. C. (2002). A constructivist approach to understanding the development of reasoning about rights and authority within cultural contexts. *Cognitive Development, 17,* 1429-1450.

Nunes, J. M. B., & Fowell, S. P. (1996). Hypermedia as an experiential tool: A theoretical model. *Information Research, 2*(1). Retrieved from http://informationr.net/ir/2-1/paper12.html

Oishi, S., Hahn J., Schimmack, U., Radhakrishan, P., Dzokoto, V., & Ahadi. S. (2005).The measurement of values across cultures: A pairwise comparison approach. *Journal of Research in Personality, 39,* 299-305.

Relan, A., & Gillani, B. B. (1997). Web-based instruction and the traditional classroom: Similarities and differences. In B. H. Khan (Ed.), *Web-based instruction* (pp. 41-46). Englewood Cliffs, NJ: Educational Technology Publications.

Ryser, G. R., Beeler, J. E., & McKenzie, C. M. (1995). Effects of a computer-supported intentional learning environment on student's self-concept, self-regulatory behavior, and critical thinking ability. *Journal of Education Computing Research, 13*(4), 375-385.

Sarles, H. (1993). *Teaching as dialogue.* Retrieved April 25, 2006, from http://ctl/stanford .edu/Tomprof/postings/718.html

Seely Brown, J., Collins, A., & Duguid, P. (1989). Situated cognition and the culture of learning. *Educational Researcher, 18*(1), 32-42.

Sherry, L., & Wilson, B. (1997). Transformative communication as a stimulus to Web innovations. In B. H. Khan (Ed.), *Web-based instruction* (pp. 67-73). Englewood Cliffs, NJ: Educational Technology Publications.

Smetana, J. G. (2002). Culture, autonomy, and personal jurisdiction in adolescent-parent relationships. In H.W. Reese & R. Kail (Eds.), *Advances in child development and behavior* (Vol. 29, pp. 51-87). New York: Academic.

Sutton, L. A. (2000). The principle of vicarious interaction in computer-mediated communications. *International Journal of Education Telecommunications, 7*(3), 223-242.

Tripathi, A. K. (2006). *Coping with innovative technology: Albert Borgman on how does technology change learning and teaching in formal and informal education.* Retrieved June 30, 2006, from http://www.acm.org/ubiquity/views/pf/v7i23_coping.pdf. Part of this essay appears in the June 20-26, 2006, edition of *Ubiquity: An ACM IT Magazine and Forum* at http://www.acm.org/ubiquity/views/a_tripathi_3.htm

Ulmer, G. (1989). *Teletheory: Grammatology in the age of video.* New York: Routledge.

Wiles, K., & Littlejohn, A. (2003). *Proceedings of Supporting Sustainable E-Learning: A UK National Forum.*

ENDNOTES

1 See Kolb (1984, 1985). See http://www.agelesslearner.com/intors/lstyleintro.html (retrieved January 26, 2006) for an additional list of bibliographic references on the subject of learning styles, as well as links to other resources dealing with multiple intelligences and models of assessment.

2 See Chariot (2002) for brief but useful tips on how to approach communication issues in cross-cultural teaching online.

3 This was an "Introduction to Literature" course, developed and administered by David Wilson at Saint Mary's University, Halifax, Canada. My thanks to David for sharing his experience and stats with me.

4 For more ideas on collaborative e-learning activities see http://www.learnativity.com. Katy Campbell's *E-ffective Writing for E-Learning Environments* (2004) is also a very good practical resource for learner-sensitive assignment design.

Chapter XVII
E-Learning and the Global Workforce:
Social and Cultural Implications for Workplace Adult Education and Training

Karim A. Remtulla
University of Toronto, Canada

ABSTRACT

Workplaces are transforming in the global age. Jobs are expanding and varying. Workers are more and more participating in a global workforce comprising people who are socially and demographically diverse, multicultural, multifaceted, and whose views on workplace priorities, accountabilities, performance, and productivity may be socially and culturally very different from one another. Ultimately, these trends infer that how workers are educated and trained in the workplace must also evolve to meet a dynamic cohort of employees with a progressively complex profile of learning needs. To make matters more interesting, one of the most noticeable trends in the workplace today is 'e-learning,' which is frequently upheld as the panacea for workplace adult education and training needs. This chapter is about e-learning, the global workforce, and their social and cultural implications for workplace adult education and training in the global age.

INTRODUCTION

Workplaces are transforming in the global age. Jobs are expanding and varying. Workers are more and more participating in a global workforce comprising people who are socially and demographically diverse, multicultural, multifaceted, and whose views on workplace priorities, accountabilities, performance, and productivity may be socially and culturally very different from one another. Ultimately, these trends infer that how workers are educated and trained in the workplace must also evolve to meet a dynamic cohort of employees with a progressively complex profile of learning needs. To make matters more interesting, one of the most noticeable trends in the workplace today is 'e-learning,' which is frequently upheld as the panacea for workplace adult education and training needs. This chapter is about e-learning, the global workforce, and their social and cultural implications for workplace adult education and training in the global age.

Including the "Introduction," this chapter contains six sections. The next section, "Background," describes some of the primary drivers of workplace transformation in the global age and its implications for workplace adult education and training. This section develops the implications that the global age, characterized by globalization, technological innovation, and the knowledge-based economy, has for workplace transformation which leads to changes in work and the workforce.

This section also points out a noticeable contradiction between workplace transformation and its implications for workplace adult education and training. Whereas work is becoming more homogenized and normalized, the workforce is becoming more socially and culturally disparate. Yet existing traditions of adult education and training in the workplace are more universal in outlook, and favor European and Western paradigms of education and training. They value workplace adult education and training economically. Such practices are unprepared to answer the socioculturally specialized learning needs of the non-European, non-Western workers who will come to constitute more and more of the global workforce. All this forms a much needed context from which e-learning can be better known and assessed.

The third section broaches the questions of "Issues, Controversies, and Problems" of e-learning. By e-learning, this chapter assumes Honey's (2001) definition that "e-learning is the process of learning from information that is delivered electronically. ... It leaves us, the learners, to identify relevant information, convert it into something meaningful and apply it appropriately" (p. 201). At this time, e-learning is heavily based on European and Western traditions of work and workplace adult education and training. It epitomizes a homogenized, normalized, and universalized solution in the workplace. It speaks to economics and technology, cost savings, hardware, and software. Still, this section argues for a more sociocultural standpoint to look at e-learning in the workplace to better inform the promise of e-learning given workplace transformation and a global workforce.

This section also introduces *constructivism* and carries out a brief overview of this set of psychological theories about knowledge and instruction. Constructivism is increasingly incorporated as an instructive foundation for adult education and training in the workplace. Again, one of the chief criticisms leveled at constructivism is its general promotion of European and Western values and ways of learning and knowing. This section examines the complex and unpredictable benefits of constructivist, adult learning theories when interpreted for e-learning. What this implies is that facing an increasingly global workforce, the theories of learning on which workplace adult education and training practices are based are primarily designed for the needs and expectations of European and Western workers.

The fourth section, "Solutions and Recommendations," posits several guiding principles on how decision makers in an environment of workplace transformation and facing a global workforce may start to think about e-learning options that use constructivist learning. In essence these principles represent an alternate, sociocultural paradigm for e-learning decision making that may be especially relevant *prior to* significant investment in e-learning solutions.

The fifth section contemplates "Future Trends." Thinking is extrapolated here around a number of uncertainties that need to be further researched. These include the future of e-learning, and the escalating necessity of a socially and culturally informed outlook on adult education and training in the workplace when faced with a global workforce.

The sixth and final section, "Conclusion," appropriately weighs up some key conclusions. This chapter is about the social and cultural issues that contextualize and complicate e-learning and workplace adult education and training in the global age. A key conclusion coming out of this discussion is the fundamental need for a more sociocultural stance to guide workplace adult education and training practitioners with e-learning given a global workforce. Such an insight is crucial for management and employees as well as researchers and practitioners to envision the future of e-learning and the global workforce with their social and cultural implications for workplace adult education and training.

BACKGROUND

Workplace Transformation in the Global Age

This section elaborates on the notions of 'workplace transformation' and 'the global age.' To prematurely engage in a conversation about e-learning and workplace adult education and training without a context for such a discussion would be misleading. It may overly simplify and intellectualize the learning needs and expectations of the global workforce in the workplace. Therefore, attending to workplace transformation is justified at this point. It reveals that workplace transformation is more closely linked to the realities of the global age and is happening outside of and independently of any e-learning intervention. This in turn establishes the transforming workplace as a context from which to embark on a more enriched discussion about e-learning and its impact on workers and learning.

This section:

1. Introduces the concepts of global age, globalization, technological innovation, and the knowledge-based economy to establish why workplaces are transforming
2. Gives further details about how workplaces are transforming with respect to work and workers
3. Talks about the implications all this holds for adult education and training in the workplace

The Global Age

How are 'workplace transformation' and 'the global age' connected? Three forces in particular stand out as characteristic of 'the global age' and key contributors to this massive, ongoing, and widespread transformation of workplaces: globalization, technological innovation, and the transition from national, industrial economies to a globally integrated, knowledge-based economy. What should not be inferred from this statement is that these three forces comprise the *only* influences behind the 'global age' or 'workplace transformation.' Globalization, technological innovation, and the knowledge-based economy are, however, widely regarded as central and essential to the questions of the global age and workplace transformation (Fenwick, 2001b; Illeris, 2003; International Monetary Fund [IMF], 2002; Jarvis, 2000; Organization for Economic Cooperation and Development [OECD], 2001). For the purposes of this chapter, 'global age' refers to the convergence of globalization, technological innovation, and the knowledge economy, where a key outcome of this confluence is workplace transformation.

Globalization, Technological Innovation, and the Knowledge Economy

What is meant by 'globalization' and what does the process of globalization do? The IMF (2002) stipulates that globalization "refers to the increasing integration of economies around the world, particularly through trade and financial flows." Earlier views on globalization, however, such as those of Robertson (1992) and Amutabi et al. (1997), take a more composite stance. For them, globalization is a process which entails social, cultural, political, and environmental dimensions as well as an economic dimension. It inculcates the migration of people, ideas, ways of life, and traditions across international borders and is not limited to just trade and financial flows. This chapter also accepts a wider angle on the process of globalization in probing workplace transformation, which by its very breadth may serve to better inform a context of workplace transformation in the global age.

What about 'technological innovation'? Castells (1996) makes the case that technological innovations in general were foundational for the vast economic restructuring that took place in the 1980s when the term 'globalization' came into common use. Jarvis (2000) too agrees that the growth and integration of a world market and the flow of capital progressed quickly as a direct result of technology. Technological innovation enables the process of globalization, which then affects nations socially, culturally, politically, environmentally, and economically.

To interpret the interaction of globalization and technological innovation as 'one way,' however, is too reductive. To advocate that globalization drives technological innovation or vice versa is insufficient when studying workplace transformation. A more appropriate characterization of the relationship between the processes of globalization and technological innovation would be one that is reciprocal and symbiotic. Not only does technological innovation expedite globalization, but globalization also makes possible the transfer and movement of ideas, cultures, traditions, and people who foster technological innovation (Amutabi et al., 1997). Again, this chapter agrees to a more composite view of technological innovation to inform a context of workplace transformation in the global age. Here, 'innovation' includes not only the imaginative 'process of invention,' but also the new products and/or services that are invented, and where 'technological innovation' is not restricted to just information and communications technologies, but evokes technology in general.

What about 'the knowledge economy' and/or 'the knowledge-based economy' (the terms 'knowledge economy' and 'knowledge-based economy' are often treated as synonymous and in this chapter are used interchangeably). What role if any do the processes of globalization and technological innovation play in the creation of 'the knowledge-based economy'? According to both the IMF (2002) and OECD (2001), technological innovation and globalization bring about the global, knowledge-based economy through their impact on production, consumption, and economic restructuring at the state level. Nevertheless, keeping in mind the wider and more composite stance of this chapter with respect to globalization and technological innovation, the social and the cultural impact at the state level cannot be excluded either.

Canada may serve as a viable example to demonstrate this phenomenon. In 2002, the Government of Canada articulated Canada's Innovation Strategy entitled, "Achieving Excellence: Investing in People, Knowledge, and Opportunity." In this plan, 'the knowledge-based economy' is framed like this:

Every part of Canada and Canadian Society has a stake in the knowledge-based economy. Barely a decade ago, it was common to equate the knowledge-based economy with specific sectors, such as information and communications technologies, or with regions, such as [the] Silicon Valley in the United States. Now the knowledge-based economy knows few, if any, industrial or geographic boundaries. (p. 6)

What is clear is that the scopes of influence of globalization, technological innovation, and the knowledge economy are not limited to nations, governments, and markets. These forces touch the local through their impact on the workplace by forcing workplaces to transform. If workplaces cannot cope and do not transform, they risk falling behind their competitors, being excluded from the global marketplace, and suffering great financial losses (Industry Canada, 2002a; IMF, 2002; OECD, 2001). Not surprisingly, workplaces are transforming in some very fundamental ways with respect to jobs and the people needed to do these jobs. As the OECD (2001) sees it, "These changes are having a significant impact on the structure of employment and on the type of labor required" (p. 102). Next, this section scrutinizes how workplace transformation is happening through changes in work and the workforce.

Workplace Transformation

Work and the Knowledge Worker

Work is becoming more homogeneous when it comes to tasks and responsibilities. One widely accepted reason for this is the power of international standards bodies that promulgate systems to instrumentalize and make uniform various job tasks across various industries. Apt examples of this are the International Standardization Organization (ISO), International Electrotechnical Commission (IEC), and International Telecommunication Union (ITU). These international standards bodies work together and construct uniformity as necessary and incontrovertible:

ISO—together with IEC ... and ITU ... —has built a strategic partnership with the WTO (World Trade Organization) ... ISO, IEC and ITU, as the three principal organizations in international standardization, have the complementary scopes, the framework, the expertise and the experience to provide this technical support for the growth of the global market. (ISO, 2006)

In turn these standards become a form of accountability on the job, forcing everyone to comply and conform to this 'international' standard (Fenwick, 2001b; Korsgaard, 1997). In more extreme cases, ISO standards are even passed into law by certain nations.

Work is also becoming normalized around certain competencies and behaviors with respect to 'high skills.' This comes from a pervasive belief that high-skilled work and competencies, based on knowledge and continuous innovation, are universally tantamount to business continuity and profitability. In terms of 'skill' this chapter refers to the OECD's (2001) definition as, "The ability to produce and use information effectively" (p. 100), further developing the belief that 'knowledge' is the outcome of these high skills. 'Competencies' according to the OECD (2001) "are viewed as complementary for participating in the knowledge economy and in new production processes ... strongly reflect the emerging dynamics of technological change and globalization" (p. 106). These definitions seem to be a *stylized* reincarnation of Drucker's (1959) originating and seminal conceptualization of 'the knowledge worker' as someone who works primarily with information and advances knowledge in the workplace.

High-skilled and competent managers are expected to *anticipate* globalization, technological innovation, and the knowledge economy, and implement pre-emptive measures through workplace transformation. The outcome is intentional and extensive flexibility in the workplace in terms of worker aptitudes, pay, and training, through interventions like job rotation, project-based assignments, teamwork, flattened organizational structures, and the mass conversion of full-time work to contract, part-time, or contingent labor (Fenwick, 2001b; Industry Canada, 2002a).

Workers and the Global Workforce

The perspective of 'workers' is, consequently, as vital as 'work' to adequately grasp workplace transformation in the global age. This movement towards the homogenization and normalization of work is not mimicked in the characteristics and composition of the global workforce.

On the contrary, workers are becoming more socially and demographically diverse. A rapidly aging workforce, a dwindling youth cohort in the workforce, and declining birth rates, especially in Europe and the West, are resulting in a smaller workforce in the future to fuel the needs of mega-corporations (Bierema, 2002; Industry Canada, 2002a). Skill shortages are becoming more severe, bringing in greater competition for well-educated and talented workers by the advanced economies (Carliner, 2003; Industry Canada, 2002a; OECD, 2001).

Workers are also becoming more multicultural. Immigration from developing countries will count for most of the labor force growth in advanced economies in the near future (Bierema, 2002; Industry Canada, 2002b). As such a dwindling workforce in Europe and the West combined with growth in global immigration will bring together greater numbers of workers of different races, classes, sexual orientations, languages, physical abilities, and religious traditions. Any incidence in the workplace of privileging certain workers over others will become swiftly apparent and much more obvious and difficult to deny (Bierema, 2002).

Finally, workers are becoming more multifaceted and bring forward an increasing variety of talents. Workers are expected to assume personal responsibility for their learning and upskilling (Korsgaard, 1997). Home life and work life are becoming intertwined. Workers are forced to learn, adapt, and merge their private lives with their work. The workforce is 'upskilling' their average educational level. One outcome of this is mass underemployment. Many workers' skills and knowledge already far exceed the career opportunities available to them and their employers' ability to use these skills despite demanding it of their workers to get work in the first place (Livingstone, 2001).

Given these highlights of workplace transformation in the global age, this section next addresses how workplace adult education and training needs to evolve to meet these challenges.

Implications for Workplace Adult Education and Training

Adult education and training in the workplace, for this chapter, means the formal and informal activities that workers undergo at work resulting in individual, social, mental, or embodied growth associated with work (Fenwick, 2001b). There are two key aspects of workplace adult education today that make it ill-prepared for workplace transformation:

1. A European and Western philosophy of learning and knowing
2. A human capital philosophy of motivation

Firstly, workplace adult education and training today relies on a foundation of generalized, universalized, and ethnocentric assumptions based in a European and Western centric worldview of knowing and learning. This, ironically, may be too narrow and confining for a global workforce undergoing workplace transformation in the global age (Alfred, 2002b; Bierema, 2002; Birden, 2003; Conceicao, 2002; Folely, 2004; Johansen & McLean, 2006; Lee & Sheared, 2002; McLean, 2006). The second trait is the attribution of economic motivation for workers to partake of workplace adult education and training, and for workplaces to provide this education and training. The roots of this lay in human capital theory, which emphasizes the economic reasons for workplace adult education and training, and assumes that learning initiatives should increase the capital value of human skills, workers are resources belonging to their employer, workers must increase their economic value to potential employers and become more employable, and employers must invest in their 'human resources' as they would in other assets for a future return on this investment through increased profits (Amutabi et al., 1997; Fenwick, 2001b; Gustavsson, 1997).

Together, these two facets of workplace adult education and training culminate into a paradigm that embodies the universalized worldview of workplace adult education and training through five implicit assumptions:

1. European and Western views and values on the purposes of learning are universal and may be unproblematically generalized to all workers.
2. The content and method of teaching must remain consistent and must get the same results.
3. 'Learning' is a 'personal' process, a solitary undertaking that is cognitive and indifferent to context, social, affective, embodied, or environmental influences.
4. The workplace sees all workers as 'equal' when it comes to access and capacity to learn.
5. The learning must align with and contribute to the economic goals of the workplace.

This paradigm defines to a great extent what the majority of practitioners of workplace adult education and training perceive when weighing up the needs of workers: all adults in the workplace learn for the same reason, they must be taught the same things, they learn in the same way, and they all have the same access and capacity to learn (Alfred, 2002b; Fenwick, 2001b; Illeris, 2003; Johansen & McLean, 2006). Carliner (2003) further encapsulates this monolithic approach to training in the workplace by noting that training is often conducted in a centralized location in many transnational corporations.

The inconsistency, when trying to understand workplace transformation and its implications for workplace adult education and training, is that current paradigms of adult education and training are more closely aligned with how workplace transformation is affecting work and, to some extent, geared towards a 'knowledge worker' of European or Western descent. Furthermore, concentrating on economic considerations alone does not translate into learning. All this seems to ignore entirely the education and training needs and motivations of the non-European, non-Western worker who will populate more and more of the global workforce in the future—a supply of workers that is ever more socially and culturally varied.

The key implication of workplace transformation for adult education and training in the workplace according to Johansen and McLean (2006), and increasingly for other practitioners, researchers, and academics of workplace adult education and training, is the mounting need for a more sociocultural bearing to assist adult education and training practitioners in contending with workplace transformation. For this chapter, a sociocultural perspective on workplace adult education and training recognizes and acknowledges that all workers are not the same. They do not all learn for the same reasons and do not all need to learn the same thing. They cannot all learn the same way and under the same circumstances. The 'sociocultural' looks beyond the individual adult learner:

Context is thus central to the social construction of meaning and the understanding of ... learning within a sociocultural environment. The environment, then, includes the individual in interactions with the culture, context, and community within which learning occurs. Indeed, it is from this sociocultural environment that the individual acquires the tools, symbols, resources, and strategies to manage the learning process. (Alfred, 2002b, p. 5)

Recent research being done to elucidate adult learning styles as well as cognitive and educational psychology predominantly refer to the necessity and importance of a social and cultural context when considering adult education and training against an increasingly complex set of adult learning needs for a global workforce (Felder & Brent, 2005; Illeris, 2003; Moore, 2005; Munro & Rice-Munro, 2004; Spencer, 2001).

Here then is the intricate context from which e-learning is borne. Globalization, technological innovation and the knowledge economy are demanding workplaces transform. Workplace transformation is resulting in work that is becoming more homogenized and normalized with an idealized sense of workers as 'knowledge workers.' The workforce is becoming more socially and culturally divergent. With repeated regularity, workers carry a multitude of talents often resulting in underemployment. And workplace adult education principles and practices are universalized in outlook, and favor European and Western ways of knowing and learning motivated by economic goals, whereas the major supply of labor in the future will be non-European and non-Western.

This chapter turns next to 'e-learning' in earnest. Does e-learning as it is perceived in the workplace today possess the sociocultural sensitivity necessary to properly educate and train a global workforce? The growing exposure of the global workforce to e-learning makes such a social and cultural enquiry essential for management and employees, as well as researchers and practitioners, to deliberate the future of e-learning and its implications for workplace adult education and training in the global age.

ISSUES, CONTROVERSIES, AND PROBLEMS

Issues

E-Learning

Despite the preconceived notions inherent in current workplace adult education and training, and the challenges of workplace transformation in the global age, e-learning is heralded as a final solution for workplace adult education and training needs for a global workforce. According to a 2003 survey conducted by the American Society for Training and Development (ASTD), 95% of the companies that responded indicated using some form of e-learning for training such that: "E-learning was estimated to account for approximately \$11.4 billion of U.S. corporate training investments in 2003" (DeRouin, Fritzsche, & Salas, 2005, p. 921). Similarly, Industry Canada reports the following information (Sirois, 2005, pp. 1-2): the global market for online learning is estimated to be US\$ 3.2 billion by 2010; the global training market for government and industry is valued at US \$300 billion; on average, over 50% of the firms in Canada describe using the Internet or related technologies for workplace training; and, in 2003, e-learning vendors generated US\$6 billion worldwide in sales. Governments, industry, and a global constituency of adult learners are demanding and anticipating online training experiences that have a *positive* impact on performance (Newton, Hase, & Ellis, 2002; Sirois, 2005).

What is it about e-learning that gives it such currency and popularity within the workplace and within a context of workplace transformation? The term 'e-learning' consists of two elements: 'e' and 'learning.' The 'e' is garnering the majority of the debate so far, generating research and discussion that focus almost exclusively on hardware and software (Ally, 2004; Bednar, Cunningham, Duffy, & Perry, 1992; DeRouin

Table 1. Comparing e-learning and the changing nature of work in the global age

Changing Nature of Work	E-Learning
Homogenization of work regarding tasks and responsibilities	*Access* by employees to content 'on demand,' anytime, and anywhere to update skills
Normalization of work around 'high-skilled work,' competencies, and behaviors	*Updates* content 'instantaneously' when necessary to reflect the changing skills and needs of work *Reduces* time needed to learn and travel to outside training facilities, and also allows employees to learn on the job and at home
Universalization of workplace adult education and training practices and outlook	*Consistent* training content delivery to all employees

et al., 2005; Duffy & Jonassen, 1992; Salas Kosarzycki, Burke, Fiore, & Stone, 2002). In this regard, e-learning speaks to a narrower economic definition of globalization. E-learning takes on a narrower outlook on technological innovation as hardware and software. These qualities align quite conveniently with workplace transformation and with workplace adult education and training (Ally, 2004; DeRouin et al., 2005; Honey, 2001; Illeris, 2003; Munro & Rice-Munro, 2004; Salas et al., 2002). Table 1 summarizes e-learning's characteristics with the changing nature of work in the global age.

An almost exclusive spotlight on hardware and software alone still does not address learning and the workplace, both of which are, as illustrated earlier, socially and culturally negotiated situations. Relatively little attention is given to 'learning' and the global workforce doing the learning. Such research is noticeably minimal when it comes to the characteristics of e-learning. In discussing the state of research into distance learning (i.e., 'DL'), Salas et al. (2002) ascertain that researchers "have made only modest inroads in understanding how the technology-based delivery of instructional programs interacts with learning outcomes" (p. 136).

Therefore, although e-learning may be popular, the basis for this popularity is, to some extent, incomplete. This overwhelming center of attention on technology in the research and study of e-learning places any review of e-learning in the workplace at somewhat of a disadvantage. All this is insufficient to answer the basic question of learning and the global workforce. This is where the need for a sociocultural framework to look at e-learning becomes paramount.

The next section presents a sociocultural framework to look at e-learning.

A Sociocultural Framework for E-Learning

A sociocultural framework is necessary to position the thinking done so far on e-learning, and reorient and reinterpret the findings towards the social and the cultural to produce a more informed depiction of e-learning, learning, and the global workforce. Such a critical framework should look not just at how e-learning influences the social and the cultural, but also at how the cultural and the social influence e-learning. To allow for a sociocultural take on e-learning in the workplace based on context, culture, and community (Alfred,

2002b), the framework below comprises three positions along a continuum: *media* at one extreme, with little or no consideration of the employee; *genre,* which considers user communities within a larger workforce; and finally, *learning,* which is directly concerned with notions of knowledge and how people learn. When taken together, these perspectives present a sociocultural perspective on current research and study on e-learning that provide better support in clarifying the social and cultural implications for workplace adult education and training and the global workforce.

A 'Media' Perspective

The idea of *media* aptly conveys the social and cultural (or lack thereof) when it comes to technology as indicated by its usage in expressions like 'computer-mediated communications.' A media perspective when analyzing e-learning looks at what channels or formats of e-learning may be made available for distribution, and ultimately, what technologies are necessary for access—in simple terms, the 'hardware.' Terms like 'computer-based instruction' and 'Web-based training' indicate the format of the content and the mode of its availability. "One can define DL as learning that is media based, remote, or asynchronous and supported by some institutional system," point out Salas et al. (2002); however, what becomes imminently clear is, "a fragmented domain consisting of many relatively new technologies" (p. 137). The rapid development and innovations of information and communication technologies makes this perspective confusing to assess the implications of e-learning, especially given the idiosyncrasies and limitations of various media.

A 'Genre' Perspective

A 'genre' perspective on e-learning primarily looks at 'software' and instructional design. 'Genre' of software involves the intended users and their relation to one another (Kress, 2003; Remtulla, 2005). This perspective shifts the focus away from the media and hardware, and onto software and applications to determine which users will ultimately benefit from e-learning and why. This approach is especially prevalent when looking at e-learning with respect to 'soft' skills training. DeRouin et al. (2005) explain:

In addition to using e-learning in the training of IT skills, a growing number of businesses have used e-learning in the training of business and soft skills ... Some of the most common business and soft skills to be taught via e-learning in organizations include management, leadership, communication, customer service, quality management, and human resources skills ... (p. 922)

A 'Learning' Perspective

A 'learning' perspective on e-learning considers pedagogy—the principles and practices of instruction that make up the e-learning. A learning approach in fact undergirds even the hardware and software decisions made as part of e-learning initiatives. Andrade et al. (2005) concur:

... technology should not be considered as a goal in itself, but as a means that facilitate learning ... it is necessary to bear in mind that during a good e-learning course, not only the technical aspects but also—and mainly—the pedagogical ones should work correctly ... (p. 658)

Implications for Workplace Adult Education and Training

The corollary to the shifting and evolving landscape of a media-based attitude on e-learning is its *homogenizing* effect on the perception of education and training needs: that greater amount of variety and functionality in media, distribution, and access of information equates to better education and training. Put another way: all education and training needs and solutions are in essence media centered.

For example, Harun (2002) discusses the continuing medical education (CME) application as part of the TeleHealth Project of the Ministry of Health of Malaysia:

In its approach to achieving the TeleHealth objectives and backed by a reliable, scalable infrastructure network, the solution proposed focuses on delivering the four applications contained within, namely, the Lifetime Health Plan (LHP), the Mass Customised Personalised Health Information and Education (MCPHIE), CME, and TeleConsultation in an integrated manner, thereby simulating real-life situations as encountered within the patient-healthcare provider environment as closely as possible ... (p. 306)

Four modalities of education were available for healthcare professionals affiliated with Malaysia's Ministry of Health: just-in-time CME, personal CME, and formal and modular distance education. Harun (2002) goes on to describe the advantages of CME which include, "the use of computer-based training for teaching information technology skills," as well as providing healthcare information available at the place and time of need, always current and tailored at the point of the need, delivered at low cost with no travel needed, and with a personalized e-learning management system to support the professional in using the system (pp. 308-309).

Telefonica, also one of the world's largest telecommunications companies, with almost 65 million clients in 41 countries (Gasco, Llopis, & Gonzalez, 2004), also revised its training program and invested in a new HR management system. The company identified and differentiated one type of training from another, "by form of communication between teachers and students" (p. 377), as either 'asynchronous technologies,' 'synchronous technologies,' 'learning management tools,' 'author's tools,' or 'learning management systems.' For Telefonica:

... it is not a question of choosing a particular technology; but wiser to use different systems or different techniques of support for its various types of formative actions. In deciding what technology to use for a specific formative action, certain general factors are important. (Gasco et al., 2004, p. 378)

Yet, neither Harun (2002) nor Gasco et al. (2004) present any significant debate about the sociocultural complexities facing professionals in the field, much of which may not be media-centered: when to retrieve information and why; the usefulness of information retrieval with respect to the values of patients/clients; the wishes of the patients/clients; the impact of motivational subjectivities, interaction, and socialization in the field between professionals and patients/clients; and the varying educational, emotional, contextual, cultural, and social experiences not only of the patients/clients, but of the professionals themselves.

A media critique on e-learning makes the homogenizing effect on adult education and training quite apparent. A media perspective reveals a perception that adult education and training in the workplace is fundamentally about (a lack of) distribution and access to information; a challenge readily addressed through expenditure on more hardware. The impetus behind the training initiatives at the TeleHealth Project and Telefonica seems to be the same, that improving information systems is synonymous with better workplace adult education and training. The focus is on access and information systems; the storage and retrieval of information. Decisions about workplace adult education and training in the workplace now become about development and/or investment in hardware, and not about the social and cultural learning needs of a socially and demographically diverse, multicultural, and multifaceted workforce.

A sociocultural critique based around a genre perspective on e-learning does provide some advantages that a media stance does not when pondering e-learning and the global workforce, and the social and cultural implications for workplace adult education and training in the workplace. A genre outlook recognizes the existence of user groups and factors in their inconsistent social and cultural needs and circumstances, whereas the media discussion reduces the entire e-learning conversation to a question of hardware. A genre outlook also addresses the fragmentation and diversity within and between communities of users in the workplace. In any organization, it is not uncommon to have many user communities, each with differing needs both within and between communities. The e-learning software in question is likely to be more standardized than is actually needed by the user groups, and may not be entirely responsive for the particular contextual, gendered, racial, social, cultural, and/or other needs of the users, especially when dealing with a global, transnational user base.

SkillSoft, an e-learning provider, conducted a study by interviewing some 200 employees from 16 participating firms located in 14 different countries across Europe, the Middle East, and Asia (Baldwin-Evans, 2004). When asked about the benefits of e-learning, interviewees named 'IT and computer literacy' as an important area of improvement. They also mentioned 'personal skills development' as another area of progress, especially "in key areas like management, leadership, coaching and mentoring, assertiveness, time management" (p. 271). Yet, the study contains some conflicting messages. When employees were questioned about barriers to e-learning, 50% indicated 'lack of time' as their major hindrance, 47% stated a fear of technology and suspicion about e-learning, 41% cited self-motivation, and 33% alluded to a lack of management support (p. 273). These conflicting messages may very likely be attributable to differences in contextual, gendered, racial, social, cultural, and/or other needs of the interviewees. In fact, Baldwin-Evans (2004) acknowledges that these responses do diverge by firm and location, though no specific details are provided in this regard (p. 273).

Hewlett-Packard also believes that employee preferences for e-learning differ around the world. According to DeRouin et al. (2005), Hewlett-Packard discovered that their employees in Asia and Europe preferred instructor-facilitated training, whereas in the United States, a self-paced format is favored. Hewlett-Packard's training motto for its global workforce is, "one size does not fit all," substantiating that "these variations in e-learning preferences by region suggest that a single type of e-learning program may not meet the needs and expectations of all employees" (p. 926).

A genre perspective on software and applications also takes into account e-learning that has been traditionally taught in classrooms. In this scenario, the debate turns to the suitability or appropriateness of the software as compared to its intended purposes (DeRouin et al., 2005); the discussion becomes about the *gap* between intentions and outcomes of applications and not about the *needs* of adult education and training in the workplace. The incongruity here is that if the 'gap' remains unidentified or intentionally omitted from software curricula, or if no standardized software solution is readily available to 'fill in the gap,' does the need cease to be 'a need'? Does this further imply that adult education and training is not warranted? Conversely, what if standardized software is available for which a workplace curriculum gap was never identified? Does a 'new' need now exist for adult education and training in the organization? Would all user communities in the workforce have this newly uncovered need? A discussion ensues about needs driven from prescribed curriculum gaps which may not necessarily answer to the complex and diverse social and cultural needs of the global workforce who are noticeably left out of the discussion.

Pollitt (2005b), for example, elaborates on the experiences of the North American retailer education and training group at the Ford Motor Company. This group is responsible for the non-technical training needs of Ford's dealer channel, which comprises some 225,000 employees (p. 639). According to Pollitt (2005b), the retailer education and training group undertook an exercise to determine what was missing from its curriculum with respect to 'soft' skills such as stress management, change management, diversity, and financial management, and then outsourced their requirements to an outside vendor. The reason to outsource was straightforward: "Building content in these areas is not a core competency of the retailer education and training group ... Therefore, the company decided to purchase the necessary skills" (p. 6.40). Over 100 online courses were made available to employees based on the curricular gap of this user group.

The mining industry in Queensland, Australia, also realized that 'organizational culture' played a major part when it came to e-learning implementation and the conflicting needs (and cultures) of differing user communities (Newton et al., 2002). Mining personnel and stakeholders across a wide range of user groups were interviewed. Management felt that standardization was beneficial to meet industry-wide competency compliance legislation and as an effective means to transfer corporate values to new employees. Industry workers instead indicted that they did not appreciate corporate policy information on screen, but preferred direct access to only the information that was most relevant to them. They further retorted that, "Viewing training as a legislative requirement in the mining industry was seen to conflict with other values in terms of providing 'good' training in terms of learners' needs of 'lifelong learning' goals" (p. 160).

Here too, the SkillSoft (Baldwin-Evans, 2004) study proves handy. Interviewees were questioned about their motivations for undertaking training through e-learning: 34% stated they did so to be more competent in their current job, and another 20% indicated that the training was compulsory and mandated by regulatory bodies or management (p. 269). However, when asked if they actually completed a course, "most of the courses being taken aren't being completed all at once," and "employees are dipping in and out of courses, skipping the sections they don't need and only learning what they need to learn at that particular time...course completion is not something they consider to be significant" (p. 271).

A genre critique on e-learning reveals a trend towards *normalizing* workplace adult education and training. This happens with universally applicable, standardized software as a prescription for filling in predetermined curricular gaps. A normalizing effect from standardized software and applications suggests that only the adult education and training needs in the workplace that are 'normal' and 'standard' are the ones that are relevant, since they are the ones acknowledged and made 'visible' by the presence of a gap in curricula and/or the availability of a standardized software solution.

The advantage that a learning perspective has over both media and genre perspectives when it comes to e-learning and adult education and training in the workplace is that it brings the social and the cultural squarely into the conversation by posing questions of epistemology (i.e., the origin and nature of knowledge) and pedagogy (i.e., the principles and practices of instruction). Such an approach to e-learning may be more relevant to understanding e-learning's implications for workplace adult education and training in the global workforce and given workplace transformation. At the core is a discussion about assumptions of knowledge, and how people come to learn things and the importance of social and cultural influences on knowledge and learning when considering a global workforce.

At the same time, a learning approach to e-learning does not completely disregard either hardware or software. The homogenizing effects of hardware and the normalizing and prescriptive trends of software imply certain views on how adults learn and need to be trained. Pedagogy does not frame adult education and training as predetermined by either a need for more hardware or finding a particular type of software, but goes beyond these perspectives to the worker and the issue of how workers learn and factors that may help or hinder learning.

The next section aims to raise some of these complications and concerns regarding constructivism, and speaks to them by delving deeper into this learning critique on e-learning. Looking closely at e-learning in this way encourages an appreciation for the repercussions of constructivist thinking on e-learning. Without an understanding of the constructivist learning theories that are increasingly penetrating e-learning, simply investing more time and money in hardware and/or software will not likely yield the gains in workplace priorities, accountabilities, performance, and/or productivity so faithfully anticipated through adult education and training in the workplace from e-learning (Gasco et al., 2004; Harun, 2002; Pollitt, 2005a, 2005b, 2005c).

In the next section, a broader and deeper questioning of controversies associated with constructivism is presented. Problems that may be anticipated with its integration with e-learning are also highlighted to further the discussion already underway with respect to the social and cultural implications of e-learning for workplace adult education and training in the global age.

Controversies

Constructivism and Constructivist Thinking

Constructivism and the constructivist view of adult learning is receiving increasing attention when it comes to e-learning (Anderson, 2004; Bednar et al., 1992; Duffy & Jonassen, 1992; Gulati, 2004). For constructivist thinkers, there are many ways to conceive and perceive the world and encounter meaning through personal experience. Constructivism sees knowledge as *constructed by the learner.* Facts are not self-evident. They are on the contrary unstable, non-neutral, and open to interpretation by the learner. As such, the goal is not to find the one correct meaning or interpretation, but to appreciate the multifaceted and active role of learners in this interpretation (Gulati, 2004; Phillips, 1995). According to Glasersfeld (1996), "It is the idea that what we call knowledge does not and cannot have the purpose of producing…an independent reality, but instead has an adaptive function" (p. 3). Fosnot (1996) maintains:

> *Constructivism ... construes learning as an interpretive, recursive, building process by active learners interacting with the physical and social world. It is a psychological theory of learning that describes how structures and deeper conceptual understanding come about, rather than one that simply characterizes the structures and stages of thought or one that isolates behaviors learned through reinforcement.* (p. 30)

The implications of this merging of constructivist pedagogy with e-learning are becoming very important for the workplace. More and more people from different regions of the world are coming into contact with each other and sharing their education and training, as well as exchanging their varied social and cultural heritage through their interactions. To promote a monolithic view of knowledge or a singular view of instruction is no longer the only alternative. This is exactly the stance of constructivism. Despite its growing appeal, however, constructivism does come with its own inherent assumptions and perceptions. These must also be weighed in tandem with e-learning and the workplace if this confluence is going to work given a global workforce that is socially and demographically diverse, multicultural, multifaceted, and undergoing workplace transformation.

There are four fundamental concerns to keep in mind about constructivism. The first concern is that constructivism is a theory about *what knowledge is and how people learn.* It ponders the entire learning process from how knowledge is initiated (i.e., epistemology) to how knowledge is acted upon and/or transferred through a means of instruction (i.e., pedagogy). Such a gamut may not be so straightforward to traverse. Fosnot (1996) sheds light on the tensions that appear along this continuum:

The theory describes knowledge as temporary, developmental, non-objective, internally constructed, and socially and culturally mediated. Learning from this perspective is viewed as a self-regulatory process of struggling with the conflict between existing personal models of the world and discrepant new insights ... (p. ix)

The second concern about constructivism is the pivotal *but pivoting* notion of 'context.' Social and cultural contexts do play a key role in knowledge, instruction, and learning according to constructivism. However, what is noteworthy is that constructivist thinkers differ in their interpretations and opinions on the *degree* of recognition that they give context and the specific role they deem appropriate for context in their particular take on constructivism. Lee (2003) provides this glimpse into the some of the particulars of context when it comes to constructivism:

Generally speaking, context is defined in terms of the total experiences and activities engaged in by members of various social groups. More than just a geographic space, it encompasses the biographic, interpersonal, political, historical, and socio-cultural settings in which individuals are socialized, shaped, and situated and in which they interact. Contextual impact usually comprises the diverse and intersecting influences that come from one's race, class, gender, nationality, communities, and the larger political and socio-cultural milieu. Context is hereby characterized as dynamic, changing, and polyrhythmic ... (p. 12)

Constructivism entails that learners accept responsibility for their own learning, implying that whatever they learn is unique to them; that symbols and tools are not sufficient to pass meaning on to learners; that concepts cannot be reduced to simpler parts and remain authentic; summarily, that learning cannot be achieved *outside* of context (Fosnot, 1996). The sociocultural is decisive (Baumgartner, 2003a; Fenwick, 2001a; Null, 2004; Phillips, 1995). Nevertheless, how context plays out in the constructivist discourse is just as widely contested among constructivists based on their differing opinions and expositions.

Third, the perception of 'experience' in knowledge and instruction is also a concern. Adults use experience as learning (Fenwick, 2001a). Nonetheless, important questions abound about *how* adults experience and the relationship between learner and experience when considering knowledge and instruction. Conflict develops when trying to understand on what basis to 'split' experiential learning from common, everyday living. Fenwick (2001a) indicates, "It seems counter-productive to separate 'experiential learning' as an evolving adult education practice from a broader consideration of learning through experience," further admitting that "attempted divisions between human experience and reflection on that experience have proved problematic for all kinds of reasons" (p. 9).

Precisely because constructivism deals with epistemology, pedagogy, experience, and context, the last and perhaps most conceptually difficult concern with constructivism is that the label 'constructivism' represents a wide array of views and stances. In fact some non-constructivist thinkers have their learning theories accepted as constructivist thinking because of the role that epistemology, pedagogy, experience, and context play in their theories of learning. In other words it is a theory about *a set of theories and/or assumptions.* This adds a whole other level of complexity with regard to *interpretation* when it comes to the tensions and concerns that already permeate constructivist thinking. 'Constructivism' is not one single notion, but a lively, vibrant, and ongoing debate of many conflicting *interpretations* that researchers and practitioners have come to associate as *constructivist thinking* over the years. Bednar et al. (1992) share their take on constructivist thinking as, "all that we know of the world are human interpretations of our experience of the world" (p. 21).

These concerns around constructivism raise a problem. Any attempt to reduce this sophistication may assist in assessing e-learning. However, such a move may inadvertently undermine a clearer understanding of constructivism and may oversimplify the implications, advantages, and disadvantages constructivism implies with respect to adult education and training in the workplace. This problem with constructivism becomes especially pertinent when applied through e-learning for a global, socially and demographically diverse, multicultural, and multifaceted workforce. The next section presents this problem more lucidly.

Problems

How to Look at Constructivism and Constructivist Thinking

Constructivism thinks about both knowledge and instruction, epistemology and pedagogy. These are not new ideas, and constructivist thinking on these notions has been around for decades (Fosnot, 1996; Glasersfeld, 1996; Phillips, 1995). Yet constructivist thinkers have to be able to justify how certain ideas remain true to their constructivist ideals—their constructivist roots—despite changing times or influences from other philosophies and disciplines. Constructivism also takes into account a multitude of interpretations. Yet all of these 'divergent' voices still maintain a common ground—a constructivist attitude—as a result of certain set of core values and beliefs that are evident despite inconsistent contexts or mixed experiences. In other words, certain thinkers can have their ideas included in any discourse on constructivist thinking and yet still not bear the label of constructivist thinker. Based on this brief exposition of constructivism and constructivist thinking, a problem arises with respect to *how* to gain some insight into constructivism from the many elucidations that comprise constructivist thinking without unduly reducing its sophistication or essentializing its attributes.

A literature review of constructivism and constructivist thinkers reveals four categories or classifications of 'distinct action' that appear comprehensive and representative of the varied interpretations of 'active' learner involvement in his or her own learning. These actions are: self-reflection, problem solving, collaboration, and, participation. Each action signifies a slightly distinct emphasis on the social context of the learner along a continuum from 'individual' to 'social.' In other words, the 'action' by which the learner constructs his or her learning changes given the social context and experiences of the learner—when the learner is alone or when the learner is a part of a community—but the active involvement of the learner in his or her own learning remains the domain of constructivism.

Next, this section lays out some prominent examples of constructivist thinking along a *continuum of learning actions* from 'individual' to 'social.' In this way, the contradictions of interpretations regarding the same action, or discrepancies between different actions, are raised to the foreground. At the same time, some of the major problems inherent in constructivist thinking regarding these actions are also shared along this same continuum.

The Act of 'Self-Reflection'

Jean Piaget (1896-1980) introduces the notions of assimilation, accommodation, adaptation, and equilibration in his thinking on knowledge and instruction. Taken from biology, 'equilibration' is the (genetic) mechanism that drives evolution. Equilibration, for Piaget, is also what drives growth and change in cognition; an active endeavor of balancing two extremes of self-regulating behavior, assimilation, and accommodation, when the learner encounters conflict, contradiction, and 'imbalance' in their environment. "Assimilation is the organization of experience with one's own logical structures and understandings," whereas accommodation involves "reflective, integrative behavior that serves to change one's own self," (Fosnot, 1996, p. 13). Fosnot (1996), however, hastens to clarify that, "Equilibration is not a sequential process of assimilation, then conflict, then accommodation. Instead it is a dynamic 'dance' of progressive equilibria, adaptation and organization, growth and change" (p. 14).

Knowledge involves the formation of mental structures—mental schema—and here again, Piaget borrows from biology. With regard to cognition, 'adaptation' occurs in this sense:

The relationship of viable biological organisms to their environment provided a means to reformulate the relationship between the cognitive subject's conceptual structures and that subject's experiential world. Knowledge, then, could be treated not as a more or less accurate representation of external things, situations, and events, but rather as a mapping of actions and conceptual operations that had proven viable in the knowing subject's experience. (Glasersfeld, 1996, p. 4)

As in biology where growth, change, and evolution happen in steps, learning for Piaget is also developmental; it happens in stages. These successive episodes of 'adaptation' and 'equilibration' comprise for Piaget what he coins his 'genetic epistemology' (Anderson, 2006; Glasersfeld, 1996).

Jerome Bruner (b. 1915) injects some thoughts on pedagogy and instruction to ground this discussion. A substantial component of Bruner's thinking in this regard revolves around the phrase "beyond the information given" (Bruner, 1973). One of Bruner's insights in this maxim is that the learner must be instructed to learn *how to learn.* Here Bruner (1973) advocates for a learner that is not a passive recipient of learning, but highly self-motivated:

... discovery in learning has precisely the effect upon the learner of leading him to be a constructionist, to organize what he is encountering in a manner not only designed to discover regularity and relatedness, but also to avoid the kind of information drift that fails to keep account of the uses to which information might have to be put. It is, if you will, a necessary condition for learning the variety of techniques of problem solving, of transforming information for better use, indeed for learning how to go about the very task of learning. (p. 406)

Bruner (1973) also invokes a highly learner-centered approach to pedagogy. He advises that the task at hand is to represent the structure of the material in terms of the way the learner sees things; in terms of the learner's mental schema. Yet any authority ascribed to thinking on pedagogical issues must be likewise critiqued on its attention to the *readiness* of the learner to learn. Learning *how to learn* is irrelevant when the learner is not predisposed or unable to learn in the first place. It is also futile if the learner does not develop this ability when facing the unknown.

Bruner's focus on the learner and on the learner's mental schema culminates in the corollary to Bruner's central passion of "going beyond the information given," and perhaps forms a central premise for what constructivist thinkers might consider a primary value and belief of constructivism: "Consider now what benefit might be derived from the experience of learning through discoveries that one makes for oneself" (Bruner,

1973, p. 403). Learning and self-discovery is its own reward.

In this regard, self-directed learning can also be included in the action of 'self-reflection.' Self-directed learning has three sides and is sometimes seen as a: (1) goal, (2) process, or (3) personal characteristic of the learner. As a goal, self-directed learning represents the objective of adult learners to be the 'masters' of their own learning. As a process, self-directed learning expresses the desire of adult learners to design how they can achieve the potential of becoming 'masters' of their own learning. Lastly, self-directed learning represents an attribute inherent in adult learners that is continuously honed as they struggle to accomplish the goal and process of self-directed learning. The ongoing refinement of this attribute is understood as a key aspect for lifelong learning.

Each side to self-directed learning also signifies *stages* of growing self-reliance. These developmental stages are experienced by the learner from heavy dependence on the instructor, to moderate self-direction, and finally, complete acceptance of responsibility for the mastery of learning, the process of learning, and the personal transformation necessary to sustain self-directed learning over a lifetime (Baumgartner, 2003c).

Implications for Workplace Adult Education and Training

When looking at 'self-reflection' pedagogies for adult education and training in the workplace, some critical questions regarding the assumptions and paradigms of self-reflection may include the following (Baumgartner, 2003c; Fenwick, 2001a):

1. **About self-reflection:**
   a. Are self-reflection, experience, and the learner separable from each other or from social, cultural, or, historical contexts?
   b. Is self-reflection always perfect and precise in every experience?
   c. Is self-reflection researchable and can it generate conclusive results?
2. **About the experience:**
   a. Is experience something concrete, tangible, and definable that can be appropriated, controlled, dissected, and analyzed for learning?
3. **About the learner:**
   a. How does the learner decide what experiences are worthy of self-reflection?
   b. Does everyone in the workplace 'self-reflect' in the same way?
   c. Is self-reflection the only way a learner can gain true meaning from learning?
   d. Is self-reflection a purely internal, rational, solitary, cognitive act for every learner?
   e. Is the learner a stable, unitary, and rational being who can be separated from social, cultural, and historical context when undertaking self-reflection?
   f. Can the learner self-reflect on experience without being influenced by non-rational effects like emotions, values, the subconscious mind, or from social, cultural, or historical contexts?
   g. Is the 'individualized' outlook for self-reflection in the workplace more representative of a specific segment of the workforce, namely those who are white, of European or Western descent, employed, better educated, and in leadership positions?
   h. Is the 'individualized' outlook for self-reflection in the workplace marginalizing groups who may not have the privilege to self-reflect in the workplace such as adults with disabilities; developmentally challenged

adults, and adults with special needs; or women, people of color, working class adults, and immigrant learners?

The Act of 'Problem Solving'

Andragogy, according to Malcolm Knowles (1913-1997), is, "the art and science of helping adults learn" (Baumgartner, 2003b, p. 6). Although Knowles is not widely accepted as a 'constructivist,' Knowles' constructivist thinking comes through in his notion of andragogy. This notion rests interestingly at the boarder between 'self-reflection' and 'problem solving' as is evident in the five assumptions he puts forward as required for adult learning through andragogy: that learners become less dependent and more self-directed over time; adults have more experiences and categorically different kinds of experiences to learn from than youth; adults time their learning for self-development; learning is problem centered; and adults are internally rather than externally motivated (Baumgartner, 2003b).

Donald Schön (1930-1997) talks about 'critical reflection' in professional practice (Fenwick, 2001a; Fenwick & Parsons, 1997) as a means for learning through solving dilemmas in the 'real' world. Schön directly addresses learning in the workplace, and the experience of professionals and practitioners as they encounter unexpected circumstances and periods of anxiety in situations that are disorganized and confused, and for which their past education and training has not readily prepared them.

Learning happens by detecting and framing problems through 'reflection-in action.' Problem framing views life as alternating between states of equilibrium and disequilibrium—between a natural and desirable state (i.e., equilibrium), a state of temporary disturbance (i.e., disequilibrium) (Fenwick & Parsons, 1997), and then repeatedly trying out solutions. Further thinking on the situation takes place after rounding out what has been learned:

When these adults meet such unique problems or situations containing some element of surprise, they are prompted to reflect-in-action by improvising an on-the-spot experimentation, thinking up and testing out and refining and retesting various solutions for the problem...professionals also often reflect-on-action in some zone of time after a problem-episode, when they examine what they did, how they did it, and what alternatives exist. (Fenwick, 2001a, p. 12)

Here, Bruner (1973) introduces 'problem-posing' as a main concept in his thinking on pedagogy and instructional method intended to support the learner and encourage their readiness to learn. Problem-posing encourages the leaner to progress to successively higher stages of intellectual development. The hard part, but the most significant, is finding adequate problems that foster this development and not undermine it:

We solve a problem or make a discovery when we impose a puzzle form on a difficulty that converts it into a problem that can be solved in such a way that it gets us to where we want to be. That is to say, we recast the difficulty into a form that we know how to work with, then work it. Much of what we speak of as discovery consists of knowing how to impose what kind of form on various kinds of difficulties. A small, but crucial part of discovery of the highest order is to invent and develop models or puzzle forms that can be imposed on difficulties with good effect. (p. 410)

In the same vein, problem-based learning purports to develop critical thinking skills through the use of 'real' problems emphasizing the importance of contextualized learning through problem solving. Problem-based learning attaches weight to 'problem solving' as a source of knowledge and the method of instruction. Ochoa and Robinson (2005) remark that problem-based learning occasionally uses small-group and case methods to convey content, but that "contextualized learning is perhaps one of the most important underpinnings" (p. 11). Problem-based learning, as Bligh (1995) sees it, teaches the learner how to frame workplace experiences as a series of 'real' problems to be solved using knowledge and skills.

Implications for Workplace Adult Education and Training

When planning 'problem solving' pedagogies for adult education and training in the workplace, some critical questions regarding the assumptions and paradigms of problem solving may include the following (Baumgartner, 2003b; Fenwick & Parsons, 1997; Lee, 2003):

1. **About problem solving:**
    a. Are problem solving, the problem, and the learner separable from each other, or from social, cultural, or, historical contexts?
    b. Is problem solving the only way to look at the workplace?
    c. Is problem solving researchable and can it generate conclusive results?
2. **About the problem:**
    a. What are the characteristics that point to something being a problem from something that is normal?
    b. What is excluded and rendered invisible by this process?
    c. Is everything that is 'different' and unique in the workplace necessarily a problem?
    d. Is a problem something fixed, concrete, tangible, and definable that may be appropriated, controlled, dissected, and analyzed for learning?
    e. Is every problem knowable and resolvable?
    f. Does every problem have to be rehabilitated?
    g. Is every problem unique or are there preconceived problems with 'cookie-cutter' solutions?
    h. Are preconceived problems authentic for the workplace?
    i. Does labeling problems reduce difference in the workplace?
3. **About the learner:**
    a. Is the learner a detached knower, separate from time, place, social position, body, gender, or personal ties, as well as social, cultural, or historical context when undertaking problem solving?
    b. Is the learner completely free, detached, and independent from other workplace constraints, impediments, or communities to solve problems?
    c. Can a free, detached problem solver solve a problem at a distance or 'from the outside'?
    d. What motivates the learner to say something is a problem and something else is not?
    e. What are the characteristics of a 'good' problem solver and who gets to decide who is going to participate in solving the problem?
    f. Who decides what problem is worthy of resolution vs. what is normal, and how do they reach this decision?
    g. Is a problem conceived in the same way by everyone in the workplace?
    h. Is problem solving purely rational and cognitive, and unaffected by the learner's non-rational effects like emotions, values, or the subconscious mind?
    i. Is the 'individualized' outlook for problem solving in the workplace more representative of a specific segment of the population, namely those who are white, of European or Western descent, employed, better educated, and in leadership positions?

j. Is the 'individualized' outlook for problem solving in the workplace marginalizing groups who may not have the privilege to 'problem solve' in the workplace such as adults with disabilities, developmentally challenged adults, and adults with special needs; or women, people of color, working class adults, and immigrant learners?

The Act of 'Collaboration'

Lev Vygotsky (1896-1934) looks at learning from a social perspective. His sociocultural theory of learning asserts that social experience through language shapes and is inseparable from cognition (Jaramillo, 1996). Vygotsky emphasizes that socialization is learning. His stance, that there can be no learning without socialization through 'words,' is highly relevant for the purposes of this chapter.

The site of learning for Vygotsky is the group. Language and discourse, written or spoken, is one mode of representation available to the learner to share meaning and that the learner uses to communicate and negotiate meaning with the group. Hirtle (1996) further underlines this message:

Words, both spoken and written in a social context, mediate meaning that helps lead to the construction of concepts or knowledge. Words function as tools which help accomplish the work of learning—mobilization of knowledge. Speech is the first tool that culture provides ... to engage in collaborative thinking with others ... . (p. 91)

To better explicate the experience of learning through group membership, Vygotsky identifies a 'zone of proximal development':

... the distance between the actual developmental level as determined by independent problem solving and the level of potential development as determined through problem solving under adult guidance, or in collaboration with more capable peers ... (Tudge, 1990, p. 157)

Vygotsky gives credence to the social, the cultural, and the historical in group contexts for knowledge and instruction.

Small group learning generally refers to sustained activities or experiences for *particular kinds of problem solving*: processes that involve reflection, information-retrieval processes, and communication in group settings. Perceiving, interpreting, and facilitating group process associated with these dynamics and issues are critical to meaning-making, learning, and change. As Dirkx and Robinson (2004) conclude, "Group process represents a powerful context that contributes to how groups and individuals in these groups understand and make sense of the work of the group" (para. 4).

Similarly, cooperative learning, as observed by Vermette and Foote (2001), uses permanent teams to complete a task and/or learn subject matter in an interactive and interdependent manner; to create meaning; to allow learners to discuss and reflect upon important information; to take risks; to see from different perspectives, to engage in provocative thinking; and to reorganize content together. Akan (2005) reasons that for cooperative learning to occur, the group must have a common goal, and there must be accountability of both the group and its members. In this way, the group members must be made keenly aware of their interdependence and must actively interact with each other; learning must be monitored to see who needs assistance; and the group must include people of all abilities.

Implications for Workplace Adult Education and Training

When deciding on 'collaboration' for adult education and training in the workplace, some critical questions regarding the assumptions and paradigms of collaboration may include the following (Dirkx & Smith, 2004; Ochoa & Robinson, 2005):

1. **About collaboration:**
   a. Are collaboration, the group, and the learner separable from each other, or from social, cultural, or, historical contexts?
   b. Is collaboration the only way to look at group and teamwork in the workplace?
   c. Is collaboration researchable and can it generate conclusive results?
2. **About the group:**
   a. Is the group as concerned about the task as it is about the members?
   b. Is group membership indicative of certain behaviors expected from its members?
   c. What are the characteristics of a 'good' group?
   d. Do hierarchies and margins form in a group?
   e. Is a group something fixed, concrete, tangible, and definable that may be appropriated, controlled, dissected, and analyzed for learning?
   f. Does the group speak for all members with one voice?
3. **About the learner:**
   a. How does the learner decide what experiences are worthy of collaboration?
   b. Is collaboration a purely internal, rational, solitary, cognitive act for every learner?
   c. Is the learner a stable, unitary, and rational being who can be separated from social, cultural, and historical context when undertaking collaboration?
   d. Can the learner collaborate without being influenced by non-rational effects like emotions, values, the subconscious mind, or from social, cultural, or historical contexts?
   e. Who decides who gets to be a member of the group?
   f. What motivates the learner to join a group?

The Act of 'Participation'

John Dewey's (1859-1952) insistence that education is learning by 'doing' places great emphasis on 'experience' as pedagogy (Fenwick, 2001a). For Dewey, education and learning are organic and inductive; learning is active, social, and builds upon past experiences. In fact, the term 'social construction' is accredited to Dewey (Hirtle, 1996, p. 91).

Dewey exactly sums up the importance of using a continuum of 'action' to capture both constructivist epistemology and pedagogy because, as Dewey sees it, "If we see that knowing is not the act of an outside spectator but of a participator inside the natural and social scene, then the true object of knowledge resides in the consequences of directed action" (Phillips, 1995, p. 6). For learning to happen, Dewey notes that an experience must include two key dimensions: continuity and interaction (Fenwick, 2001a). Dewey values the mind as an active participant and creator of knowledge, not a passive recipient where learning results of meaningful reflection upon active experiences in social climates.

In the same way, situated learning is essentially a matter of learning from everyday life. What this means is that learning is contained *in* the experience; in a specific time and place; involving other learners, the environment, and to create meaning in a particular milieu. The kind of learning that is developed is then only applicable to similar situations as it is the unique result of a highly contextual social process encompassing ways of thinking, perceiving, and problem solving. Here, learning is not and cannot be separated from the world of action, but exists in complex, social situations and activities. Stein (1998) elaborates that "learning becomes a social process dependent upon transactions with others placed within a context that resembles as closely as possible the practice environment" (p. 3). Meaningful interaction comprises learning that emerges out of the experience.

Lave and Wenger (1991) introduce a more anthropological, community-centered focal point on learning: legitimate peripheral participation. Really, the discussion is no longer about 'participation' but 'co-participation,' because the cognitive boundaries that separate the individual from the community/environment in other constructivist interpretations of epistemologies and pedagogies are completely overridden by Lave and Wenger's (1991) social co-participation within 'community.' Co-participation in practice encompasses learning and understanding, "Thus knowing is interminably inventive and entwined with doing…The objective is to become a full participant in the community of practice, not to learn about the practice. The community itself defines what constitutes legitimate practice" (Fenwick, 2001a, p. 26).

The complete emersion of the individual in the community is a marked departure from the more individuated foci of other constructivist thinking. Whether reflection, problem solving, or collaborating, the locus of control for action remained with the individual and the individual mind. Here the individual becomes indistinguishable from community through co-participation with fellow members of the community, and the learning that transpires as a result of this emersion goes so far as to even shape the identity of the individual. Knowledge cannot be centered within individual learners according to legitimate peripheral participation. What Lave and Wenger (1991) propose is a very different bearing on constructivism:

Learning is a process that takes place in a participation framework, not in an individual mind. This means, among other things, that it is mediated by differences of perspective among the co-participants. It is the community, or at least those participating in the learning context, who 'learn' under this definition. Learning is, as it were, distributed among co-participants, not a one-person act. (Hanks, 1991, p. 15)

Through legitimate peripheral participation, Lave and Wenger (1991) offer a constitutive role for learning, learning for improvisation, emergent processes from actual interaction, the improbability of behavior based on prefabricated mental schema, learning as negotiated and strategized, and a socially constructed contribution to the community (Barab & Duffy, 2000; Hanks, 1991; Lave & Wenger, 1991). The co-participant is legitimized by the community, and in turn legitimizes and perpetuates the community. "The challenge, it would seem," says Hanks (1991), "is to rethink action in such a way that structure and process, mental representation and skillful execution, interpenetrate one another profoundly" (p. 16).

Implications in the Workplace for Adult Education and Training

When planning 'participation' pedagogies for adult education and training in the workplace, some critical questions regarding the assumptions and paradigms of participation may include the following (Fenwick, 2001a):

1. **About participation:**
    a. Are participation, community, and the learner inseparable from each other or from social, cultural, or historical contexts?

  b. Is participation so dependent on community and context that whatever is learned is impossible to transfer to other communities and contexts?
  c. Is participation so concerned with context that it ignores the kind of knowledge being acquired and the ways the material needs to be is engaged?
  d. Is participation in community researchable and can it generate conclusive results?
2. **About the community:**
  a. Is the 'universal' outlook of participating in a community good for everyone?
  b. Is the 'universal' outlook, that every community has a center, representative of a specific segment of the population, namely those who are white, of European or Western descent, employed, better educated, and in leadership positions?
  c. Does the 'universal' outlook of communities with centers ignore the existence of resistance in communities that may be unjust?
  d. Does the 'universal' outlook of communities with centers create marginalized groups which include adults with disabilities, developmentally challenged adults, and adults with special needs; or women, people of color, working class adults, and immigrant learners?
  e. Who decides which learners get to participate in the community and what is knowledge in the community?
  f. Is the learner's identity solely defined by the community; what about multiple identities; what if learners belong to more than one community?
  g. What are the characteristics of a 'good' community?

Based on this 'continuum of learning,' a framework in the next section can be developed in a much more comprehensive manner.

SOLUTIONS AND RECOMMENDATIONS

This section makes recommendations on how decision makers may start to think about and plan for the challenges that lie ahead for workplace adult education and training in their organizations when contemplating e-learning solutions that use constructivist pedagogy. The following guiding principles formulate an alternate, sociocultural paradigm for decision making which goes beyond cost, hardware, and software, and might prove handy for decision makers when evaluating their e-learning options:

1. A global workforce undergoing workplace transformation with differing workplace adult education and training needs will not be solved through technology alone. A 'homogenized' solution based on technology offered by vendors will likely exclude the social and cultural needs of a global workforce in the process. *E-learning decision makers must include e-learning solutions as part of a broader strategic vision of workplace adult education and training given the global age.*
2. Constructivism and constructivist thinking deal with epistemology, pedagogy, experience, context, and a wide range of views and stances and interpretations on each of these notions. A 'normalized' software solution offered by vendors presumes that such negotiation and interpretation are no longer characteristic of constructivism and that a divergence of opinion no longer exists. Such a solution will likely alienate certain parts of the workforce while privileging others.

E-learning decision makers must be able to identify the four constructivist learning actions to lay out a mapping of possible issues, controversies, and problems that lie ahead of any implementation of constructivist-based e-learning. They must also be able to tell the difference between solutions that are truly constructivist from those that only 'claim' to be.

3. Workplace transformation provides the context for e-learning. A 'universalized' solution offered by vendors is likely to have limited use and restricted relevance to the complexity of workplace transformation and will not likely to be scalable or customizable to meet the future of changing workforce composition. *E-learning decision makers must facilitate an integrative 'snapshot' of the social and cultural implications e-learning holds for workplace transformation to make it relevant and useful in the long term. They must go beyond cost, hardware, and software to look at culture, context, and community.*

A new plan for e-learning interventions comes to fruition to truly assess whether learning needs in the workplace will be met to accomplish workplace priorities, accountabilities, performance, and productivity.

FUTURE TRENDS

More research is needed on e-learning from a sociocultural perspective. A sociocultural critique remains imperative as the complexity of the workplace multiplies exponentially on a global scale, and the social and cultural begin to play out more prominently through a growing, global workforce that is socially and culturally heterogeneous. If the focus in the future continues towards a greater appreciation of learning in the workplace, and not just the technology, the sociocultural deficiencies of e-learning leave much to be desired when it comes to a global workforce and workplace adult education and training.

If the dominant point of view for research and decision making in the future remains a technological one (i.e., hardware and software), then decision makers may be satisfied as long as the technology performs well. The sociocultural criticisms of e-learning may go unnoticed. The divergence of the technological and sociocultural positions on e-learning and the workplace, however, will factor significantly in the future as the dollars of investment continue to mount and workplaces continue to grapple with an increasingly global workforce that may not be meeting workplace priorities, accountabilities, performance, and productivity despite heavy investment in workplace adult education and training through e-learning technology.

An additional focus of research in the future must also be the role of adult education and training in the workplace. Future trends in the workforce point to greater workplace transformation, an aging workforce, greater immigration, the changing roles of women, greater diversity, pluralism, multilingualism, the presence of workers with differing ability, and a greater prominence of minorities (Bierema, 2002). Still, workplace adult education and training seems ill-prepared. Other ways of knowing that are not cognitively or psychologically focused are equally legitimate, but are made invisible with the unproblematic adoption of European and Western theories of learning on the international stage for workplace adult education and training (Alfred, 2002a; Birden, 2003; Flowers, 2003; Folely, 2004; Lee & Sheared, 2002; McLean, 2006). What will become urgent from a sociocultural position is the role of workplace adult education and training for a global workforce and how this will all play out through the burgeoning dominance of e-learning.

CONCLUSION

This chapter examines the impact of e-learning on the global workforce, and the social and cultural implications for workplace adult education and training. The focal point here is on e-learning in the workplace given a complex and rich context consisting of: globalization, technological innovation, the knowledge-based economy, workplace transformation, the altering nature or work, the shifting composition of the workforce, and constructivist-based knowledge and instruction. Yet present-day research on e-learning emphasizes hardware, software, and the mechanics of costs and technology.

Workplace transformation is resulting in work that is becoming more homogenized and normalized with an idealized, stylized, and reincarnated sense of employees as 'knowledge workers.' On the contrary, the workforce is becoming more socially and culturally diverse where workers are faced with underemployment, flexibility in the workplace, and the mass conversion of full-time work to contract, part-time, or contingent labor. Yet workplace adult education principles and practices are universalized in attitude, and favor European and Western ways of knowing and learning, with economic motivators and outcomes, despite the major supply of labor in the future being non-European and non-Western.

For e-learning to be valuable to a global workforce that is socially and culturally diverse, a social and cultural standpoint is vital to counter the current tendencies in e-learning and the workplace towards homogenization, normalization, and universalization. Practitioners in workplace adult education and training can influence this goal by acknowledging, identifying, and applying instructional interventions that evolve and authentically speak to a dynamic cohort of employees with a progressively complex profile of learning needs.

REFERENCES

Akan, O. H. (2005). Concrescent conversations: Generating a cooperative learning experience in principles of management—a postmodern analysis. *Journal of Education for Business, 80*(4), 214-217.

Alfred, M. V. (2002a). Linking the personal and the social for a more critical democratic adult education. *New Directions for Adult and Continuing Education, 96*, 89-95.

Alfred, M. V. (2002b). The promise of sociocultural theory in democratizing adult education. *New Directions for Adult and Continuing Education, 96*, 3-13.

Ally, M. (2004). Foundations of educational theory for online learning. In T. Anderson & F. Elloumi (Eds.), *Theory and practice of online learning* (pp. 3-31). Athabasca, AB: Athabasca University.

Amutabi, M., Jackson, K., Korsgaard, O., Murphy, P., Quiroz Martin, T., & Walters, S. (1997). Introduction. In S. Walters (Ed.), *Globalization, adult education and training: Impacts and issues* (pp. 1-12). London: Zed Books.

Anderson, M. D. (2006). *Overview of Piaget's theory of development.* Retrieved January 25, 2006, from http://facultyWeb.cortland.edu/andersmd/PIAGET/PIAGET.HTML

Anderson, T. (2004). Toward a theory of online learning. In T. Anderson & F. Elloumi (Eds.), *Theory and practice of online learning* (pp. 3-31). Athabasca, AB: Athabasca University.

Andrade, J., Ares, J., García, R., Rodríguez, S., Seoane, M., & Suárez, S. (2005). A pedagogical overview on e-learning. In R. Khosla (Ed.), *KES 2005* (pp. 658-664). Berlin: Springer-Verlag Berlin Heidelberg.

Baldwin-Evans, K. (2004). Employees and e-learning: What do the end-users think? *Industrial and Commercial Training, 36*(7), 269-274.

Barab, S. A., & Duffy, T. M. (2000). From practice fields to communities of practice. In D. H. Jonassen & S. M. Land (Eds.), *Theoretical foundations of learning environments* (pp. 25-55). Mahwah, NJ: Lawrence Erlbaum.

Baumgartner, L. M. (2003a). Adult learning theory: The basics. In L.M. Baumgartner, M.-Y. Lee, S. Birden, & D. Flowers (Eds.), *Adult learning theory: A primer* (pp. 1-4). Columbus, OH: Ohio State University.

Baumgartner, L. M. (2003b). Andragogy: A foundational theory/set of assumptions. In L. M. Baumgartner, M.-Y. Lee, S. Birden, & D. Flowers (Eds.), *Adult learning theory: A primer* (pp. 5-10). Columbus, OH: Ohio State University.

Baumgartner, L. M. (2003c). Self-directed learning: A goal, process, and personal attribute. In L. M. Baumgartner, M.-Y. Lee, S. Birden, & D. Flowers (Eds.), *Adult learning theory: A primer* (pp. 23-28). Columbus, OH: Ohio State University.

Bednar, A. K., Cunningham, D., Duffy, T. M., & Perry, D. J. (1992). Theory into practice: How do we link? In T. M. Duffy & D. H. Jonassen (Eds.), *Constructivism and the technology of instruction: A conversation* (pp. 17-34). Mahwah, NJ: Lawrence Erlbaum.

Bierema, L. L. (2002). The sociocultural contexts of learning in the workplace. *New Directions for Adult and Continuing Education, 96*, 69-78.

Birden, S. (2003). Critical and postmodern challenges for education. In L. M. Baumgartner, M.-Y. Lee, S. Birden, & D. Flowers (Eds.), *Adult learning theory: A primer* (pp. 29-34). Columbus, OH: Ohio State University.

Bligh, J. (1995). Problem-based learning in medicine: An introduction. *Postgraduate Medical Journal, 71*(836), 323-326.

Bruner, J. S. (1973). Beyond the information given: Studies in the psychology of knowing. In J. M. Anglin (Ed.), *Education* (pp. 401-425). New York: W. W. Norton & Company.

Carliner, S. (2003). The economy-training op connection. *T + D, 57*(8), 65-66.

Castells, M. (1996). *The rise of the network society. The information age: Economy, society, and culture* (Vol. 1). Oxford, UK: Blackwell.

Conceicao, S. (2002). The sociocultural implications of learning and teaching in cyberspace. *New Directions for Adult and Continuing Education, 96*, 37-45.

DeRouin, R. E., Fritzsche, B. A., & Salas, E. (2005). E-learning in organizations. *Journal of Management, 31*(6), 920-940.

Dirkx, J. M., & Smith, R. O. (2004, October 6-8). *Learning to see through the invisible: The problem of process in online* collaborative learning. In *Proceedings of the Midwest Research-to Practice Conference in Adult, Continuing, and Community Education,* Indianapolis, IN.

Drucker, P. F. (1959). *Landmarks of tomorrow: A report on the new 'post-modern' world.* New York: Harper.

Duffy, T. M., & Jonassen, D. H. (1992). Constructivism: New implications for instructional technology. In T. M. Duffy & D. H. Jonassen (Eds.), *Constructivism and the technology of instruction: A conversation* (pp. 1-16). Mahwah, NJ: Lawrence Erlbaum.

Felder, R. M., & Brent, R. (2005). Understanding student differences. *Journal of Engineering Education, 94*(1), 57-72

Fenwick, T. (2001a). Experiential learning: A theoretical critique explored through five perspectives. *Information series no. 385.* Retrieved March 4, 2006, from http://www.eric.ed.gov/ERICDocs/data/ericdocs2/content_storage_01/0000000b/80/26/30/a1.pdf

Fenwick, T. (2001b). Tides of change: New themes and questions in workplace learning. *New Directions for Adult and Continuing Education, 92*, 3-17.

Fenwick, T., & Parsons, J. (1997). *A critical investigation of the problems with problem-based learning.* Retrieved March 4, 2006, from http://www.eric.ed.gov/ERICDocs/data/ericdocs2/content_storage_01/0000000b/80/26/d2/a2.pdf

Flowers, D. (2003). An afrocentric view of adult learning theory. In L. M. Baumgartner, M.-Y. Lee, S. Birden, & D. Flowers (Eds.), *Adult learning theory: A primer* (pp. 1-4). Columbus, OH: Ohio State University.

Folely, G. (Ed.). (2004). *Dimensions of adult learning: Adult education and training in a global era.* Berkshire, UK: Open University Press.

Fosnot, C. T. (1996). Constructivism: A psychological theory of learning. In C. T. Fosnot (Ed.), *Constructivism: Theory, perspectives, and practice* (pp. 8-33). New York: Teachers College Press.

Gasco, J. L., Llopis, J., & Gonzalez, M. R. (2004). The use of information technology in training human resources: An e-learning case study. *Journal of European Industrial Training, 28*(5), 370-382.

Glasersfeld, E. V. (1996). Introduction: Aspects of constructivism. In C. T. Fosnot (Ed.), *Constructivism: Theory, perspectives, and practice* (pp. 3-7). New York: Teachers College Press.

Gulati, S. (2004, April). Constructivism and emerging online learning pedagogy: A discussion for formal to acknowledge and promote the informal. In *Proceedings of the Conference of the Universities Association for Continuing Education,* London.

Gustavsson, B. (1997). Life-long learning reconsidered. In S. Walters (Ed.), *Globalization, adult education and training: Impacts and issues* (pp. 237-249). London: Zed Books.

Hanks, W. F. (1991). Forward. In J. Lave & E. Wenger (Eds.), *Situated learning: Legitimate peripheral participation* (pp. 13-24). New York: Cambridge University Press.

Harun, M. H. (2002). Integrating e-learning into the workplace. *Internet and Higher Education, 4*, 301-310.

Hirtle, J. S. P. (1996). Social constructivism. *English Journal, 85*(1), 91-92.

Honey, P. (2001). E-learning: A performance appraisal and some suggestions for improvement. *The Learning Organization, 8*(5), 200-202.

Illeris, K. (2003). Workplace learning and learning theory. *Journal of Workplace Learning, 15*(4), 167-178.

Industry Canada. (2002a). *Achieving excellence: Investing in people, knowledge and opportunity.* Retrieved June 25, 2004, from http://www.innovationstrategy.gc.ca/gol/innovation/site.nsf/vDownload/Page_PDF/$file/achieving.pdf.

Industry Canada. (2002b). *Knowledge matters: Skills and learning for Canadians.* Retrieved March 2, 2004, from http://www11.sdc.gc.ca/sl-ca/doc/knowledge.pdf

International Monetary Fund (IMF). (2002). What is globalization? *Globalization: Threat or opportunity?* Retrieved July 28, 2006, from http://www.imf.org/external/np/exr/ib/2000/041200.htm

International Organization for Standardization. (2006). ISO and world trade. *Overview of the ISO system.* Retrieved August 15, 2006, from http://www.iso.org/iso/en/aboutiso/introduction/index.html#two

Jaramillo, J. A. (1996). Vygotsky's sociocultural theory and contributions to the development of constructivist curricula. *Education, 117*(1), 133-140.

Jarvis, P. (2000). Globalisation, the learning society and comparative education. *Comparative Education, 36*(3), 343-355.

Johansen, B.-C. P., & McLean, G. N. (2006). Worldviews of adult learning in the workplace: A core concept in human resource development. *Advances in Developing Human Resources, 8*(3), 321-328.

Korsgaard, O. (1997). The impact of globalization on adult education. In S. Walters (Ed.), *Globalization, adult education and training: Impacts and issues* (pp. 15-26). London: Zed Books.

Kress, G. (2003). *Literacy in the new media age*. New York: Routledge.

Lave, J., & Wenger, E. (1991). *Situated learning: Legitimate peripheral participation*. New York: Cambridge University Press.

Lee, M.-Y. (2003). A critical analysis of andragogy: The perspective of foreign-born adult learners. In L. M. Baumgartner, M.-Y. Lee, S. Birden, & D. Flowers (Eds.), *Adult learning theory: A primer* (pp. 11-16). Columbus, OH: Ohio State University.

Lee, M.-Y., & Sheared, V. (2002). Socialization and immigrant students' learning in adult education programs. *New Directions for Adult and Continuing Education, 96*, 27-36.

Livingstone, D. W. (2001). Expanding notions of work and learning: Profiles of latent power. *New Directions for Adult and Continuing Education, 92*, 19-30.

McLean, G. N. (2006). Rethinking adult learning in the workplace. *Advances in Developing Human Resources, 8*(3), 416-423.

Moore, J. (2005). Recognising and questioning the epistemological basis of educational psychology practice. *Educational Psychology in Practice, 21*(2), 103-116.

Munro, R. A., & Rice-Munro, E. J. (2004). Learning styles, teaching approaches, and technology. *The Journal for Quality and Participation, 27*(1), 26-32.

Newton, D., Hase, S., & Ellis, A. (2002). Effective implementation of online learning: A case study of the Queensland mining industry. *Journal of Workplace Learning, 14*(4), 156-165.

Null, J. W. (2004). Is constructivism traditional? Historical and practical perspectives on a popular advocacy. *The Educational Forum, 68*(2), 180-188.

Ochoa, T. A., & Robinson, J. M. (2005). Revisiting group consensus: Collaborative learning dynamics during problem-based learning activity in education. *Teacher Education and Special Education, 28*(1), 10-20.

Organization for Economic Cooperation and Development (OECD). (2001). *Education policy analysis 2001—competencies for the knowledge economy*. Retrieved August 21, 2004, from http://www.oecd.org/dataoecd/42/25/1842070.pdf

Phillips, D. C. (1995). The good, the bad, and the ugly: The many faces of constructivism. *Educational Researcher, 24*(7), 5-12.

Pollitt, D. (2005a). E-learning connects Cable & Wireless with big cost savings. *Human Resource Management International Digest, 13*(1), 19-20.

Pollitt, D. (2005b). E-learning delivers management skills to Ford's North American dealers. *Training & Management Development Methods, 19*, 6.39-36.42.

Pollitt, D. (2005c). ScottishPower goes DIY with e-learning. *Training & Management Development Methods, 19*, 6.33-36.37.

Remtulla, K. (2005). *Literacy in the new media age*. Retrieved January 3, 2006, from http://www.com.washington.edu/rccs/bookinfo.asp?ReviewID=366&BookID=295

Robertson, R. (1992). *Globalization*. London: Sage.

Salas, E., Kosarzycki, M. P., Burke, C. S., Fiore, S. M., & Stone, D. L. (2002). Emerging themes in distance learning research and practice: Some food for thought. *International Journal of Management Reviews, 4*(2), 135-153.

Sirois, V. (2005). *The Canadian education and training in*dustry. Retrieved September 8, 2005, from http://strategis.ic.gc.ca/epic/internet/incet-ecf.nsf/en/ok01770e.html

Spencer, B. (2001). Changing questions of workplace learning researchers. *New Directions for Adult and Continuing Education, 92*, 31-40.

Stein, D. (1998). *Situated learning in adult education*. Retrieved March 4, 2006, from http://www.eric.ed.gov/ERICDocs/data/ericdocs2/content_storage_01/0000000b/80/2a/2b/04.pdf

Tudge, J. (1990). Vygotsky, the zone of proximal development, and peer collaboration: Implications for classroom practice. In L. C. Moll (Ed.), *Vygotsky and education: Instructional implications and applications of socio-historical psychology* (pp. 155-172). New York: Cambridge University Press.

Vermette, P., & Foote, C. (2001). Constructivist philosophy and cooperative learning practice: Toward integration and reconciliation in secondary classrooms. *American Secondary Education, 30*(1), 26-37.

Chapter XVIII
The Impact of Sociocultural Factors in Multicultural Communication Environments:
A Case Example from an Australian University's Provision of Distance Education in the Global Classroom

Angela T. Ragusa
Charles Sturt University, Australia

ABSTRACT

Changes in the availability and quality of communication technology have revolutionized, and fundamentally altered, learning environments. As citizens of the "Information Age," the breadth and impact of global communication are triggering unprecedented transformation of social structures and institutions. This chapter explores the impact of commodification on education when institutions of higher education sell knowledge as a commercial good. The contemporary phenomenon of distance education is increasingly offered and purchased by an international market which experiences heightened pressure for standardization from the global citizens it serves. It is argued here that technological changes necessitate reevaluation of communication processes, discursive practices, and organizational policies. To stay competitive and produce quality products for increasingly international audiences, institutions must create well-articulated policies. By providing insight on the impact multiple sociocultural and communicative norms have on virtual communication, this research uses qualitative discursive analysis of case examples to examine how variance in the structure and delivery of virtual communication environments at a leading distance education university in Australia affects student satisfaction, perception, and learning outcomes. Whereas previous research fails to include a theoretical or conceptual framework, this work draws upon interdisciplinary work from the fields of sociology, education, and science and technology studies. How "cyberspace" changes interaction rituals, masks cultural norms, and alters entrenched social expectations by creating new sensitivities is discussed, along with the ramifications of variation in technological availability, competence, and expectations in global classrooms. In sum, ideas for informing change in policy, administration, and the delivery of distance education and virtual communication in global environments are discussed to equip leaders and participants with skills to foster effective communicative and interaction strategies.

INTRODUCTION

Analyzing the "Australian experience" with distance education[1] (DE), this chapter uses qualitative data to show how the type and design of virtual learning environments influence both the process and purpose of communication outcomes in global learning environments. DE is a global phenomenon with "over 130 countries developing or offering distance courses, many of them based on new information and communication technologies (ICTs)" (Shields, Gil-Egui, & Stewart, 2004, p. 120). However, the "global marketplace for e-learning products and services" varies widely among and within countries, courses offered, and technologies (Bowles, 2004). As multiple sectors (education, corporate, government) are involved in its development and delivery, inclusion of culture as a primary foci of analysis remains limited (Bowles, 2004; Monolescu, Schifter, & Greenwood, 2004; Palloff & Pratt, 2001; Brooks, Nolan, & Gallagher, 2001). Failing to prioritize cultural analysis may be costly and ineffective in communicative environments with multinational participants. Applying secondary data analysis techniques using a sociological and science and technology studies (STS) lens, this chapter informs existing literature not only by prioritizing culture, but moreover overcomes the methodological challenge facing international studies (Au-Yeung, Ha, & Au, 2004; Hawkey, 2004) of isolating "pure e-learning" (Bowles, 2004) environments in higher education to study virtual interaction without the added presence of face-to-face interaction. According to situated learning theory, knowledge cannot be removed from learning context and environment. Hence, not only are local norms set within a larger, global community, but a host of interaction-based, communicative norms are called into question as participants negotiate the boundaries of electronic communication dynamics.

Critical analysis of such dynamics provides insight into the symbiotic relationship between culture and technology in affecting communication. As such, the principal objective of this chapter is to enable better-informed decision making and the refinement of virtual communication (VC), in learning and other environments, by offering in-depth examples of how cultural norms and interactions can, and do, impact VC. By exploring how structure impacts virtual dialogue, the chapter asks how normative assumptions affect the negotiation of communication rituals based on situational meanings. Anticipated to be of particular interest to academic researchers, teachers, managers, and administrators of cross-cultural online learning environments, as well as business leaders working in international VC environments, this chapter uses case examples to demonstrate the strategic value of effective management; offers pragmatic steps that managers, teachers, and participants can take to maximize the virtual communicative experience; and asks all to reflect upon how sociocultural experiences interact with technologies to affect successful VC delivery.

BACKGROUND

Education as a Setting for Virtual Communication

Education, as an institution, is a field complete with theories, methodologies, and practices which can lead to the dichotomization of knowledge production into either positivism or social constructivism. Classical educational theorists such as Dewey (1938) and Piaget (1927) have argued that learning can be understood as a process of *praxis,* or doing. *Praxis* stands in contrast with the ancient *tabula rasa*/ "blank slate" (Brooks & Brooks, 1993, in Palloff & Pratt, 2001) notion of students being empty blackboards waiting to be written upon. In assessing the use of multimedia tools to build collaborative virtual[2] communities, Goldman-Segall (1992, p. 260) offers a study contradictory to "the traditional school curriculum [that] is based on the attendance of empiricism, rationalism, and pragmatism" (Scheffler, 1965). Goldman-Segall's ethnographic research demonstrates how multimedia educational tools not only are mechanisms for cultural reflection/taking "the other" perspective, but moreover reveal what "the virtual community feeds back into the real community" (p. 293). The outcome is a negotiated understanding reflexive of participants' intentions and actions as collaborators in an educational project. Knowledge is constructed as an interaction between learner and environment, which subsequently reconfigures both (Semple, 2000). Applying a constructivist epistemology (Vygotsky, 1986), learning can be understood as a dialectical process infused by culture where meaning is constructed from a negotiation of cognitive thoughts within specific environments. Termed "electronic pedagogy" (Palloff & Pratt, 2001), this alternative view argues that when individuals work independently on subject matter and then engage in reflexive discussions about what fellow participants contributed, critical thinking skills are exercised. This chapter presents research supportive of Goldman-Segall's observation that multimedia, and particularly VC, enable individuals to reflect and view world events from others' perspectives in ways unique and different from face-to-face interactions.

In the past decade, much has been written about the pros and cons of DE from an educational framework. The main contributors are from education, technology, and administration disciplines. Additional disciplines beginning to address core cultural dimensions include human-computer interaction, usability, and technical communication, and the field of rhetoric, which now includes discussions of culture and online issues. However, each admittedly has been slow to contribute in meaningful ways. Although researchers in these paradigms are well equipped to analyze the effectiveness of DE, discussion of cultural issues is less rigorous and sparse, even when the imperative for including culture is acknowledged (Jorgensen, 2002). Jorgensen's (2002) review suggests that issues surrounding successful "community building" are a crucial component of culture in online communities (Borthick & Jones, 2000; Wegerif, 1998; Davis, 1997). Culture has also attributed great importance to the building of social capital within sociological theory (Putnam, 2005; Coleman, 1988) and society's general functioning (Durkheim, 1951; Merton, 1968). To understand how communities[3] function successfully, some traditional functionalists (Durkheim, 1951; Parsons, 1954) emphasize the establishment of common social norms. While functionalism may be useful to understand macrostructural organization of communities, those connecting macro- and micro-level analysis (Bourdieu, 1977; Giddens, 1979), and particularly some feminist analyses (Collins, 1990; Smith, 1990), reveal that power dynamics differ by unit of analysis. Within the education literature, questions of how to create the best virtual community reflect the challenge of how to create a social presence in an online environment (Lynch, 2002).

Australian market research attributes variation in cultural attitudes, communication, infrastructure, and government policy (Bowles, 2004) with responsibility for the current nature and climate of DE. Despite the vast majority of research being imported to Australia from the U.S. and the UK, there exists an absence of reliable Australian data on electronic learning (otherwise known as "e-learning") trends, as governmental indicators and Bowles (2004) claim. These implications challenge the establishment of benchmarks and expectations as "the exponential growth of contacts and networks among educators in an increasingly multinational market exists in a context of local and global stratification" (Ragusa & Atweh, 2004). As the example ethnomathematics reveals, "Research has failed to develop an ability to produce knowledge about people from within. Moreover, the knowledges generated have failed to assist in the transformation of reality, leading not to social change in justice but rather confirmation of the status quo" (Ragusa & Atweh, 2003). According to theories within STS, "Developments in transportation communication as well as the crisis of world ecology have created the so-called global society" (Hess, 1995, p. vii). In turn, the "people of diverse nationalities find themselves in increasing contact with each other. In many countries women, underrepresented ethnic groups, gays and lesbians, and other previously excluded groups have gained a greater voice" which results in "public debates on diversity, pluralism, oppression, exclusion, inclusion, colonialisms, identity politics and other issues that can be glossed as multicultural" (Hess, 1995, p. vii).

Social-psychological research shows variation exists in comfort and preference for VC (Lynch, 2004). Introverts tend to be more comfortable with electronic communication and reflection than extroverts (Palloff & Pratt, 1999). By applying this knowledge to a cultural critique, a Western/non-Western divide emerges. For example, research in Hong Kong shows students prefer to passively receive instruction rather than partake in active pedagogy heralded by new technology and distance learning communication strategies of the 21st century (Au-Yeung et al., 2004). It is crucial to be cognizant of such cultural variation, especially with increasing student numbers and class size.

The "Australian experience" with DE shows the limitation of research and recommendations which fail to include (Monolescu et al., 2004; Palloff & Pratt, 2001; Brooks et al., 2001) or give cursory attention to (Albalooshi, 2003; Lynch, 2002, 2004), culture as a fundamental variable of focus. This is not acceptable in increasingly global educational environments, of which DE perhaps leads the way by containing potentially the highest levels of multinational participants. Lynch (2002) cites the impact of cultural values which "are reflected in informal rules and reward structure of the organization" (p. 71) at the bottom of the list of Sherry and Wilson's (1997) indicators of transformative communication. However, as this chapter reveals, ignoring and/or sidelining the impact of culture on VC environments may be both costly and damaging.

Distance Education in Australia

In 2003, the Australian National Training Authority published two volumes detailing the expectations and delivery of online/computer-based education in Australia. As part of a strategic plan to meet national goals for delivering flexible learning educational products, the ability to meet policy objectives remains hampered by the lack of procedures for data collection, institutional variation, and limited coordination between states and the federal government. Changes in the delivery of education are forewarned by the former Minister for Education, Science and Training, Brendan Nelson MP, as crucial:

Globalization, massification of higher education, a revolution in communications and the need for lifelong learning, leave Australian universities nowhere to hide from the winds of change. (Nelson, 2005)

However, this call for change is at least 13 years old. In 1992, the Minister for Employment, Education and Training, Kim Beasley, Member of Parliament, commissioned a report which provided "formal advise from the Higher Education Council [HEC] on DE in Australia," according to Peter Laver's letter to the Minister. HEC wrote: "The potential is there...both for use of DE materials and for alternative modes of delivery ... the council's preferred position is consistent with the consultants: mixed mode should refer to flexible learning arrangements that can be made available to all students in all institutions" (HEC, 1992, p. 5). However, although "the potential exists technically in Australia for every higher education student to work in mixed mode format...for DE ... usage is still not great. This wider use can be attributed more to enterprising teachers then to the influence of DECs [DE Centers]" (HEC, 1992, p. 6). As articulated by the federal government, the problem lies not within technological limitations but within cultural and business norms. From the cultural side:

One major problem is cultural. Australian material has generally been produced for Australian cultural conditions and for use in Australian universities ... simply making courseware available to an overseas user on a 'take it or leave it' basis, leaving unattended a host of difficulties ranging from aptness of content to the cultural appropriateness of particular forms of examination questioning, could not suffice as an approach in the medium or long-term. (HEC, 1992, p. 7)

Clearly, cultural and business norms overlap. This is evident by the government's following statement:

The other problem is the current fragmentary approach to the business enterprise of delivering higher education overseas. Although the start-up costs are likely to be heavy, the prospective return is great for institutions and for Australia, both in terms of fees won and in terms of generating an ethos on campus towards more flexible approaches to learning. Yet, one of the certainties that Australian institutions must face is competition from other (Asian) nations already active in developing higher education services for delivery overseas. (HEC, 1992, p. 8)

Although the report may seem philanthropic in its discussion of culture, further elaboration negates construing the Australian government's interest as anything but economic:

To be sensitive and responsive to the cultural needs of overseas students, who are to be engaged through DE materials and modes of delivery, may make the difference between developing an overseas market worth a few million dollars annually and a truly significant enterprise. (p. 8)

Hence, technological solutions alone are problematic.[4] For example, research on global collaborations among mathematics educators reveals:

Increased technological capability does not translate into changes in the equitability of global knowledge sharing practices. Even as computer networks facilitate collaborative research, those from the periphery 'will experience pressure against working in native languages, or on questions different from those attracting attention in the main centers. They will be measured against their peers in the centers, not against those in their own institution or region' (Gibbons et al., p. 131). Ultimately, 'globalization destroys local cultures and organizations' (Gibbons et al., p. 131) and fosters standardization over plurality, contributing towards Western imperialism. (Ragusa & Atweh, 2003)

Increasingly, when it comes to knowledge production, global interdependence among developed countries is total, while developing countries are left with "starvation and handouts from the developed parts of the world" (Winchester, 2005, p. 1). Furthermore, although we may like to think the Internet enables us to live in a "global village," nation-state boundaries and ethnic prejudices, not to mention global disparities in world indicator statistics on health, mortality, education, wealth, and so forth (The World Bank, 2005), can bring firsthand accounts of some very large social problems right into your "own backyard." This is particularly likely when global citizens come together to learn, as "globalization of the world's economies is leading to increased emphasis on internationalization of the curriculum" (Barjis, 2003).

Increasingly mobile and transient societies, fuelled by knowledge-overload in the Information Age, require new methods of communication. International competition, technological proficiency, multiple careers in renewed emphasis on productivity and quality are realities industries and businesses worldwide face (Barjis, 2003). Technology has irrevocably altered business models and policies, and higher education is no exception, in Australia and worldwide (Stein, 2003). VC presents an unprecedented opportunity for learning, dialogue, and social change. As we journey into the virtual horizon, however, it is necessary to take practical steps as businessmen/women, politicians, educators, technocrats, and general voyagers, to prepare for the journey with knowledge and cultural sensitivities of our fellow travelers, our virtual environment, and ourselves.

Case Study: Secondary Data Analysis of Virtual Communication

In Western countries, such as America and Australia, student evaluation typically takes the form of quantitative student questionnaires distributed either electronically or via the postal service. In Asia (Au-Yeung et al., 2004) and the UK (Hawkey, 2004), even when case study methodology is applied to assess collaborative online learning communities, analysis remains quantitative. Critical analysis of quantitative studies reveals quantitative surveys' subjective recall of past events can be problematic.[5] Another problematic methodological issue of international studies (Au-Yeung et al., 2004; Hawkey, 2004) of VC environments in higher education is their impossibility to analyze communication that never entails face-to-face interaction. This complexity is also evident within Australia:

'Pure' e-learning delivery—that is, learning that relies entirely on information and communication technologies—can be hard to isolate, as the technology is often used to enable other forms of communication, such as face-to-face meetings between learners and facilitators, video or telephonic conferencing or exchange of materials.' The National Centre for Vocational Education and Research believes that pure e-learning is rare in Australia. (Bowles, 2004, pp. 25-26)

This situation makes the current study unique because the only formal interactions students had were virtual, mediated by a computer. The present study enhances the literature by addressing this situational challenge, in addition to foregrounding culture. Qualitative, secondary data/discourse analysis conducted via a case study[6] approach is used to analyze communicative situations in my teaching experiences with two large DE subjects over two consecutive six-month periods in 2005-2006 at a rural Australian university[7] known for its leadership as a DE provider. This research adds to the understanding of how structure impacts the effectiveness of virtual dialogue. Moreover, it provides insight and strategic ideas regarding the development of successful VC in global environments.

Environment and Demographics

The structure of the two learning environments was similar in the provision of material (everyone received a hardcopy DE "package" containing a subject outline, reading book, and study guide), yet differed greatly in level of VC required. "VG1" (Virtual Group 1) contained nearly 300 participants and "VG2" approximately 140. Both contained individuals from across Australia and abroad, representing a range of ethnicities, nationalities, ages, disabilities, academic aptitudes, and life situations.

However, unlike VG2, which was led using a constructivist pedagogy, understanding the process of education as a socially constructed, reflexive practice characterized by collaboration, group projects, and praxis (Johnson & Johnson, 1994; Schon, 1991), VG1 contained only one asynchronous[8] general virtual forum which participants were asked to visit weekly for announcements. Students were left to be individual distance learners, with the added dimension of having access to the voluntary, general forum. In contrast, VG2 participated in multiple asynchronous electronic forums: a general forum for announcements, working group sub-forums for group virtual presentation preparation, and topic sub-forums for student discussion leadership. All students in VG2 were required to participate weekly in virtual discussions run by group leaders.

Outcomes

The constructivist and collaborative learning environment of VG2 is upheld by some as a key to virtual success. Jorgensen's (2002) literature review lends support to this conception. "When students are actively involved in collaborative (group) learning online, the outcomes can be as good as or better than those for traditional classes, but when individuals are simply receiving posted material and sending back individual work, the results are poorer then [sic] in traditional classrooms" (Jorgensen, 2002, p. 9). This quote epitomizes the difference between the two subjects. While both are defined equally as "DE," the correspondence[9] learning experience of VG1, supplemented by virtual chat and announcements, revealed minimal student satisfaction. In contrast, those in VG2 reported much learning of both subject content and life skills, such as team management.

Despite variation in the quantity and quality of VC, both cohorts expressed learning resulted from their electronic dialogue. Although VG1 was active on the general forum (1,540 postings for 335 members and 103,802 "distinct reads"), its designation as an auxiliary learning tool resulted in little dialogue relating to subject content. Despite formal encouragement, rather than use the electronic forum to ask specific content-related questions, VG1 member primarily used it to vent grievances and assuage discontent/frustration with their learning experiences. This does not mean learning did not ensue. To the contrary, students learned much about the politics of communication using (post)modern technology. In contrast, VG2 produced 765 postings from 81 users, and 25,360 "distinct reads" on the general forum and multiple other communications not reviewed in the 11 topical discussion forums. To highlight some challenges and nuances involved in VC, this chapter proceeds with qualitative discourse analysis of virtual communicative text.

Qualitative analysis reveals three key characteristics resulting from multicultural VC experiences:

1. Intergenerational Negotiation
2. 'Real-World' Impact of Ethnic Culture and Racism
3. The Impact of Technology and Systemic Practices

Intergenerational Negotiation

Although both VGs contained mature-age and traditional students, perhaps because VG1 was not structured as task-focused/topical (like VG2), introductions differed in their revelation of personal details. This impacted future communications in three ways. First, as those in VG1 disclosed their age and geographical location, allegiances formed. Participants self-segregated into "oldies," "crusties," "dinosaurs," and "late starters," with several 40-something-year-olds sharing their fears and excitement about starting university at this point in their life. As middle-agers became more active in the virtual dialogue, some younger participants, particularly those in their teens, expressed shock. For example, a student born in 1985 found it incredulous to study with others who left high school that same year. Soon, it became a competition among the young not to be the youngest and to be the oldest among the old. Some simply looked forward to graduating in the same year as one of their children.

Second, although initially a topic of much conversation, age quickly took a backseat to the establishment of commonality, particularly by level of role-overload, which continued throughout the semester. All ages lamented the struggle to balance work, family, and study. Self-initiated disclosure of life situations fostered impromptu allegiances, most commonly family status (mothers/fathers, grandparents, etc.), school leavers, and geographical location.

A third outcome of informally structured virtual dialogue among a group of 288 "strangers" was the formation of informal mentoring. Younger participants reading descriptions of life circumstances of older participants actively sought out mentoring from those "more mature" and "handling the stress better." Younger ones perceived older generations to have the advantage of wisdom, along with better-developed stress management techniques. However, even though the requests were met with encouragement, older participants perceived younger ones to be advantaged because of their familiarity with technology, recent studying skills, and belief that stress is not correlated with age. Regardless of any extrinsic "truth," the experience of being in this VC situation together led to the development of a collective consciousness expressed by such sentiments as "being in the same boat" and able to "fall in a heap together." These conversations raised awareness of subjectivity, as discussions made it apparent that what constitutes maturity, old/young, privileged, and so forth, largely depends on one's perspective and position relative to others.

In contrast, although an equally varied range of student ages, experiences, and locations existed in VG2, the formality of the structure seemed to inhibit the formation of bonds based on demographic data. While some disclosed personal details and chatted occasionally, being assigned to a specific discussion group and work tasks had the effect of focusing virtual conversations. This led to discussions of team/task organization and learning, rather than primarily emotional support. Furthermore, VG2 infrequently left the structure of the "virtual classroom," whereas in VG1 some asked others to continue the discussion via personal e-mail.

What can be learned from this comparison, among other things, is the impact structure and management can have on the quantity and type of VC. Providing a group-work structure, facilitated by electronic sub-forums,[10] resulted in role clarity and alleviated anxiety associated with the uncertainties of DE. As the exclamation of two in VG2 reveal, combining effective technology and "human" management can relieve stress associated with working in virtual groups: "Thank god for Angela! I think this will make things heaps easier rather than using our own personal e-mails, as that way no one misses out on whats going on etc." "I agree with you and will go straight to our subforum once I have finished here. Great work!"

It is important to realize although similar group dynamics exist in face-to-face and virtual groups (Lynch, 2004), virtual groups remain significantly different in structure. VC restricted to written form may be more tedious, time consuming, interrupted by unanticipated events, and subject to a host of considerations that differ from face-to-face interactions. Consequently, although some research indicates the effectiveness of electronic discussions is greater than for face-to-face discussions (Brooks et al., 2001), attenuating an exhaustive list of structural issues identified in the literature (Bowles, 2004; Lynch, 2002; Brooks et al., 2001), such as resolving issues of time management, discussion leadership, technological complications, authoritative control, and dialogue ground rules, is essential for effective VC.

'Real-World' Impact of Ethnic Culture and Racism

The second important feature of VC is anticipating and managing the impact of 'real-world' events. The use of reflection in virtual learning environments, which is facilitated by the structure and reality of electronic learning technologies and VC practices, has caused a shift in pedagogy:

Constructive knowledge involves the opportunity to critically analyze information, converse with others about its meaning, reflect upon how information fits within your personal belief in value structures, and arrive at a meaningful understanding of that information ... this method is projected important, and lends itself well, to the online learning environment because learning online supports dialogue and the collaborative development of understandings. (Lynch, 2004, p. 109)

As it becomes common practice for DE to bridge individuals' roles as students and other life roles (Lynch, 2004), external factors may become increasingly influential. When dialogue can occur 24 hours/7 days a week from the comfort of one's own preferred environment, under the influence of substances or not, within a structure that does not require communications to be pre-reviewed, boundaries that usually govern face-to-face interaction are removed. This creates heightened potential for unmediated conflict, posing a greater risk to VC breakdown.

It is imperative managers and administrators conduct risk assessment prior to commencement and build coping mechanisms within the structure to ensure that high-quality service delivery is not undermined. To minimize undesirable outcomes and maximize communicative action, risk assessment cannot be an entirely top-down strategy. VC requires negotiation on the part of each contributor. If equity in legitimacy of voice is granted to uphold democratic/pluralist ideology, be aware it will engender diversity, and thus conflict resolution/minimization may require equal plurality in approach.

Knowledge production and true communication necessitates "we have to learn from the painful experiences and the irreparable suffering of those who have been humiliated, insulted, injured, and brutalized that nobody may be excluded in the name of moral universalism—neither underprivileged classes nor exploited nations, neither domesticated women nor marginalized minorities"(Habermas, 1993, p. 15). When the public partakes in a VC discussion, traditional statues (gender, class, race, etc.) responsible for enforcing social norms remain covert. This is simultaneously liberating and constricting, as those accustomed to social privilege may find themselves lacking power they may ordinarily take for granted. In contrast, VCs grant traditionally marginalized social groups' power to participate without the restrictions master statuses typically impose in face-to-face interactions.

Analysis of virtual dialogue reveals it is not so easy to remove social identity and status that years of socialization have firmly entrenched via ideology. Even if social cleavages are not immediately recognizable through written prose, expressed value systems and the cultural relativity of social standpoints (Harding, 1991; Smith, 1987) become readily exposed when debate ensues, as the following case example illustrates. The result is twofold. On one hand, VC traverses boundaries (geographical, social, and economic), facilitating intimate dialogue among sectors of the public which otherwise may isolated. This poses a unique opportunity for education, as individual life stories and experiences work to dispel (or uphold) stereotypes. On the other hand, the threat exists for larger groups to divide along traditional nation-state, age, and other boundaries in the quest for validation of moral superiority. For instance, as Butler (1990) uses the category of gender to argue, gender is not some intrinsic state of being, but rather a negotiated performance reproduced and re-enacted daily. So, too, is culture a socially constructed product reinvented via VC.

The informality of VC fosters an increasingly more intimate type of dialogue than face-to-face classroom interactions. This encourages deeper reflection on cultural norms and beliefs. The disclosure of "taboo" prejudices too politically incorrect to reveal beyond the safe environment of one's computer terminal becomes commonplace as individuals' biases are identified by others. Asynchronous VC is unique because it provides a written account of one's own thoughts and dialogue which continue to exist even when one's current opinions may no longer be reflected in past writings. In this study, the institutionalized VC structure made viewing all prior correspondence continuously available. The impact re-reading past communications had is noticeable. Perhaps due to the fractured nature of asynchronous VC, the expression of self-identities, while seeming reflexive of "real-life" realities, metamorphosed over the six months of virtual dialogue. Several who entered the dialogue espousing rigid, clear boundaries and belief systems revealed in later communication how their interactions, both within and outside the learning environment, changed. For some, this caused a discrepancy between their "old" and "new" selves, while for others virtual interactions served to preserve preconceived notions. To explore this phenomenon further, two case examples will be drawn upon.

- ***Case Example 1.* The impact of multiculturalism on VC:** VG1 dialogue regarding racism and immigrants emerged in response to the Macquarie Fields[11] Riots in Sydney. What ended up being a nine-day virtual conversation among 35 participants was initiated by a 34-year-old who wrote how she was disgusted and appalled at the rape of a pregnant woman abducted by three men in Macquarie, Sydney. She reveals the incident caused her mental anguish and conflict because although she attended a highly multicultural school in inner-city Sydney, such incidents as racial riots in the media cause her to feel she's becoming racist. She identifies that this is at odds with her identity as a "typical" Australian and notes how infuriating it is that the more Australia lets immigrants into the country, the more ethnic fights are imported from immigrants' country of origin.

  The ensuing VC reveals three challenges technology poses to discussions of social problems. First, initial responses highlighted the complexity of dealing with heated emotional and sociocultural issues. A primary concern was how to communicate and define national/cultural identity in an environment where value systems and beliefs varied widely, and where individual characteristics remained unknown. For example, the very first response used the well-known history of Australia's rape and decimation of thousands of aboriginals, and the years of aboriginal assimilation into white British culture, to question who counts as immigrants. Yet, because another student failed to read the discussion closely, s/he challenged the respondent's justification of aboriginal men raping and murdering white women, arguing many ethnicities share a country and need to stop blaming each other for past actions.

The discussion that followed reveals the second challenge virtual reality poses to effective communication—the complexity of articulating any universal meaning/intention using subjective cultural symbols (words). When the original initiator reflected/re-read her initial words, she articulated how she did not explain herself properly, tried to qualify what she meant by "typical," and backpedaled about her intended meaning. Responses to her clarification shows effort to clarify VC sometimes does little to change/retract original statements and may not assuage others' discontent. For example, a fourth contributor challenged her "typical" Australian comment, noting that indigenous Australians are as typical as you can get and that noted accusations and judgments make matters worse.

This virtual dialogue continued, as individuals debated concepts of race, racism, nationality/nationalism, immigration, hate, prejudice, Eastern vs. Western cultures, terrorism, rape and violence, power, control, and dominance. Arguments about ability to transcend culture led to standpoint/point-of-view discussions, escalating until the physical reality of seeing such dialogue alive as text on the computer screen inspired one to reflect how the very act of dialogue occurring among multicultural individuals was in and of itself inspiring. This participant found it empowering to note as a cultural artifact, a product of democratic societies. Although the reflection offset spiraling cynicism, uncertainty of being heard among competing cyberspace voices remained a reality, as the preface "whoever else cares to read this response" implied. Perhaps luckily for this communicator, the "reality" of the conversation itself was used by others to evidence cultural pride as the tide of conversation flowed to embrace multiculturalism.

Haraway's (1991) concept of cyborgs captures the symbiotic relationship between humans and machines, and is increasingly manifested through social analysis of VC. "Technology is beginning to mediate our social relationships, our self-identities and our wider sense of social life to an extent we are only just beginning to grasp" (Featerstone & Burrows, 1995, p. 13). Although the racism example confirms the similarity of virtual asynchronous and synchronous face-to-face communication whereby inclusion of an alternative point of view shifts both focus and climate, it also reveals difference. What differs between the two communication styles (perhaps more synchronous vs. asynchronous) remains the quantity of time to devote to reflection. Synchronous dialogue demands an immediate response, often with little time to analyze the evidence previously provided. Yet, in asynchronous environments, a communicator can do considerable research before responding, as another contributor did when s/he noted the paucity of information provided to evidence claims of the influence culture has on events and ideas. Hence, the examples reveal asynchronous dialogue using computers as a medium alter the meaning, flow, and outcome of communication.

This brings us to the third challenge VC poses: negotiating definitions of authority. In the classical Weberian tradition, power can be defined as the ability to achieve desired ends in the face of resistance. In this case, the power both the rapists and rioters exerted in Sydney demonstrates traditional power. Traditional power often lacks legitimization, as reflected in the conversation. The difference between power and authority lies within perception. Authority is power others perceive as legitimate, not coercive. When authoritative voices enter a dialogue, opposition is minimized. This occurred in this case when two police officers, one in Sydney, the other in Melbourne, added their perspectives. However, for VC, authority poses unique challenges: (a) there may be several authoritative voices competing for power in the dialogue, (b) it may be harder to verify whether one's identity/position is indeed true, and (c) control over the direction of communication may be relinquished to persuasive writers, due to written dialogue's uni-dimensionality.

In this example, the Sydney police officer decided he had heard enough from what he defined as "armchair critics," telling them it was time to "get real and listen to how it really is!!" As he described leading a riot team of police during the Macquarie Fields Riots, his authoritative prose left little room for further discussion. Emotionally charged, and backed by years of experience and hardships, he made a charismatic plea for hard-core justice, where all individuals are treated the same under the law, irrespective of ethnicity or politics.

These insights were met with cries of joy from many, with thanks for the officer "set[ting] us straight on the subject." The authority commanded by the police officer and role as expert were repeatedly asserted. Another added, "you have expressed how a lot of my family, friends & I feel about the Macq fields riots but you have the evidence to back it up." The disadvantage of such discussions is the dichotomy that quickly is established between the knowers (those with the right to speak, who are identified as "absolutely right and are coming from a position of experience and knowledge") vs. those with less worthy ideas. The impact of position to grant communicative authority is notable, and was acknowledged by many who stated they were glad to "have someone in a position of experience who can shed light on this subject" and "always find it hard to form and express an opinion on these types of matters because I never know the real situation."

Who has the right to speak (and be heard) has been a topic of interest in feminist epistemology (Roof & Wiegman, 1995; Garry & Pearsall, 1989) and of critical political theorists, such as Antonio Gramsci. However, deciding whose perspective deserves legitimization is a contentious issue, and in this case, continues to fuel deep-seated prejudices. "Those who have developed these lovely theories do so from an educated persons perspective. These people have little knowledge of the type of individual that has no regard for the laws we are all expected to live by and for the safety, property and well-being of others." Such was the sentiment of a self-identified Victorian police officer, who added he worked in disadvantaged and culturally diverse areas. He proceeded to use his status to express dissatisfaction with Australia's welfare state "where we make excuses for poor (and even criminal) behavior."

Once again, this led to a flood of bigoted, culturally insensitive exchanges, such as a response from a participant who has a police officer as her partner. What began as agreement with the Victorian officer, regarding those who see situations firsthand, and everyday having more right to speak than others without such knowledge, digressed into disclosure of how, although she could "respect and appreciate poverty, youth issues and disadvantaged groups," she was "tired of systems protecting these people to the detriment of others." Her tirade ended with a revelation about how much better off her sister on "government handouts" is than she who owns a business.

VG participants, equipped with all their identities and subsequent bodies of knowledge (as professionals in various fields, occupants of various neighborhoods and countries, situational knowledge from interpersonal relationships, etc.) acted out real-world battles in cyberspace. With each passing communication, participants divided more firmly along political lines, varying in expressions of sympathy vs. critique of police officers and level of open-mindedness. After nine days, interest subsided and new issues were up taken. However, unlike verbal dialogue, which lives in our memories alone, months later every detail of the communication remained intact—as a "real" artifact for all to view, if desired.

Unlike face-to-face communications, VCs appear to foster a unique combination of emotive and rationalized thoughts about world events. Although it may be impossible to anticipate every response in a debate, written responses tend to be more personal and reflexive than face-to-face discussions and exhibit less inhibition. In large groups, the geospacial/physical constraints imposed by size, where individuals may not get to discuss ideas one on one, is vastly altered by VC. VC enables multiple one-on-one conversations, which may result in a ratio of one to hundreds. However, it is questionable whether individuals consciously think about what it means to express deeply personal thoughts in a public forum.

- ***Case Example 2.* The subjectivity of electronic communication:** Research exploring communication norms in virtual environments in education (Palloff & Pratt, 2001) has drawn particularly from stage and systems theory to explain the organization and development of electronic group dynamics, including conflict and resolution (McClure, 1998; Tuckman & Jensen, 1977). The second case example provides support for the need of task clarity, focused collaboration, and quick leadership intervention when conflict arises for successful group functioning (McClure, 1998). It also gives credence to Palloff and Pratt's (2001) warning of the heightened potential for misinterpreted communication, due to lack of face-to-face interaction rituals and less adherence to social norms based on the work of Schopler, Abell, and Galinsky (1998). Adding to Thomas' (1923) "definition of the situation," as articulated in classic symbolic interactionism theory, the case illustrates how subjective meanings individuals attach to a situation have real-world consequences, which may be independent of any "objective" reality. As Lynch (2004) articulates, "when reading a typed message, there is a strong tendency to project—consciously or not—your own expectations, wishes, anxieties, and fears into what the person wrote. This may lead to further conflict and, because it is written, sometimes builds a feeling of resentment that lasts longer" (p. 13).

Long before the instance occurred, during the first weeks of communicating virtually, a participant described the complexities a later example would reveal: "Sorry if the 'tone' of my post to the forum provoked you. Thing about cyberspace communication is that it lacks tone—or one has to work harder at communicating—something I'm still adjusting too." Although this instance does not reveal miscommunication, there is agreement VC is indeed subject to misinterpretation: "No, didn't provoke me as such and I agree with the lack of tone thing. Perhaps my messages are being misread also...pros and cons of technology!"

Applying the once groundbreaking insights of Cooley (1922), whose "looking-glass self"[12] concept revolutionized classical social-psychological theory, the computer may be understood as a mirror reflecting our self-image back to us. As we view this reflection, communicative uncertainty prevails: Is this indeed who I am? Have I expressed myself in a way that enables others to understand me as I wish them to? When an artifact—a computer—is added to the complex process of communication among humans, situational dynamics change. Let us consider the following example.[13]

In my role as lecturer, it is common practice to produce a final grade distribution chart so students can see their performance relative to others. To convey this data virtually, I wrote the following message:

While I cannot offer you confirmation of your final grades, I have created a chart for you to see how the class did as a whole, which I am attaching as a word file. I think you will be happy to see that out of the xx of you who completed all 3 assessment items, only xx of you failed, which is just x%.

While initially this received expressions of thanks, such was not the case for one failing student in VG1. This participant wrote how my note expressed an insensitivity to students' feelings, contained "unsupportive comments" by confirming some failed, and would make those who failed feel worse. The correspondence ended with exclamations of how I do not know what damage I might be causing and how she found "this behavior quite disgusting, egotistical, and unethical." This spurred 23 further communications, including a handful charging to my defense, telling her not to "tough talk behind a computer," how failing is a reality of life, how her response demonstrates "unacceptable behavior from an adult," and since they "are no longer children" university is preparing them for the "real world." Others pointed to the "danger of technology" where instant thoughts can be "fired off" in the heat of the moment, whereas in the past, emotions would subside prior to mailing a letter. Noting the difference VC makes. One asks: "Would you have spoken to the person that way, face to face, in front of the class of 170 people? You might say 'Yes', but I suggest embarrassment would result." In this case, older students took on mentoring roles, outlining strategies to effectively deal with disturbing electronic information.

With painstaking effort, I responded by clarifying my intentions. However, no amount of preparation could have prepared me for her reply. From the apology and confession, I and the other approximately 200 virtual community members learned she just suffered the death of her daughter. Concluding communications displayed a range of emotions—empathy, sorrow for actions based upon imperfect knowledge, even resentment, as one noted although she "felt" for her situation, the group member only contributed once to the virtual dialogue. The armchair philosophers implored others to be nonjudgmental and reflected how "the answers to problems are not always as apparent as they appear on the surface." In sum, all seemed to learn an invaluable life lesson that may not have been possible without participating in VC.

The Impact of Technology and Systemic Practices

The third broad lesson to be learned from this case study is: when using technology to communicate, all aspects of communication can (and will!) be affected—from the hardware making it possible, or impossible, to the even slight confusion over definitions.[14] VC depends on high-quality technical delivery for success, which is contingent on geographical location,[15] resources, politics, and user ability—all beyond the scope of this work. However, it is worth noting that technological failure impacts communication. When technology fails, for any reason, confidence in the system wanes and anxiety rises. As the following experiences reveal, technological failure itself does not often cause the most customer dissatisfaction. Humans seem to expect machines, and technological systems, to fail. Many exhibit great distrust of technology. As the next example reveals, what humans are least forgiving of is other humans.

A useful, yet fallible, electronic assessment submission system is the institutional means available to students to electronically submit their work in this case. At times, students submit work only to find out later it was received as unreadable due to computer format. When their work is returned as "unreadable," some students get incited. In VG2, one student found that even though the Web site noted the submission as successful, her hard copy returned via the postal service saying "unreadable" led to her feeling "disappointed" to see hours of work done for naught. Rightly so, she wished to receive such advice weeks earlier, upon submission. Yet, when computers are not set up to properly "talk" to humans, errors in the system remain invisible.

This single experience was exacerbated by another's inflated perception of the event. Noting "what a horror story I really feel for you," another student extrapolated the instance to her fear of taking the virtual exam. Student panic, resulting in questions about what happens if the Internet service provider "decides to disconnect during the exam," led to continued fret and pleads for help to "eliminate this extreme worry." If such communication was conducted by telephone or e-mail, the result would have been a one-on-one conversation where each concern was subsequently addressed. However, because it occurred in a public arena, it provided fuel for the entire virtual community to read and reflect upon. Hence, the very structure of "public" VC—whereby individuals post messages for all to read—appears to encourage not only disclosure of individual fears, but moreover generates chaos among others[16] potentially impervious to such ideas.

Perceptions, and perception management, are often all that exists in VC. Therefore, it is crucial that strong management exists to intervene and curtail escalated fears where necessary, taking proactive steps to negotiate positive outcomes. In VG2, far greater usage of educational technology (CD-ROM readings, sub-forums, electronic lectures, and student presentations in addition to electronic exams and assignment submission) existed. Hence, it is unsurprising that more technological difficulties, including elevated apprehension and student stress relating to inadequate technological knowledge, manifested. When two participants vented frustration at the inconveniences caused (babysitters planned, work schedules rearranged) due to Web site failure making their exam available, another participant mediated discontent by articulating faith in management. After confirming how everyone held such concerns about balancing "work, children, home and study," she expressed faith in the lecturer's management of the situation and ability "to give us the time we need to re-organize our schedules."

For global virtual communities, time is relative. The technical problem above occurred outside standard business hours, leaving management unaware until the next day. Thus, mechanisms for dealing with failure are best built into the system. Although every technological failure cannot be anticipated, it is most important to create a "humanistic side," through consistent dialogue to encourage confidence that crises will be attended as quickly and professionally as possible.

Specifically for DE, the establishment of reliable technological systems is synonymous with quality performance. For all its imperfections, technology is arguably a more viable alternative than solving geographical inadequacies of rural infrastructure and communication issues relating to distance. Both groups identified discrepancies in the privileges Australian urban residents experience compared to rural and international residents. Compounded by outdated policies, such as mandated use of the postal service to return assignments and grades, institutional rules can be inconsistent with efforts at technological leadership. Systemic inconsistencies lead to customer frustration. As new students in Saudi Arabia and Pipalyatjara communicated via e-mail, the delivery of mail varied widely (i.e., in Pipalyatjara, because of no rain in two years, "the mail plane [was] having some trouble landing," and in Sudan, mail delivery by camel was said to be slow). Hence, university policies can add to student frustration, such as when individuals receive material for one subject, yet not others.[17]

CONCLUSION

From the many examples discussed, three broad future recommendations can be put forth to encourage successful VC. VC should be: (a) culturally responsible, (b) flexible in time and style, and (c) mandatory, if group cohesion is desirable.

When electronic communication is actively managed and made routine, it becomes part of the skill set—another tool in the toolkit of the global learner. Participants learn a new set of social and communication norms that differ in type and kind from face-to-face interaction. Without physical and verbal cues, interaction—indeed communication—becomes subject to hyper-analysis by participants. As analysis reveals, this can lead to heightened sensitivity and subjectivity of interpretation for content and intent. When participation is optional, the "free rider" phenomenon (Olson, 1965), commonly associated with social movements and collective behavior, ensues. It becomes all too easy, indeed even rational, for many to sit back and benefit from the contributions of a few hard workers. Hence, crucial to successful creation of an active virtual community is the participation of members. To achieve this, and simultaneously preserve the flexibility that makes DE so popular among mature-age and international students, it is recommended that learning materials be structured to enable participants to self-select into a set number of virtual discussions. Not only does this teach effective time management skills, but moreover it gives individuals the option to not partake in discussions that, at least in the social sciences, may be cultural hot buttons. For example, although lecturers might be eager to hear the opinions and reflections on personal experiences regarding issues such as abortion, refugees, terrorism, and so forth, as distance educators we must realize our capacity to intervene/counsel/educate is mediated by our *distance*. The realities of our students' lives (and the impact a disturbing virtual conversation may have) must take precedence.

The virtual community analyzed in this chapter makes explicit cultural norms operating when individuals construct meaning and make sense of their everyday world—wherever that is. Fostering dialogue among individuals holding varied statuses within the social hierarchy (that is, our global society) challenges cultural (and class, gender, age, ethnic, etc.) norms "...from different standpoints different aspects of the ruling apparatus and of class come into view" (Smith, 1987, p. 107). Applied as a methodological tool to VC, critical analysis—infused with phenomenological sensibilities that lend insight into how micro-level internalization of sociocultural values and norms (Berger & Luckmann, 1966) impact communication—provides knowledge of how cultural awareness and individual belief systems result from systemic socialization. This subsequently creates a foundation for future analyses of global systems composed of nation-state citizens. Societies are not merely aggregations of individuals. As Hess (1995) theorizes, multicultural spaces can be understood as stages whose very existence predispose its participants to conflict as they negotiate meanings of nationalities, ethnicities, classes, genders, histories, and so forth.

REFERENCES

Albalooshi, F. (Ed.). (2003). *Virtual education: Cases in learning and teaching technologies.* Hershey, PA: IRM Press.

Australian National Training Authority. (2005). *Our universities: Backing Australia's future.* Retrieved February 15, 2006, from http://www.backingaustraliasfuture.gov.au/ministers_message.htm

Au-Yeung, L. H., Ha, T., & Au, G. (2004). The experience of new WBI-adopters in Hong Kong. *Journal of Educational Technology Systems, 31,* 411-422.

Barjis, J. (2003). An overview of virtual university studies: Issues, concepts, trends. In F. Albalooshi (Ed), *Virtual education: Cases in learning and teaching technologies* (pp. 1-20). Hershey, PA: IRM Press.

Berger, P., & Luckmann, T. (1966). *The social construction of reality.* New York: Doubleday.

Borthick, A. F., & Jones, D. R. (2000). The motivation for collaborative discovery learning online and its application in an information systems assurance course. *Issues in Accounting Education, 15,* 181-211.

Bourdieu, P. (1977). *Outline of a theory of practice.* Cambridge, MA: Cambridge University Press.

Bowles, M. S. (2004). *Relearning to e- learn: Strategies for electronic learning and knowledge.* Carlton, Victoria: Melbourne University Press.

Brooks, D. W., Nolan, D. E., & Gallagher, S. M. (2001). *Web-teaching: A guide to designing interactive teaching for the World Wide Web* (2nd ed.). New York: Kluwer Academic/Plenum.

Brooks, J., & Brooks, M. (1993). *In search of understanding: The case for constructivist classrooms.* Alexandria, VA: Association for Supervision and Curriculum Development.

Butler, J. (1990). *Gender trouble: Feminism and the subversion of identity.* New York: Routledge.

Coleman, J. S. (1988). Social capital in the creation of human capital. *American Journal of Sociology, 94,* 95-120.

Collins, P. (1990). *Black feminist thought: Knowledge, consciousness and the politics of empowerment.* Boston: Unwin Hyman.

Cooley, C. H. (1922). *Human nature and the social order.* New York: Scribner.

Davis, M. (1997). Fragmented by technologies: A community in cyberspace. *Interpersonal Computing and Technology, 5*(1-2), 7-18.

De Freitas, S., & Oliver, M. (2005). Does e-learning policy drive change in higher education?: A case study relating models of organizational change to e-learning implementation. *Journal of Higher Education Policy and Management, 27*(1), 81-95.

Dewey, J. (1938). *Experience and education.* New York: Macmillan.

Dukheim, E. (1951). *Suicide.* Glencoe, IL: The Free Press.

Falk, I., Grady, N., Ruscoe, J., & Wallace, R. (2003). *Designing effective e-learning interventions, learning to e-learn research.* Center for Teaching and Learning in Diverse Educational Contexts. Hobart, TAS: Northern Territory University & Unitas Company, Ltd.

Featherstone, M., & Burrows, R. (Eds.). (1995). *Cyberspace, cyberbodies, cyberpunk.* London: Sage.

Gary, A., & Pearsall, M. (Eds.). (1989). *Women, knowledge and reality: Explorations in feminist philosophy.* Boston: Unwin Hyman.

Gibbons, M., Limoges, C., Nowotny, H., Schwartzman, S., Scott, P., & Trow, M. (Eds.). (1996). *The new production of knowledge.* London: Sage.

Giddens, A. (1979). *Central problems in social theory.* Berkeley, CA: University of California Press.

Goldman-Segall, R. (1992). Collaborative virtual communication: Using learning-constellations, a multimedia ethnographic research tool. In E. Barrett (Ed.), *Sociomedia* (pp. 257-296). Cambridge, MA: The MIT Press.

Habermas, J. (1993). *Justification and application: Remarks on discourse ethics.* Cambridge, MA: The MIT Press.

Haraway, D. (1991). *Simians, cyborgs, and women.* New York: Routledge.

Harding, S. (1991). *Whose science? whose knowledge?* New York: Cornell University Press.

Hawkey, K. (2004). Assessing online discussions working 'along the grain' of current technology and educational culture. *Education and Information Technologies, 9*(4), 377-386.

Herring, S. C. (2004). Slouching towards the ordinary: Current trends in computer-mediated communication. *New Media and Society, 6*(1), 26-36.

Hess, D. (1995). *Science and technology in a multicultural world.* New York: Columbia University Press.

Higher Education Council. (1992). *Response to the ministerial reference on distance education.* Canberra, ACT: Higher Education Council.

Johnson, D. W., & Johnson, R. T. (1994). Cooperative learning in the culturally diverse classroom. In R. A. DeVillar, C. T. Faltis, & J. P. Cummings (Eds.), *Cultural diversity in schools* (pp. 57-73). Albany, NY: State University of New York Press.

Jorgensen, D. (2002). The challenges and benefits of asynchronous learning networks. In H. Iyer (Ed.), *Distance learning: Information access and services for virtual users* (pp. 3-17). New York: Haworth Information Press.

Kreuger, L. W., & Neuman, W. L. (2006). *Social work research methods: Qualitative and quantitative applications.* Boston: Pearson Education.

Lash, S. (2002). *Critique of information.* London: Sage.

Leebron, E. J. (2004). Media entrepreneurship as an online course: A case study. In D. Monolescu, C. C. Schifter, & L. Greenwood (Eds.), *The distance education evolution: Issues and case studies* (pp. 240-257). Hershey, PA: Information Science Publishing.

Little, A. (2004). *The politics of community: Theory and practice.* Edinburgh: Edinburgh University Press.

Lynch, M. (2002). *The online educator: A guide creating to virtual classroom.* London: Routledge Falmer.

Lynch, M. (2004). *Learning online: A guide to success in the virtual classroom.* New York: Routledge Falmer.

Merton, R. (1968). *Social theory and social structure.* New York: The Free Press.

McClure, B. (1998). *Putting a new spin on groups.* Hillsdale, NJ: Lawrence Erlbaum.

Monolescu, D., Schifter, C.C., & Greenwood, L. (Eds.). (2004). *The distance education evolution: Issues and case studies.* Hershey, PA: Information Science Publishing.

Olson, M. (1965). *The logic of collective action: Public goods and the theory of groups.* Cambridge, MA: Harvard University Press.

Palloff, R. M., & Pratt, K. (1999). *Building learning communities in cyberspace: Effective strategies for the online classroom.* San Francisco: Jossey-Bass.

Palloff, R. M., & Pratt, K. (2001). *Lessons from the cyberspace classroom: The realities of online teaching.* San Francisco: Jossey-Bass.

Parsons, T. (1954). *Essays in sociological theory.* New York: The Free Press.

Piaget, J. (1927). *The child's conception of time.* New York: Ballantine Books.

Putnam, R. D. (Ed.). (2005). *Democracies in flux: The evolution of social capital in contemporary societies.* Oxford: Oxford University Press.

Ragusa, A., & Atweh, B. (2003, November). Analysing global collaborations using a post-structuralist model of social justice and social change. In *The Centre for Social Change Research 2003 Conference Proceedings.* Brisbane: Queensland University of Technology.

Ragusa, A., & Atweh, B. (2004, August). Social justice, social change and higher education: Analyzing global collaborations among higher education researchers. In *Proceedings of the Society for the Study of Social Problems Conference,* San Francisco.

Ritzer, G. (2004). *The globalization of nothing.* Thousand Oakes, CA: Pine Forge Press.

Robinson, J. (1999). The time diary method: Structure and uses. In W. Pentland, A. Harvey, M. Lawton, & M. McColl (Eds.), *Time use research in the social sciences* (pp. 47-88). New York: Kluwer Academic/Plenum.

Roof, J., & Wiegman, R. (Eds.). (1995). *Who can speak? Authority and critical identity.* Chicago: University of Illinois Press.

Royse, D. (1999). *Research methods in social work* (3rd ed.). Chicago: Nelson-Hall.

Scheffler, I. (1965). *Conditions of knowledge.* Chicago: Scott, Foresman & Company. Cited in Goldman-Segall (1992).

Schifter, C. (2004). Faculty participation in DE programs: Practices and plans. In D. Monolescu, C. C. Schifter, & L. Greenwood (Eds.), *The distance education evolution: Issues and case studies* (pp. 1-21). Hershey, PA: Information Science Publishing.

Schon, D. (1991). *The reflective turn.* New York: Teachers College Columbia University.

Schopler, J., Abell, M., & Galinsky, M. (1998). Technology-based groups: A reviewed conceptual framework for practice. *Social Work, 4*(3), 254-269.

Semple, A. (2000). Learning series and the influence on the development and use of educational technologies. *Australian Science Teachers Journal, 46*(3), 21-28.

Sherry, L., & Wilson, B. (1997). Transformative communication as a stimulus to Web innovations. In B. Khan (Ed.), *Web-based instruction.* Englewood Cliffs, NJ: Educational Technology Publications.

Shields, S. F., Gil-Egui, G., & Stewart, C. M. (2004). Certain about uncertainty: Strategies and practices for virtual teamwork in online classrooms. In D. Monolescu, C.C. Schifter, &

L. Greenwood (Eds.), *The distance education evolution: Issues and case studies* (pp. 116-141). Hershey, PA: Information Science Publishing.

Smith, D. (1987). *The everyday world as problematic: A feminist sociology.* Boston: Northeastern University Press.

Smith, D. (1990). *The conceptual practices of power: A feminist sociology of knowledge.* Boston: Northeastern University Press.

Stein, A. (2001). Preparation for e-learning: An Australian study. In F. Albalooshi (Ed.), *Virtual education: Cases in learning and teaching technologies* (pp. 140-155). Hershey, PA: IRM Press.

Sydney Morning Herald. (2005a, March 1). *Fatal crash: Driver's arrest imminent.* Retrieved February 8, 2006, from http://www.smh.com.au/news/National/Fatal-crash-drivers-arrest-imminent/2005/03/01/1109546829118.html

Sydney Morning Herald. (2005b, March 1). *Fourth night of riots: Nine charged.* Retrieved February 8, 2006, from http://www.smh.com.au/news/National/Fourth-night-of-riots-nine-charged/2005/03/01/1109546826854.html

Thomas, W. (1923). *The unadjusted girl.* Boston: Little, Brown.

Tuckman, B., & Jensen, M. (1977). Stages of small group development revisited. *Group and Organizational Studies, 2*(4), 419-427.

Vygotsky, L. (1986). *Thought and language.* Cambridge, MA: Massachusetts Institute of Technology.

Wegerif, R. (1998). The social dimension of asynchronous learning networks. *Journal of Asynchronous Learning Networks, 2*(1), 34-49.

Wilson, J. M. (2002, January 11). Successful distance-education spinoffs. *Chronicle of Higher Education,* 48, 18.

Winchester, I. (2005). Globalization, diversity and education for a democratic 21st century. *Journal of Educational Thought, 39*(1), 1-5.

World Bank. (2005). *World development indicators 2005.* Retrieved February 13, 2006, from://Web.worldbank.org/WBSITE/EXTERNAL/DATASTATISTICS/0,,contentMDK:20523710~hlPK:1365919~menuPK:64133159~pagePK:64133150~piPK:64133175~theSitePK:239419,00.html

ENDNOTES

1. Precisely defining what constitutes "DE" remains a contested issue. Adjectives such as "flexible delivery," "virtual," "correspondence," and "online" are found intermittently dispersed throughout the literature to describe non-face-to-face education. Although DE is not a new phenomenon, with correspondence programs existing during the 19th century, the addition of technologically driven communication systems is recent (Schifter, 2004).
2. The concept of "virtual" learning spaces is defined by Graves (2000) "as those not constricted by time or place" (Leebron, 2004, p. 241).
3. Politics surrounding the concept "community" reveals it to be highly contested terrain (Little, 2002). For the scope of this research, community encapsulates neither the classical political theories of Aristotle, Tonnies, or Paine, nor entertains arguments over communitarianism vs. individualism. In this study I conceptualize community as a system of shared commitments and social values "expressed and promoted symbolically and culturally" (Frazer, 1999, p. 209 in Little, 2004, p. 131) as Frazer articulates in discussions of the welfare state. In sum, what I termed "virtual community" draws upon recent conceptualizations, due to

societal changes brought on by modernity and globalization which no longer require geographical connection to establish "community." As operationalized, community here is informed by identity politics and increasingly limited by mental and infrastructure limitations rather than geography.

4 Prior to 2005, academic policy governing student computer access made computer requirements minimal. This meant, for social equity reasons, educators could not require students to participate in electronic forums. When policy in 2005 changed to reflect new technological minimum standards, making computer access mandatory, this did little to eradicate existing disparity among students. Despite shifting policy, variation in student access continued, with some having ready access via employment and/or home, while others traveled to public spaces with public terminals. Hence, practical implications of social inequality remain and continue to need addressing.

5 For example, when trying to understand how and why individuals in specific communities spend their time, Robinson's (1999) research reveals survey respondents' estimations are grossly unrepresentative of reality and tend to exhibit bias towards over-reporting workload commitments. Although time use varies in kind from VC participation, the activity of cognition associated with memory recall is similar. Surveys produce a qualitatively different type of reflection and selection of content than analysis of primary texts yields. While the cohort for this study completed both qualitative and quantitative surveys, only discourse analysis is used as the basis for case study.

6 As with every methodology, the inevitability of limitations must be noted. Using case study as an analytical tool means the findings are necessarily situational, contextual, and non-generalizable (Kruger & Neuman, 2006; Royse, 1999).

7 In this environment, student forums were largely informal and participation optional.

8 Discussions of DE provisions described as "asynchronous learning network (ALN)" (Wilson, 2002) and "computer-mediated communication (CMC)" (Herring, 2004) have been attributed with increasing popularity (Herring, 2004) among educators and described as "a canonical model of online education that has received wide acceptance" (Wilson, 2002). Although the ALN model, as described by the CEO of UMassOnline, contains "threaded discussions and live chat," for the Australian case study, live chat is not currently possible.

9 Traditional DE refers to learning by correspondence, whereby students are sent reading material and assessment items to complete and return by mail. The university in this study sought to enhance DE by simply creating electronic discussion forums. This approach is inadequate as unstructured electronic discussions do not necessarily translate into better learning. Even with a study guide, without the guidance of lectures, many students were unable to grasp the subject material.

10 As one student defined, "sub groups are the groups we are in for our discussion groups ... what every topic you are discussing is the sub group you are in."

11 The "Macquarie Fields Riots" occurred in Sydney's southwest for four nights following a fatal motor vehicle accident on February 25, 2005, when two young men, passengers in a stolen motor vehicle, died after being pursued by Sydney police, while the driver fled the scene (*Sydney Morning Herald*, 2005a, 2005b).

12 Cooley's research explored how we imagine our appearance is to others, our imagination of their judgments of our appearance, and any subsequent modification.

13 In contrast, no such situation arose in VG2. As the size, virtual structure, and assignment design differed, it is impossible to speculate why. Future studies may wish to analyze why, and under what conditions, such differences in communication arise.

14 For example, although I thought asking students to send me an e-mail "so I can figure out if it's individual accounts or the whole system which isn't working," would clearly indicate I was referring to one's e-mail account, not the subject forum, this proved to be an incorrect assumption. Distinguishing between electronic communication technologies was problematic for some who incorrectly attempted to clarify my instructions by saying by "e-mailing Angela meant this actual post on the forum she has made" (for which I responded "Actually I meant real e-mail ...")Although a minor point, had students used the wrong communication device, the technological problem would have been unsolved.

15 Consider the following virtual correspondence received: "I am currently working in Sudan with the United Nations, in somewhat remote areas ... My issue is that to date, I have not received the prescribed reading material. I wanted to test the best method of receiving goods from Australia, and decided to put 'camel mail' to the test ... To date—non arrival!"

16 Such sentiments continue to be put forth in telephone conversations with students, some who ultimately decide to only read lecturers' forum postings to minimize distress.

17 This is due to the university's practice of mailing out DE material by subject, instead of by student. In this case, printing failure caused delay in the publication of some subject material, yet not others.

About the Authors

Kirk St.Amant (kirk.st-amant@ttu.edu) is an assistant professor in the Department of English at Texas Tech University (USA). He has a background in anthropology, international government, and technical communication, and his research interests include intercultural communication via online media. He has taught online and conventional courses in technical and professional communication and in intercultural communication for Texas Tech University, Mercer University, and James Madison University. He has taught courses in e-commerce, distance education, and business communication in the Ukraine as a part of the USAID-sponsored Consortium for the Enhancement of Ukrainian Management Education (CEUME).

* * *

Vinita Agarwal is a PhD candidate in the Department of Communication at Purdue University (USA). Her research interests center on information technology, healthcare, privacy, patient safety, and policy. She received her Master's in Communication from the University of Illinois at Chicago, and is also an alumnus of the Mass Communication Research Center (MA, India) with an undergraduate degree in Physics (Delhi University, India). Ms. Agarwal's paper offering a critique of autonomy and consent in international privacy regulation recently won a top four student paper award at NCA (2005).

Edgar Barrantes is originally from Costa Rica. He is currently a doctoral student in rhetoric and professional communication at New Mexico State University (USA). His research interests include outsourcing technical communications in Latin America and U.S.-Latin American intercultural professional communication.

Pamela Estes Brewer lectures in the English & Philosophy Department at Murray State University (USA), and coordinates the professional writing program. She is completing her doctorate in Technical Communication and Rhetoric at Texas Tech University. Her research interests include intercultural virtual workplaces and related studies. She has presented at the Society for Technical Communication, the Association of Teachers of Technical Writing, and Computers & Writing conferences. In addition, she was part of a research team that won the $10,000 STC grant for its work concerning authors' preferences in editing, and served on the organizing committee for the 2006 Computers & Writing Online Conference.

Sejung Marina Choi is an assistant professor of advertising at the University of Texas at Austin (USA). Her research interests are in the areas of source credibility, cause-related marketing, advertising on the Internet, and cross-cultural consumer behavior. Her work has appeared in *Journal of Advertising, Journal of Consumer Affairs, Journal of Popular Culture, Journal of Current Issues and Research in Advertising, Journal of Computer-Mediated Communication, International Journal of Advertising*, and various conference proceedings. Prior to her graduate degrees, she worked in the advertising agency business, particularly in account management. Her past clients include various multinational advertisers and local advertisers.

Tatjana Chorney emigrated to Canada from Belgrade (Yugoslavia, 1992). She studied English linguistics and literature at the University of Belgrade, and then literature and the humanities at York University in Toronto. She holds a PhD in English from the University of Toronto. She has taught literature, composition, and technical writing in community colleges and universities. She now teaches English Literature and Cultural Studies at Saint Mary's University in Halifax (Canada). She has published essays on literature, reading, teaching, and computers, and now works on a federally funded project involving interactive reading habits in the hypermedia.

Amy Dalzell is a doctoral student in rhetoric and professional communication at New Mexico State University (USA). Her research focuses on the relations of visual, verbal, and aesthetic rhetorics.

Qiumin Dong is originally from China. She is currently a doctoral student in rhetoric and professional communication at New Mexico State University (USA). Her research interests include teaching professional communication in China, and U.S.-Chinese intercultural communication.

Anthony Faiola is a researcher and educator with an interest in the cognitive aspects of human-computer interaction. He is an associate professor in the Indiana University School of Informatics at IUPUI (USA). Dr. Faiola is a three-time Fulbright scholar to Russia in new media and communication technology; he has also worked and guest lectured throughout Europe on the subject of media usability theory and practice. His research focuses on the influence of cultural cognition on interactive media, as well as design theory as it applies to interaction design and new media education.

Christopher Ford is a doctoral student in rhetoric and professional communication at New Mexico State University (USA). His research interests include online communication and gaming narratives.

Rachel Gallagher is an MA student in rhetoric and professional communication at New Mexico State University (USA).

Shaoyi He is an associate professor of information systems in the School of Business Administration at California State University San Marcos (USA). He received his PhD from the University of North Carolina at Chapel Hill (1998). Since then, he has worked as a faculty member at Long Island University, The Pennsylvania State University, and California State University San Bernardino. Dr. He also worked at IBM for three-and-a-half years. His current research interests include interplay of technology, culture, and language in global e-business; multilingual issues in e-commerce Web site localization; and language barriers in marketing across cultures.

Marc Hermeking teaches and researches in international and cross-cultural marketing at the Institute for Intercultural Communications at the Ludwig-Maximilians University Munich, Germany. He studied business administration (diploma) and marketing psychology, European ethnology, and intercultural communications (doctor). During his study he worked in the media industry. His doctoral thesis deals with influences of culture on the international transfer and usage of technology (Kulturen and Technik, 2001). He established the "Seminar fuer Interkulturelles Marketing" at the Institute for Intercultural Communications (2000); since then he has been lecturer at Munich University, a trainer for industrial corporations, and has published several contributions for textbooks and journals.

Yasmin Ibrahim is a senior lecturer in the Division of Information and Media Studies at the University of Brighton (UK), where she lectures on globalization and the media and political communication. Her main research interests include the use of the Internet for empowerment and political communication in repressed polities and diasporic communities, global governance, and the development of alternative media theories in non-Western contexts.

Jennifer Jones is an assistant professor of business law at the University of the Incarnate Word. She holds a JD from the University of Houston and an LLM in International Law from American University, Washington College of Law. Before joining the faculty at UIW, she served as an international attorney in the Office of Chief Counsel for Import Administration and as advisor to the Deputy Assistant Secretary of Import Administration at the U.S. Department of Commerce.

Kendall Kelly teaches technical communication at Texas State University in San Marcos (USA). She holds an MA in English from Boston College and an MLIS from the University of Texas at Austin. She is currently pursuing a PhD in technical communication and rhetoric from Texas Tech University.

Sipai Klein is originally from Israel. He is currently a doctoral student in rhetoric and professional communication at New Mexico State University (USA). His research interests include Web-based and distance pedagogies in rhetoric and professional communication.

Wei-Na Lee is a professor of advertising at the University of Texas at Austin. Her research interests include cross-cultural consumer behavior, multicultural marketing communication, and consumers in a technology-mediated environment. Her work has been published in *Journal of Advertising, Journal of Advertising Research, Journal of Business Research, Psychology & Marketing, Journal of International Marketing, International Journal of Advertising,* and various book chapters and conference proceedings.

She co-edited the 2005 book *Diversity in Advertising*. She was the recipient of the American Academy of Advertising Research Fellowship (1992, 2002, and 2004). Lee was executive director of the Office of Survey Research at the University of Texas at Austin (1998-2001).

Marohang Limbu is originally from Nepal. He completed his MA in Rhetoric at New Mexico State University (USA), and is currently a doctoral student in rhetoric and composition at the University of Texas-El Paso. His research focuses on intercultural rhetoric in indigenous populations.

Sorin Adam Matei is a researcher and educator with an interest in social uses of communication technologies. He is an assistant professor in the Department of Communication at Purdue University (USA). Dr. Matei has a variety of research and instructional interests, including sociospatial shaping of communication technology; online social interaction and communities; and spatial analysis applied to communication systems, and international communication processes and flows.

Bolanle A. Olaniran earned his PhD from the University of Oklahoma in 1991. He is a professor in the Department of Communication Studies at Texas Tech University (USA). His research areas include computer-mediated communication, cross-cultural communication, and organization communication. His works have appeared in international, national, and regional journals, and he has written several book chapters.

Angela T. Ragusa earned her PhD and MS degrees in sociology, and her MS in science and technology studies from Virginia Polytechnic Institute (USA). She is currently a sociology lecturer at Charles Sturt University (Australia), with 10 years of university teaching experience. Dr. Ragusa's interest in virtual communication and distance education stems from her work to improve students' educational experiences. While her overarching area of research interest is social change, recent areas of publication include media analysis and representation; higher education, social justice, and globalization; materialism, time use, and social values; and work stress and occupations. Past workplaces include the University of New England, Queensland University of Technology, Griffith University, Virginia Tech, the United Nations, and JP Morgan.

Lisa Ramirez is originally from Belize. She is currently an MA student in rhetoric and professional communication at New Mexico State University (USA). Her research interests include the rhetoric of science in Latin America and ESL writing.

Thomas C. Reeves (Web site: http://it.coe.uga.edu/~treeves/ e-mail: treeves@uga.edu) is a professor in the Department of Educational Psychology and Instructional Technology at the University of Georgia (USA), where he teaches evaluation, research, and instructional design. His research interests include evaluation and assessment methods, cognitive tools and mental models, socially responsible research, authentic tasks for online learning, and instructional technology in developing countries. He has been an invited speaker in 22 countries. Reeves was the first person to receive the AACE Fellowship Award from the Association for the Advancement of Computing in Education (2003).

Karim A. Remtulla is a PhD candidate in the Department of Adult Education and Counselling Psychology at the University of Toronto. His research focuses on the theoretical and practical implications of e-learning on adult education. He received his master's in business administration from the Schulich School of Business at York University and his master's in science in information management from the School of Information Studies at Syracuse University. He has published in the *International Journal of Baudrillard Studies* and presented at the World Conference on Educational Multimedia, Hypermedia & Telecommunications, as well as the World Forum Proceedings of the International Research Foundation for Development.

Simeon J. Simoff is associate professor in information technology and computing science and head of the E-Markets Research Group at the University of Technology, Sydney (Australia). He is also co-director of the Institute of Analytic Professionals of Australia. He is known for the unique blend of interdisciplinary scholarship which integrates the areas of data mining, design computing, virtual worlds, and digital media with application in the area of electronic trading environments. His work in these fields has resulted in nine co-authored/edited books and more than 170 research papers, and a number of cross-disciplinary educational programs in information technology and computing. He is co-editor of the CRPIT series *Conferences of Research and Practice in Information Technology.* He has initiated and co-chaired several international conference series in the area of data mining.

Nicole St. Germaine-Madison earned her MA in scientific and technical communication at Bowling Green State University (USA), and is a PhD candidate in technical communication and rhetoric at Texas Tech University. Her research interests include translation and localization and medical documentation, particularly as they apply to the Tejano and Mexican-American population.

Charles J. Stewart is a distinguished professor of communication at Purdue University (USA). He teaches undergraduate courses in persuasion and interviewing, and graduate courses in persuasion and social protest, extremist rhetoric on the Internet, and the rhetoric of conspiracy. He is a founding fellow of Purdue's Teaching Academy, is inscribed in *Purdue's Book of Great Teacher's*, and is the recipient of the National Communication Association's Ecroyd Award for Outstanding Teaching in Higher Education. He is the author of nine books and numerous articles and chapters in the areas of interviewing, persuasion, social movements, rhetorical criticism, and counseling.

Fay Sudweeks is a senior lecturer in information systems at Murdoch University (Australia). She has a Bachelor of Arts in Psychology and Sociology, a Master of Cognitive Science and Doctor of Philosophy (Communication Studies), and is a member of the Australian Computer Society. Her current research interests are social, cultural, and economic aspects of CMC, group dynamics, and e-learning. She has published four edited books, 14 edited proceedings, and more than 70 papers in journals, books, and conference proceedings. She is on the editorial board of the *Journal of Computer-Mediated Communication, Human Communication Research, New Media and Society, International Journal of E-Learning, Journal of Electronic Commerce Research*, and *Journal of Electronic Commerce in Organizations.* With Charles Ess, she has co-chaired five international and interdisciplinary conferences on *Cultural Attitudes Towards Technology and Communication.*

Rosanna Tarsiero is an eclectic knowledge management consultant and facilitator for Gionnethics (Italy), where she combines a solid background in medical science with more than six years of experience in moderating and facilitating online communities. Her academic publications, conference participations, and practical expertise cover a broad spectrum of topics. Of particular interest are her works on online self-help communities, communities of practice, experiential and social learning in facilitation settings, human-computer interaction, online volunteering, and management of volunteers with mental illness.

Barry Thatcher is an associate professor of rhetoric and professional communication at New Mexico State University (USA). His research interests include intercultural professional communication, border rhetoric, border health literacy, history of rhetoric in Latin America, and U.S.-Latin American professional communication.

Chun-Min Wang (e-mail: flat@uga.edu) is a doctoral candidate in the Department of Educational Psychology and Instructional Technology in the College of Education at the University of Georgia (USA). A native of Taiwan, his research interests include cultural issues in online education, design-based research, evaluation of instructional technology, and creativity in instructional design. His PhD research is focused on the intersection of cultural dimensions in online learning environments with a special interest in Chinese international students studying in the United States. He is also involved in the evaluation of digital libraries funded by the National Science Foundation.

Yun Xia earned his PhD from Southern Illinois University (USA, 2002). He is currently an assistant professor in the Department of Communication at Rider University (USA). He teaches print media and digital media design courses. His research interests are on the social impact of computer-mediated communication, educational applications of communication technologies, visual intelligence of graphic communications in new media, and semiotic analysis of communication signs in new media. His current studies include the exploration of logic forms in analog and digital communication and technology use in different cultures. His works have appeared in journals such as *Human Communication, China Media Research*, and *The American Journal of Semiotics*.

Index

I

K

L

M

N

O

P

R

S

T

U

V

W